Lecture Notes in Computer Sci

T0237930

Commenced Publication in 1973
Founding and Former Series Editors:
Gerhard Goos, Juris Hartmanis, and Jan van Leeuwen

Michael H.F. Wilkinson
Jos B.T.M. Roerdink (Eds.)

Mathematical Morphology and Its Application to Signal and Image Processing

9th International Symposium, ISMM 2009
Groningen, The Netherlands, August 24-27, 2009
Proceedings

 Springer

Volume Editors

Michael H.F. Wilkinson
Jos B.T.M. Roerdink
University of Groningen
Institute of Mathematics and Computing Science
P.O. Box 407, 9700 AK Groningen, The Netherlands
E-mail: {m.h.f.wilkinson, j.b.t.m.roerdink}@rug.nl

Library of Congress Control Number: Applied for

CR Subject Classification (1998): I.4.10, I.5.4, C.3, J.3, G.2, I.2.8

LNCS Sublibrary: SL 6 – Image Processing, Computer Vision, Pattern Recognition, and Graphics

ISSN 0302-9743

ISBN 978-3-642-03612-5 Springer Berlin Heidelberg New York

springer.com

© Springer-Verlag Berlin Heidelberg 2009

Typesetting: Camera-ready by author, data conversion by Scientific Publishing Services, Chennai, India
Printed on acid-free paper   SPIN: 12732043   06/3180   5 4 3 2 1 0

Preface

The 9th ISMM conference covered a very diverse collection of papers, bound together by the central themes of mathematical morphology, namely, the treatment of images in terms of set and lattice theory. Notwithstanding this central theme, this ISMM showed increasing interaction with other fields of image and signal processing, and several hybrid methods were presented, which combine the strengths of traditional morphological methods with those of, for example, linear filtering. This trend is particularly strong in the emerging field of adaptive morphological filtering, where the local shape of structuring elements is determined by non-morphological techniques. This builds on previous developments of PDE-based methods in morphology and amoebas. In segmentation we see similar advancements, in the development of morphological active contours.

Even within morphology itself, diversification is great, and many new areas of research are being opened up. In particular, morphology of graph-based and complex-based image representations are being explored. Likewise, in the well-established area of connected filtering we find new theory and new algorithms, but also expansion into the direction of hyperconnected filters. New advances in morphological machine learning, multi-valued and fuzzy morphology are also presented.

Notwithstanding the often highly theoretical reputation of mathematical morphology, practitioners in this field have always had an eye for the practical. Most new theoretical and algorithmic developments are driven by urgent needs from practical applications. Thus many application areas are covered in this volume, ranging from biomedical imaging, through materials science to satellite imaging and traffic analysis. The development of standardized software packages which make all the latest algorithms easily available to the image processing professional is very much a part of this practical side of morphology.

Finally, we wish to thank all the members of the Program Committee for their efforts in reviewing all submissions and giving extensive feedback, essential to the quality and success of this conference.

June 2009

Michael Wilkinson
Jos Roerdink

Organization

ISMM 2009 was organized by the Institute for Mathematics and Computing Science, University of Groningen, Groningen, The Netherlands.

Chair

Michael H.F. Wilkinson	University of Groningen, The Netherlands
Jos B.T.M. Roerdink	University of Groningen, The Netherlands

Steering Committee

Junior Barrera	Universidade de São Paolo, Brazil
Isabelle Bloch	Télécom ParisTech (ENST), Paris, France
Renato Keshet	HP Labs, Haifa, Israel
Petros Maragos	National Technical University of Athens, Greece
Jos B.T.M. Roerdink	University of Groningen, The Netherlands
Christian Ronse	Université de Strasbourg, France
Philippe Salembier	Universita Polytècnica de Catalunya, Barcelona, Spain
Pierre Soille	European Commission, Joint Research Centre, Ispra, Italy
Hugues Talbot	Université Paris-Est, ESIEE Paris, France

Local Organizing Committee

Desiree Hansen
Alphons Navest
Jos Roerdink
Marike Vos
Michael Wilkinson

Invited Speakers

Bernhard Burgeth	Saarland University, Saarbrücken, Germany
Luc Florack	Eindhoven University of Technology, The Netherlands
Reinhard Klette	University of Auckland, New Zealand

Program Committee

J. Angulo	T. Geraud	P. Salembier
A. Asano	A. Hanbury	J. Serra
R. Audigier	A.C. Jalba	P. Soille
S. Beucher	D. Jeulin	H. Talbot
I. Bloch	S. Lefevre	I. Terol-Villalobos
G. Borgefors	R. Lotufo	E.R. Urbach
M. Buckley	P. Maragos	C. Vachier
B. Burgeth	B. Marcotegui	R. van den Boomgaard
M. Couprie	L. Najman	M. Van Droogenbroeck
J. Cousty	G.K. Ouzounis	M. Welk
J. Crespo	N. Passat	M.A. Westenberg
E. Decencière	J.B.T.M. Roerdink	M.H.F. Wilkinson
A.N. Evans	C. Ronse	

Sponsoring Institution

Institute for Mathematics and Computing Science, University of Groningen, Groningen, The Netherlands.

Table of Contents

Algorithms

Discrete Driver Assistance

Reinhard Klette[1], Ruyi Jiang[2], Sandino Morales[1], and Tobi Vaudrey[1]

[1] The University of Auckland, Auckland, New Zealand
[2] Shanghai Jiao Tong University, Shanghai, China

Abstract. Applying computer technology, such as computer vision in driver assistance, implies that processes and data are modeled as being discretized rather than being continuous. The area of stereo vision provides various examples how concepts known in discrete mathematics (e.g., pixel adjacency graphs, belief propagation, dynamic programming, max-flow/min-cut, or digital straight lines) are applied when aiming for efficient and accurate pixel correspondence solutions. The paper reviews such developments for a reader in discrete mathematics who is interested in applied research (in particular, in vision-based driver assistance). As a second subject, the paper also discusses lane detection and tracking, which is a particular task in driver assistance; recently the Euclidean distance transform proved to be a very appropriate tool for obtaining a fairly robust solution.

Keywords: Discrete mathematics, driver assistance, stereo analysis, lane detection, distance transform.

1 Vision-Based Driver Assistance

Driver assistance systems (DAS) are developed to (i) *predict* traffic situations, (ii) *adapt* driving and car to current traffic situations, and (iii) *optimize* for safety. Vision-based DAS applies one or multiple cameras for understanding the environment, to help achieve goals (i-iii).

After specifying a processing model, possibly in continuous space, any specification for its algorithmic use will depend on discrete mathematical models, such as numerical algorithms [13], or concepts in discrete mathematics such as adjacency sets $A(p)$ of pixels p, digital straight lines, or distance transforms, which are examples from digital geometry [14]. Typically, continuous models are used in motion analysis up to the moment when mapping those concepts into algorithms, but matching techniques for multi-ocular vision typically already start with a discrete model.

This paper is organized as follows: Section 2 describes techniques applied in binocular correspondence analysis, followed by Section 3 with (further) illustrations of matching results in vision-based DAS. Lane detection via distance transform is the subject of Section 4. A few conclusions are given in Section 5.

M.H.F. Wilkinson and J.B.T.M. Roerdink (Eds.): ISMM 2009, LNCS 5720, pp. 1–12, 2009.

2   Stereo Algorithms

Stereo algorithms are designed for calculating pairs of corresponding pixels in concurrently recorded images. After calibration and rectification [6], images L (left) and R (right) are in *standard stereo geometry* (i.e., parallel optical axes, coplanar image planes, aligned image rows), defined on pixels of an $M \times N$ grid Ω. See Figure 1; the third view has been used in [17] for prediction error analysis.

Thus, stereo pixel correspondence is basically a 1D search problem, compared to motion pixel correspondence (e.g., [10]) which is a continuous 2D search problem. Two corresponding pixels $p_L = (x, y)$ and $p_R(x - \Delta(x, y), y)$ identify a *disparity* $\Delta(x, y)$ which defines the *depth* $bf/\Delta(x, y)$, where b is the *base distance* between both focal points, and f is the uniform focal length of both cameras (after rectification). However, the search should also account for disparity consistency between adjacent scan lines (e.g., between rows y, $y - 1$, and $y + 1$).

2.1   Data and Continuity Terms

This *stereo matching problem* is an instance of a general *pixel labeling problem*: given is a finite set $\mathcal{L}$ of labels $l, h, \ldots$; define a labeling Δ which assigns to each pixel $p \in \Omega$ (in the *base image*; we assume L to be the base image) a label $\Delta_p \in \mathcal{L}$. Consider a *data term* of penalties $D_p(\Delta_p)$ for assigning label Δ_p to pixel p. The simplest data term is given by $D_x(l) = |L(x, y) - R(x - l, y)|^b$, for a fixed row y, $1 \leq x \leq M$, and b either 1 or 2, assuming that image pairs are *photo-consistent* (i.e., corresponding pixels have about the same value).

[7] compares various data terms within a *the-winner-takes-all strategy*: for each pixel $p = (x, y)$ in the left image, a selected data term is applied for all potential matches $q = (x - l, y)$ in the right image, for $l \geq 0$; that l is taken as disparity which defines a unique (within the whole row) minimum for this cost function; if there is no such unique global minimum then the disparity at p remains undefined. – For example, results in [7] indicate that the census cost function seems to be very robust (w.r.t. image data variations) in general.

Fig. 1. One time frame of image sequences taken with three cameras (called: *third, left,* and *right camera* - from left to right) installed in HAKA1, test vehicle of the *.enpeda..* project. Note the reflections on the windscreen, and differences in lightness (e.g., image of right camera is brighter than the other two). Left and right views are rectified.

The minimization of the following *energy* (or: *cost*) *functional* E defines a basic approach for solving the stereo matching problem:

$$E(\Delta) = \sum_{p \in \Omega} \left(D_p(\Delta_p) + \sum_{q \in A(p)} C(\Delta_p, \Delta_q) \right) \tag{1}$$

This functional combines a data term with a *continuity term* $C(\Delta_p, \Delta_q)$, which is often simplified to a unary symmetric function $C(|\Delta_p - \Delta_q|)$, for assigning labels Δ_q to adjacent pixels $q \in A(p)$. Further terms may be added (e.g., for occlusion, or ordering constraint). The continuity term assumes that projected surfaces are *piecewise smooth* (i.e., neighboring pixels represent surface points which are at about the same distance to the cameras). A convex function C supports efficient global optimization, but leads to oversmoothed results [11].

Common choices for a unary continuity function are either a simple step function (Pott's model), a linear function, or a quadratic function, where the latter two need to be truncated for avoiding oversmoothing. A simple choice is also a two-step function, which penalizes small disparity changes at adjacent pixels with a rather low weight (to allow for slanted surfaces), but penalizes larger disparity changes with a higher weight.

Finding a global minimum Δ, which minimizes the energy in Equation (1), assuming a continuity function which is not enforcing some kind of over-smoothing, is an NP-hard problem; see [15]. Purely local matching strategies (e.g., hierarchical correlation based methods) failed to provide reasonable approximate solutions. Strategies favored recently follow some semi-global optimization scheme.

2.2 Semi-Global Paradigms for Sub-Optimal Solutions

Basically, current stereo algorithms follow one of the following three paradigms: scanline optimization often implemented by (DP) *dynamic programming* [18], (BP) *belief propagation* [3], or (GC) *graph-cut* [15]. These paradigms aim at finding a sub-optimal solution to the stereo matching problem.

SGM using Scanline Optimization. *Semi-global matching* (SGM) is commonly identified with applying scanline optimization along several digital rays, all incident with the start pixel p in the base image [9]. Original dynamic programming stereo [18] was defined for energy minimization along a single scan line. Assume that row y remains constant; matching aims at minimizing

$$E_m(\Delta) = \sum_{x=1}^{m} \left(D_x(\Delta_x) + \sum_{\hat{x} \in A(x)} C(\Delta_x, \Delta_{\hat{x}}) \right) \quad \text{with} \quad E(\Delta) = E_M(\Delta) \tag{2}$$

Value m defines the *stage* of the dynamic optimization process; when arriving at stage m we have assignments of labels Δ_x, for all x with $1 \le x < m$ (possibly excluding pixels close to the left border of the left image), and we have not yet assignments for $x \ge m$; we select Δ_m by taking that $l \in \mathcal{L}$ which minimizes

$$D_m(l) + C(l, \Delta_{m-1}) + E_{m-1}(\Delta) \tag{3}$$

Fig. 2. Left: streaks in a single-line DP result (for left and right image as shown in Figure 1). Right: visible search line patterns in the calculated depth map for 8-ray DP using mutual information as cost function (also known as SGM MI).

(Obviously, the term $E_{m-1}(\Delta)$ can be deleted for the minimization task.) At $m = 1$ we only have $\Delta_1 = 0$, for $m = 2$ we may decide between $l = 0$ or $l = 1$, and so forth. When arriving at stage $x = M$, we have an optimized value $E(\Delta)$ (modulo the applied DP strategy); we identify the used labels for arriving at this value by backtracking, from $x = M$ to $x = M - 1$, and so forth.

Dynamic programming propagates errors along the used digital line; here, this occurs along image rows, from left to right, resulting in horizontal streaks in the calculated depth map. Disparities in adjacent pixels of the same line, or in adjacent rows may be used to define a continuity term in the used energy function for reducing this streak effect. A further option is to combine forward DP also with the following backward DP strategy:

$$E^m(\Delta) = \sum_{x=m}^{M} \left(D_x(\Delta_x) + \sum_{\hat{x} \in A(x)} C(\Delta_x, \Delta_{\hat{x}}) \right) \quad \text{with} \quad E(\Delta) = E^1(\Delta) \quad (4)$$

where Δ_m is selected as that $l \in \mathcal{L}$ which minimizes

$$D_m(l) + C(l, \Delta_{m-1}) + E_{m-1}(\Delta) + C(l, \Delta_{m+1}) + E^{m+1}(\Delta) \quad (5)$$

($E_{m-1}(\Delta)$ and $E^{m+1}(\Delta)$ can be ignored again.) Obviously, this requires to proceed up to $x = m$ with 'normal' DP both from left and from right, then combining values along both digital rays into one optimized value Δ_m at $x = m$ following Equation (5). This increases the time complexity, compared to the simple approach in Equation (3), and time-optimization is an interesting subject.

This double-ray approach was generalized to optimization along multiple digital rays [9], thus approximating the global (NP-complete; see above) solution. For a digital ray in direction **a**, processed between image border and pixel p, consider the segment $p_0 p_1 \ldots p_{n_a}$ of that digital ray, with p_0 on the image border, and $p_{n_a} = p$; the energy contribution along that ray at pixel p is defined via scanline optimization as in Equation (3). All used digital rays **a** (ending at p) are assumed to have identical impact; the label at pixel p is obtained by generalizing Equation (5); we assign that disparity l which minimizes

$$D_p(l) + \sum_{\mathbf{a}} [C(l, \Delta_{n_{\mathbf{a}}-1}) + E_{n_{\mathbf{a}}-1}(\Delta)] \qquad (6)$$

Labeling Δ in Equation (6) obtains thus a further value Δ_p at pixel p. Again, all those $E_{n_{\mathbf{a}}-1}(\Delta)$ can be deleted for minimization, and algorithmic time optimization (say, scanline optimization along all lines at first, and then combining results) leads to a feasible solution. [8] also includes a second-order prior into the used energy function.

Belief Propagation. Belief propagation is a very general way to perform probabilistic inference; the BP stereo matching algorithm in [3] passes messages (the "belief" which is a weight vector for all labels) around in a 4-adjacency image grid. Message updates are in iterations; messages are passed on in parallel, from a pixel to all of its 4-adjacent pixels. At one iteration step, each pixel of the adjacency graph computes its message based on the information it had at the end of the previous iteration step, and sends its (new) message to all the adjacent pixels in parallel.

Let $m_{q \rightarrow p}^i$ denote the message send from pixel q to adjacent pixel p at iteration i, defined for all $l \in \mathcal{L}$ as follows:

$$m_{q \rightarrow p}^i(l) = \min_{h \in \mathcal{L}} \left(C(h, l) + D_q(h) + \sum_{r \in A(q) \backslash p} m_{r \rightarrow q}^{i-1}(h) \right) \qquad (7)$$

l is just one of the $|\mathcal{L}|$ possible labels at p, and h runs through $\mathcal{L}$ and is again just a possible label at q. We accumulate at p a vector of length $|\mathcal{L}|$ of all messages received from all $q \in A(p)$, and this contains at its position $l \in \mathcal{L}$ the following:

$$D_p(l) + \sum_{q \in A(p)} m_{q \rightarrow p}^i(l) \qquad (8)$$

Besides $D_p(l)$, we also have the sum of all the received message values for $l \in \mathcal{L}$. Instead of passing on vectors of length $|\mathcal{L}|$, a belief propagation algorithm typically uses $|\mathcal{L}|$ *message boards* of the size of the images, one board for each label l. At the end of an iteration t, that disparity with minimum cost is selected as being the result for pixel $p \in \Omega$.

BP fails in cases of photometric inconsistencies between left and right image; [5] showed that some edge preprocessing of both images is of benefit, and [19] performed a systematic study which shows the residual images (rather than original input images) carry the important information for correspondence analysis.

Graph-Cut. Consider 4-connected pixels of the base image; this defines an undirected graph (Ω, A) with nodes Ω and edges A. Assume two additional nodes s and t, called *source* and *sink*, respectively, with directed edges from s

Fig. 3. Left: BP result (for left and right image as shown in Figure 1). Right: GC result.

to all the nodes in Ω, and from those nodes to t. This defines altogether an undirected graph $G = (\Omega \cup \{s,t\}, A \cup \{s\} \times \Omega \cup \Omega \times \{t\})$. Edges in this graph are weighted by $w(p,q)$ (continuity values to undirected edges and data values to undirected edges), also called *capacities*.

An (s,t)-*cut* of G is a partition of $\Omega \cup \{s,t\}$ into subsets S and $\overline{S}$, with $s \in S$ and $t \in \overline{S}$. The energy $E(S)$ of such an (s,t)-cut is the sum of all weights of edges connecting S with $\overline{S}$:

$$E(S) = \sum_{p \in S, q \in \overline{S}, \{p,q\} \in A} w(p,q) \tag{9}$$

A *minimum* (s,t)-*cut* is an (s,t)-cut with minimum energy. Ford and Fulkerson (see [4]) proved that the calculation of a minimum cut is equivalent to the calculation of a maximum flow; the calculation of a min-cut is commonly implemented via calculating a max-flow. Used algorithms have about $\mathcal{O}(n^4)$ worst case run time, but run in practice in about $\mathcal{O}(n^3)$ expected time or better.

Let Δ be a labeling of Ω. Any α-*expansion* of Δ into Δ' satisfies that

$$\Delta'_p \neq \Delta_p \implies \Delta'_p = \alpha$$

for every pixel $p \in \Omega$. The following *expansion-move algorithm* is a greedy algorithm which runs in practice in near-linear time:

    start with an arbitrary labelling Δ on Ω;
    do { success := false;
        for each label $\alpha \in \mathcal{L}$ {
            calculate the minimum-energy α-expansion Δ' of Δ;
            if $E(\Delta') < E(\Delta)$ **then** { $\Delta := \Delta'$;   success := true} } }
    until success = false;
    return Δ

The calculation of the minimum-energy α-expansion is performed by applying a min-cut (meaning, via max-flow) algorithm. An α-extension either keeps an old label Δ_p or assigns the new label α; this defines a partition of the graph into set S (old label) and $\overline{S}$ (new label). Equation (9) defines the energy for this partitioning (labeling). See, for example, [1] for more details.

3   Improving Stereo Results by Preprocessing

Obviously, the illustrated resulting depth maps are not satisfactory for DAS. Errors are often due to varying illumination conditions or other real world imaging effects, and this is different to studies using only ideal images taken indoors or under controlled conditions. The paper [19] identified residual images as a promising type of input data for stereo or motion correspondence algorithms.

3.1   Edge Operators

Earlier than [19], the paper [5] studied the effect of edge-preprocessing on stereo matching, showing that edge-preprocessed input data improve resulting depth maps in general, especially when applying BP.

Fig. 4. Top: Sobel edge maps of the left-right stereo pair in Figure 1, used as stereo input pair. Middle: depth maps of single line DP (left) and Birchfield-Tomasi cost function in 8-ray DP (also known as SGM BT). Bottom: BP (left) and GC results.

Figure 4 shows four resulting depth maps for the left-right stereo pair in Figure 1, but after applying the (3 × 3) Sobel edge operator. In case of 8-path DP, the use of the Birchfield-Tomasi cost function (SGM BT) shows better results compared to the use of mutual information (SGM MI). – Edge images are high-frequency components of images, and the same is true for residual images.

3.2   Residual Images

We consider an image I as being a composition $I(p) = \mathbf{s}(p) + \mathbf{r}(p)$, for $p \in \Omega$, where $\mathbf{s} = S(I)$ denotes the *smooth component* and $\mathbf{r} = I - \mathbf{s}$ the *residual*. We use the straightforward iteration scheme:

$$\mathbf{s}^{(0)} = I, \quad \mathbf{s}^{(n+1)} = S(\mathbf{s}^{(n)}), \quad \mathbf{r}^{(n+1)} = I - \mathbf{s}^{(n+1)}, \quad \text{for } n \geq 0.$$

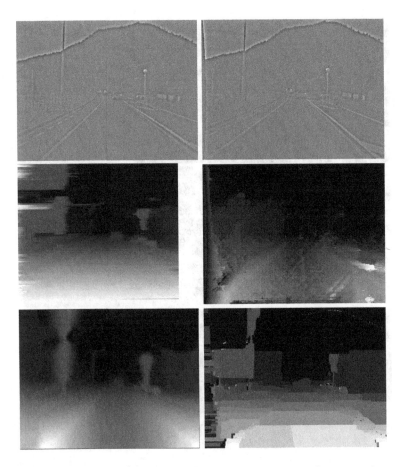

Fig. 5. Top: residual images using 40 iteration of 3×3 mean, used as stereo input pair. Middle: depth maps of single line DP (left) and Birchfield-Tomasi cost function in 8-ray DP (also known as SGM BT). Bottom: BP (left) and GC results.

Figure 4 shows four resulting depth maps for the left-right stereo pair in Figure 1, but on the residual images defined by a 3×3 mean operator and $n = 40$ iterations.

As a general conclusion, single-line DP is quite robust and provides a fast and approximate depth map ("a good draft", and some kind of temporal or spatial propagation of results might be useful [16]), SGM-BT fails absolutely on original data but seems to perform better than SGM-MI on preprocessed images, BP is highly sensitive to illumination changes, and improves very nicely when using optimized parameters on preprocessed input data, and GC also improves on preprocessed input data.

4   Lane Detection and Tracking

Lane detection and tracking has been a successful research subject in DAS. [12] reviews briefly related work in vision-based DAS and discusses a new lane model, also providing two algorithms for either time-efficient or robust lane tracking (to be chosen in dependency of current road situation).

4.1   Bird's-Eye View and Edge Detection

The proposed lane detection and tracking algorithms work on a single image sequence. However, the figures show results for both stereo sequences in parallel, illustrating this way the robustness of the method with respect to different camera positions.

The process starts with mapping a given image into a bird's-eye view (i.e., a homography mapping a perspective image into an orthographic top-down view), based on calibrated projections of corners of a rectangle in front of the car; see Figure 6 for two resulting bird's-eye views. This is followed by an edge detection method which aims at detecting vertical step edges (such as lane marks) rather than horizontal edges; small artifacts are eliminated from these binarized edge images which would otherwise disturb the subsequent distance transform.

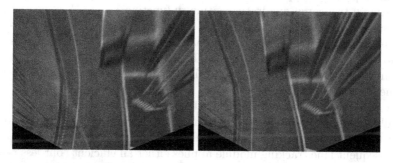

Fig. 6. Bird's-eye views (for left and right image as shown in Figure 1)

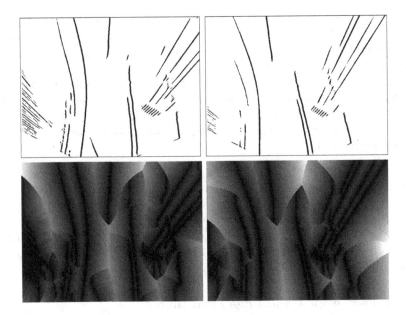

Fig. 7. Detected edges (top) and RODT results (bottom). Both for bird's-eye views as shown in Figure 6. In RODT, dark = low distance, white = high distance.

4.2 Distance Transform and Lane Tracking

Consider a distance transform, applied to a binary edge map, which labels every pixel $p \in \Omega$ by its shortest distance to any edge pixel; see, for example, [14], for distance transforms in general. Experiments have been performed with various kinds of distance transforms, and preference was given to the Euclidean distance transform (EDT). For example, [2] proved that a 2D EDT can efficiently be calculated by two subsequent 1D EDT. (The developed procedure for calculating lower envelopes is also applicable for calculating lower envelopes in a BP algorithm while using a truncated quadratic continuity function.)

[20] suggested the *orientation distance transform* (ODT) which separates EDT values into a row and a column component, represents as complex numbers. We use the real (i.e., row) part of the ODT. See Figure 7 for binary input edge maps and resulting RODT (i.e., real ODT) maps, where gray values increase with measured distance.

In a predefined start row (near to the image's bottom) we identify a left and right boundary point for the current lane based on the calculated RODT values; these two boundary points initialize a particle filter for lane detection. The subsequent lane tracking module applies either an efficient (but less robust; designed for good road conditions), or a robust (but less efficient) algorithm.

Figure 8 shows final results of lane detection (using robust method), for left and right sequence. Some kind of unification might be considered; however, our experience shows that the method performs very robust on a single image

Fig. 8. Detected lanes (for left and right image as shown in Figure 1), illustrating robustness of the technique (i.e., independence of camera position)

sequence. Both tracking algorithms are operating in the bird's-eye views, and both are using results of the RODT for evaluating possibilities of finding lane boundaries.

The RODT not only provides information about the expected centerline of a lane but also about lane boundaries. However, it takes slightly more computation time than the total for generating the bird's-eye view, edge detection, and removal of artifacts.

5 Conclusions

This paper informs the reader about a few subjects where discrete mathematics have met program development in recent vision-based DAS, and proved to be very useful for defining fairly efficient, accurate or robust techniques. The discussed stereo techniques have been proposed elsewhere, and we provided a brief and uniform presentation, together with experimental illustration. The lane detection and recognition solution was reported in [12]. Vision-based DAS is expected to move further ahead, from low-level stereo and motion analysis into advanced subjects for understanding complex traffic scenes, and further interactions with discrete mathematics are certainly coming this way.

References

1. Boykov, Y., Veksler, O., Zabih, R.: Fast approximate energy minimization via graph cuts. IEEE Trans. Pattern Analysis Machine Intelligence 23, 1222–1239 (2001)
2. Felzenszwalb, P.F., Huttenlocher, D.P.: Distance transform of sampled functions. Cornell Computing and Information Science, TR 2004-1963 (September 2004)
3. Felzenszwalb, P.F., Huttenlocher, D.P.: Efficient belief propagation for early vision. Int. J. Computer Vision 70, 41–54 (2006)
4. Ford, L.R., Fulkerson, D.R.: Flows in networks. Technical report R-375-PR, US Air Force Project RAND (1962)

5. Guan, S., Klette, R., Woo, Y.W.: Belief propagation for stereo analysis of night-vision sequences. In: Wada, T., et al. (eds.) PSIVT 2009. LNCS, vol. 5414, pp. 932–943. Springer, Heidelberg (2009)
6. Hartley, R.I., Zisserman, A.: Multiple View Geometry in Computer Vision. Cambridge University Press, Cambridge (2000)
7. Herman, S., Klette, R.: The naked truth about cost functions for stereo matching. MI-tech TR 33, University of Auckland (2009)
8. Hermann, S., Klette, R., Destefanis, E.: Inclusion of a second-order prior into semi-global matching. In: Wada, T., et al. (eds.) PSIVT 2009. LNCS, vol. 5414, pp. 633–644. Springer, Heidelberg (2009)
9. Hirschmüller, H.: Accurate and efficient stereo processing by semi-global matching and mutual information. In: Proc. CVPR, vol. 2, pp. 807–814 (2005)
10. Horn, B.K.P., Schunck, B.G.: Determining optical flow. Artificial Intelligence 17, 185–203 (1981)
11. Ishikawa, H.: Exact optimization for Markov random fields with convex priors. IEEE Trans. Pattern Analysis Machine Intelligence 25, 1333–1336 (2003)
12. Jiang, R., Klette, R., Wang, S., Vaudrey, T.: New lane model and distance transform for lane detection and tracking. In: Proceedings of CAIP (to appear, 2009)
13. Kimmel, R.: Numerical Geometry of Images. Springer, New York (2004)
14. Klette, R., Rosenfeld, A.: Digital Geometry. Morgan Kaufmann, San Francisco (2004)
15. Kolmogorov, V., Zabih, R.: Multi-camera scene reconstruction via graph cuts. In: Heyden, A., Sparr, G., Nielsen, M., Johansen, P. (eds.) ECCV 2002. LNCS, vol. 2352, pp. 82–96. Springer, Heidelberg (2002)
16. Liu, Z., Klette, R.: Dynamic programming stereo on real-world sequences. In: Köppen, M., et al. (eds.) ICONIP 2009, Part I. LNCS, vol. 5506, pp. 527–534. Springer, Heidelberg (2009)
17. Morales, S., Klette, R.: Prediction error evaluation of various stereo matching algorithms on long stereo sequences. MI-tech TR 38, University of Auckland (2009)
18. Ohta, Y., Kanade, T.: Stereo by two-level dynamic programming. In: Proc. IJCAI, pp. 1120–1126 (1985)
19. Vaudrey, T., Klette, R.: Residual images remove illumination artifacts for correspondence algorithms! In: Proc. DAGM (to appear, 2009)
20. Wu, T., Ding, X.Q., Wang, S.J., Wang, K.Q.: Video object tracking using improved chamfer matching and condensation particle filter. In: Proc. SPIE-IS & T Electronic Imaging, vol. 6813, pp. 04.1–04.10 (2008)

The "False Colour" Problem

Jean Serra

Université Paris-Est, Laboratoire d'Informatique Gaspard Monge, Equipe A3SI,
ESIEE Paris, France
j.serra@esiee.fr

Abstract. The emergence of new data in multidimensional function lattices is studied. A typical example is the apparition of false colours when (R,G,B) images are processed. Two lattice models are specially analysed. Firstly, one considers a mixture of total and marginal orderings where the variations of some components are governed by other ones. This constraint yields the "pilot lattices". The second model is a cylindrical polar representation in n dimensions. In this model, data that are distributed on the unit sphere of $n - 1$ dimensions need to be ordered. The proposed orders, and lattices are specific to each image. They are obtained from Voronoi tesselation of the unit sphere The case of four dimensions is treated in detail and illustrated.

1 Introduction

When one takes the supremum of two numerical functions f and g, the resulting function $f \vee g$ may largely differ from the two operands, though at each point x the supremum $(f \vee g)(x)$ equals either $f(x)$ or $g(x)$. In multidimensional cases, the situation becomes worse. If our two functions now represent colour vectors, their supremum at point x may be neither $f(x)$ nor $g(x)$. A false colour is generated. This parasite phenomenon, due to the multidimensionality of our working space, appears therefore when dealing with satellite data, or with the composite data of the geographical information systems.

The present study aims to analyze, and, if possible, to control the phenomenon. In such matter one rarely finds a unique good solution, but usually several attempts, more or less convincing. In this respect, the case of colour imagery, and the associated lattices, turns out to be an excellent, and very visual, paradigm for multidimensional situations. It has motivated many approaches, from which one can extract the three following themes.

The first theme deals with the advantages and disadvantages of a total ordering for multivariate data. For some authors, this ordering seems to be an absolute requirement, which should be satisfied by lexicographic means [5] [14]. By so doing, one favours the priority variable to the detriment of the second, and so on. By reaction, other authors observe that total ordering in $\mathbb{R}^1$ has the two different finalities of i) locating the extrema, and ii) defining distances between grey levels. When passing from grey to colour, i.e. from $\mathbb{R}^1$ to $\mathbb{R}^3$, it may be advantageous to dissociate the two roles. Numbers of markers can replace

M.H.F. Wilkinson and J.B.T.M. Roerdink (Eds.): ISMM 2009, LNCS 5720, pp. 13–23, 2009.
© Springer-Verlag Berlin Heidelberg 2009

extrema, and numbers of distances can describe vector proximity. This point of view, clearly explicit in [8], led A.Evans and D.Gimenez to remarkable connected filters for colour images, and F. Meyer to a watershed algorithm for colour images [11].

However, a total ordering is sometimes necessary, e.g. when we work in a space for which hue is not a coordinate. As we saw, in the usual product lattice $R \times G \times B$, the supremum of two triplets (r, g, b) may have a hue different from that of the two operands. One could argue that it suffices to take cylindrical coordinates, i.e. to replace R, G, B by L, S, H (for luminance, saturation, hue), equipped with a correct norm, such as L_1 or $max - min$ [1], to bring down the objection, since the hue is then directly under control. But this solution ignores that the higher is saturation, the more hue is significant. If the pixel with the highest hue has also a low saturation, the former is practically invisible, though it can be strongly amplified by a product of suprema, just as if it was a false colour. In conclusion, let us say that the matter is less keeping or not a total ordering than ensuring that some variables must not be treated separately. For example, we could demand that the hue of the supremum be that of the pixel with the highest saturation. Then the latter "pilots" the hue, hence the name of this ordering, and of the associated lattices. They are studied below in section 3.

Now, when passing from the Euclidean representation R, G, B to a polar system such as L, S, H, we introduce new drawback. On the unit circle, the hue ordering is not only arbitrary, but also the cause of a strong discontinuity between violet and red. How to master this parasite effect? One can imagine several strategies. For example, we may restrict ourselves to use increments only [9], or we may take several origins for the hue, according to its histogram, as proposed by E.Aptoula and S.Lefèvre in [3], [4]. We may make the hue depend on some neighborhood around each pixel, and not on the whole image, etc.. For multivariate data in n dimensions, the same problem reappears, now on the unit sphere in $\mathbb{R}^n$. Section 4 below proposes to segment this unit sphere by means of Voronoi polyhedra, which generalizes the method of Aptoula and Lefèvre .

Finally the third theme, typical of multivariate data, consists in using some variables for mixing the segmentations of other ones. For example, P. Soille considers a hierarchy of partitions and extracts at each level the classes that satisfy a constraining criterion [13]. The quaternions algebra offers another way, investigated by T.A. Ell and S.J. Sangwine [7] for Fourier transform of colour images, where the real axis is particularized. In [1] J.Angulo and J. Serra segment separately the three grey images H, L, and S, and then keep either hue or luminance segmentation, according as saturation is high or low. This mode of classification is extended to more dimensions in the example of remote sensing presented in section 6.

The three above themes lie on the implicit assumption of finiteness of the data sets, as they resort to Proposition 2 below. But the usual working spaces $\mathbb{R}^2$ and $\mathbb{Z}^2$ are not finite. How to reconcile the two points of view? This initial mathematical step will be the matter of the next section.

2   Lattices of Finite Parts

Proposition 2 imposes we restrict ourselves to *finite* families of points only. But if we take for framework a given finite part of $\mathbb{R}^n$ or $\mathbb{Z}^n$, we have to renounce to translation invariance, hence to Minkowsli operations. Observe however that the problem is not working with the subsets of a finite set, but working with finite sets that remain finite under the operations that transform them.

Set case. A good way for expressing this idea consists in starting from an arbitrary set E, possibly finite, countable, or even continuous, and focusing on its finite parts exclusively. As they are closed under intersection, with $\varnothing$ as smallest element, we just need to provide them with a universal upper-bound, namely E itself, for obtaining a complete lattice. Hence we can state [12].

Proposition 1. Set lattice of finite parts (LFP): *Let E be a set, and let $\mathcal{X}'$ be the class of its finite parts. The set $\mathcal{X} = \mathcal{X}' \cup E$ forms a complete lattice for inclusion ordering, where, for every family $\{X_i, X \in \mathcal{X}, i \in I\}$, possibly infinite, the infimum $\wedge X_i$ and the supremum $\vee X_i$ are given by*

$$\wedge X_i = \cap X_i,$$
$$\vee X_i = \cup X_i \text{ when } \cup X_i \text{ is upper-bounded by an element of } \mathcal{X}',$$
$$\vee X_i = E \text{ when not.}$$

Class $\mathcal{X}$ is closed under infimum, since even when I is the empty family, we have that $\wedge\{X_i, i \in \varnothing\} = E \in \mathcal{X}$. Proposition1 applies for sets of points with integer coordinates in $\mathbb{Z}^n$ or in $\mathbb{R}^n$, as well as for infinite graphs.

Function case. Before extending Proposition 1 to numerical functions, we recall a classical result on ordered sets.

Proposition 2. *Let T be an ordered set. Every finite family $\{t_i, i \in I\} \in T$ admits a supremum $\vee t_i$ and an infimum $\wedge t_i$ which are themselves elements of the family iff the ordering of T is total.*

The proposition cannot be generalized to countable families, even neither to finite families when the ordering is partial only (e.g. $R \times G \times B$ colour space). In the following, the set T of Proposition 2 is a numerical lattice, and corresponds to the arrival space of the functions under study. It may be $\bar{\mathbb{R}}$, or a closed part of $\bar{\mathbb{R}}$, or $\bar{\mathbb{Z}}$, or any subset of $\bar{\mathbb{Z}}$. The two universal bounds of T are denoted by M_0 and M_1. As we did for $\mathcal{P}(E)$ with Proposition 1, we can associate with T a finite lattice $\mathcal{T}$. It suffices to put, for any family $\{t_j, j \in J\}$ in T, that

$$\curlywedge\{t_j, j \in J\} = \wedge t_j \text{ when } \mathrm{Card}(J) < \infty, \text{ and } \curlywedge\{t_j, j \in J\} = M_0 \text{ when not,}$$
$$\curlyvee\{t_j, j \in J\} = \vee t_j \text{ when } \mathrm{Card}(J) < \infty, \text{ and } \curlyvee\{t_j, j \in J\} = M_1 \text{ when not.}$$

Lattice $\mathcal{T}$ is made of all finite families of numbers, plus M_0 and M_1. Consider now the class $\mathcal{F}$ of all functions $f : E \to T$ with a finite support. Finiteness must hold not only on the support of f and on $f(x)$, but also on the number of values taken at point x by any family $f_j \in \mathcal{F}$. This constraints lead to the following function lattice [12].

Proposition 3. Function lattice of finite parts : *the class $\mathcal{F}$ of functions f : $E \to \mathcal{T}$ with finite extrema and finite support forms a complete lattice for the pointwise numerical ordering. At point $x \in E$ the infimum $\curlywedge$ and the supremum $\curlyvee$ of a finite or not family $\{f_j, j \in J\}$ in $\mathcal{F}$, are given by the expressions*

$$(\curlywedge \{f_j, j \in J\})(x) = \wedge f_j(x) \quad \text{when } x \in \cap X_j, \text{ and card } J \text{ are finite,}$$
$$(\curlywedge \{f_j, j \in J\})(x) = M_0 \quad \text{when not;}$$
$$(\curlyvee \{f_j, j \in J\})(x) = \vee f_j(x) \quad \text{when } x \in \cup X_j \subseteq \mathcal{X}', \text{ and card } J \text{ are finite,}$$
$$(\curlyvee \{f_j, j \in J\})(x) = M_1 \quad \text{when not.}$$

Lattice $\mathcal{F}$, and its multidimensional versions, are shared by all models we develop from now on, even when it is not explicitly recalled.

3   Pilot Lattices

This section is devoted to the first theme met in introduction, and to its consequence on "false colours". We now work in the n dimensional space $T^{(n)}$, $n < \infty$. Its elements are the ordered sequences of n real numbers called components. The space $T^{(n)}$ can welcome many ordering relations leading to complete lattices, two extreme representatives of which being the marginal ordering and the lexicographic one. The first one is the product ordering of each component, and the second one describes a route over the whole space, where the first component is prioritary, then the second, the third, etc..If $t^1, t^2 \in T^{(n)}$, with $t^1 = (t_1^1, t_2^1, ... t_n^1)$, and $t^2 = (t_1^2, t_2^2, ... t_n^2)$, we get

– for the marginal ordering:

$$t^1 \leq t^2 \quad \text{iff} \quad t_i^1 \leq t_i^2, \quad 1 \leq i \leq n$$

hence
$$\vee \{t^j, j \in J\} = (\vee t_1^j, \vee t_2^j, .. \vee t_n^j), \quad j \in J.$$

When family J is finite, each component of the supremum is a t_i^j, (Proposition 2), but taken from a point j that may be different for each component.

– for the lexicographic ordering

$$t^1 \leq t^2 \quad \text{iff} \quad \exists i \text{ such that } t_i^1 < t_i^2 \text{ and } j < i \text{ imply } t_j^1 = t_j^2 \tag{1}$$

In other words, the order is obtained by comparing the leftmost coordinate on which the vectors differ. The ordering being now total, the supremum $\vee \{t^j, j \in J\}$ of any finite family J is one of its elements $t^j = (t_1^j, t_2^j, .. t_n^j)$ with all its components (Proposition 2).

The pilot structures take place between these two extremes. $T^{(n)}$ be a n-numerical space, and let a partition of $T^{(n)}$ into k complementary sub-spaces.

Definition 1. . *Let $\{T_s^{(n)}, 1 \leq s \leq k\}$ be a family of function lattices of finite parts that are totally ordered. Their direct product, endowed with the marginal ordering is called* Pilot lattice.

The pure marginal case is obtained for $k = n$, and the total ordering for $k = 1$. Except for marginal ordering, the supremum and infimum of any family always involve several components of some same elements of the family. Therefore they do not ensure us to completely preserve the initial data, but some of their components only. The next section gives examples of pilot lattices.

4 Polar Ordering in $\mathbb{R}^n$

We now develop the second theme pointed out in introduction. The idea is now to build a pilot lattice for cylindrical coordinates in $\mathbb{R}^n$, where the unit sphere be equipped with a significative total ordering [12].

Luminance and saturation. Colour polar representations are of cylindric type in $\mathbb{R}^3$. The main diagonal stands for the cylinder axis, and its basis is given by the chromatic disc. The colour point (r, g, b) is projected in x^l on the axis, and in x^s on the base, and the so-called polar representations consist in various quantizations of these two projections. The generalization to $\mathbb{R}^n$ is straightforward. Let $(x_1...x_n)$ be the multispectral coordinates of point x, and x^l (resp. x^s) be its projection on the main diagonal D (resp. on the plane Π orthogonal to D passing by the origin O). Introduce the mean $m = \frac{1}{n}\sum x_i$. Point x^l has all its coordinates equal to m. As for point x^s, its coordinates satisfy the equation $\sum x_i^s = 0$, because vectors Ox^l and Ox^s are orthogonal, and also the $n - 1$ equations

$$x_1 - x_1^s = x_2 - x_2^s = \cdots = x_n - x_n^s,$$

telling that x is projected parallel to the main diagonal. Hence,

$$x_i^s = \frac{1}{n}[(n-1)x_i - \sum_1^{n-1} x_i^s] = x_i - m \qquad 1 \leq i \leq n. \tag{2}$$

According to the chosen norm, such as L_1 or L_2, the "luminance" (resp. the "saturation") is given by the average of the absolute values, or the quadratic average of the coordinates of x^l (resp. x^s). In the "chromatic" plane Π, vector x^s is expressed in spherical coordinates, i.e. by one module (the saturation) and by $n - 2$ directions (the hues), since $n - 2$ angles $\alpha_1...\alpha_{n-2}$ are needed for locating a point on the unit sphere S_{n-1} in $n - 1$ dimensions.

Hues. We purpose to construct a pilot lattice from the product of three total orderings on luminance, saturation and hues. For the first two ones, which are energies, a usual numerical lattice is convenient. It remains to model the hues. Aptoula-Lefèvre ordering [3], [4] can be generalized as follows.

   Let $\{c^j, 1 \leq j \leq k\}$ be a finite family of poles on S_{n-1}, of coordinates c_i^j , $1 \leq i \leq n - 1$. Just as in two dimensions, we use the notation $c \div c^j$ to indicate the value of the acute angle cOc^j (i.e. $\leq \pi$) between point c and pole c^j. Take, on the unit sphere S_{n-1}, the Voronoï polygons w.r. to poles c^j, and assign for each point of the sphere the distance to its closest pole. In case of several equidistant poles, a priority rule allows to decide between them: conventionally c_1 prevails over c_2, which prevails over c_3 etc.. Finally, in case of two points equidistant from a same pole α, we iterate the process in the unit sphere S_{n-2} orthogonal axis $O\alpha$, and possibly in S_{n-3} etc., until we find an angle inequality. Then we say that c is closer to its pole than c' is closer to its own one, and we write $c \sqsupseteq c'$ when

$$either \quad \min_{j}\{c \div c^j\} < \min_{p}\{c' \div c^p\} \quad 1 \leq j, p \leq k, \tag{3}$$

$$or \quad \min_{j}\{c \div c^j\} = c \div c^{j_0} = \min_{p}\{c' \div c^p\} = c' \div c^{p_0} \quad \text{and} \quad j_0 > p_0, \tag{4}$$

$$or \quad \min_{j}\{c \div c^j\} = c \div c^{j_0} = \min_{p}\{c' \div c^p\} = c' \div c^{j_0} \quad \text{and iteration in } S_{n-2}. \tag{5}$$

The last condition may seem complicated, but it considerably simplifies in the useful case or $\mathbb{R}^4$. The three conditions classify points on the unit sphere according a total ordering based on angular interval, which can be replaced by any angular distance. The physical meaning here is the same as the resemblance to reference hues in the colour case.

Finally, we have in hand three finite total orderings: two numerical ones for luminance and saturation, plus ordering $\sqsupseteq$ for the hues. Propositions 1 and 3 apply, and provide the multivariate data with a pilot lattice.

Choice of the initial data. Formally speaking, this pilot lattice allows us to segment in a space with more than 100 dimensions as those that occur in satellite imagery [6]. However, the strong redundancy of the bands makes that processing cumbersome. Indeed, one rarely finds, in literature on remote sensing, image processing involving more than 4 principal components. On the other hand, the situation is now different from the colour case, as the first component is by construction the most important one. Therefore, it seems more appropriate to consider it as the main diagonal (the "grey tones").

Case of four components. We now develop in detail the four dimensional case, when the first components of a multi-spectral image are w, x, y, z. The first principal component w is chosen as the luminance, and describes the main diagonal in $\mathbb{R}^4$. The coordinates of the luminance and saturation vectors are given by

$$x^l = (w, 0, 0, 0) \quad \text{and} \quad x^s = (0, x, y, z)$$

with $x, y, z \geq 0$. Hyperplane Π is nothing but the space $\mathbb{R}^3$, and the polar cylindric coordinates in Π nothing but the usual spherical ones of $\mathbb{R}^3$. This context suggests to adopt the L_2 norm, since then the expressions of the saturation and of the two hues are those of the module ρ, the colatitude θ and the longitude ψ of the usual spherical coordinates in $\mathbb{R}^3$, i.e.

$$\rho = \sqrt{x^2 + y^2 + z^2} \tag{6}$$

$$\cos\theta = \frac{z}{\rho}, \quad \cos\psi = \frac{x}{\rho \sin\theta}, \quad \sin\psi = \frac{y}{\rho \sin\theta}. \tag{7}$$

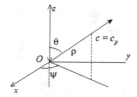

Fig. 1. Polar coordinates (ρ, θ, ψ) of point c, itself projection c_p of a point in R^4

Both colatitude θ and longitude ψ vary from 0 to $\frac{\pi}{2}$ since $x, y, z \geq 0$. These angles are depicted in Fig.1. Let $\{c^j, 1 \leq j \leq k\}$ be k poles on the unit sphere, ordered by decreasing priorities, and with coordinates $c^j = (x^j, y^j, z^j)$.

The angle cOc^j between point c and pole c^j is bounded by 0 and $\pi/2$. One obtains it from the scalar product of the two vectors c and c^j

$$\cos(cOc^j) = \langle c, c^j \rangle = \frac{xx^j + yy^j + zz^j}{\rho\rho^j}. \tag{8}$$

This relation allows us to re-formulate the first two relations (3) and (4) of the hue ordering in a simpler manner. We have $c \sqsupseteq c'$ when

$$\text{either } \min_j \{\langle c, c^j \rangle\} > \min_p \{\langle c', c^p \rangle\} \qquad\qquad 1 \leq j, p \leq k, \tag{9}$$

$$\text{or } \min_j \{\langle c, c^j \rangle\} = \langle c, c^{j_0} \rangle = \min_p \{\langle c', c^p \rangle\} = \langle c', c^{p_0} \rangle \quad \text{and } j_0 < p_0. \tag{10}$$

The third relation (5) corresponds to the case when c and c' are equidistant to their closest pole c^{j_0}. Then they can be ordered by increasing longitudes ψ. Finally, if $\psi(c) = \psi(c')$, what happens when c, c', and c^{j_0} lie in a same vertical plane of passing by $0z$, the ordering is completed by increasing colatitudes:

$$\min_j \{\langle c, c^j \rangle\} = \langle c, c^{j_0} \rangle = \min_p \{\langle c', c^p \rangle\} = \langle c', c^{j_0} \rangle \text{ and}$$
$$\text{either } \psi(c) < \psi(c') \tag{11}$$
$$\text{or } \psi(c) = \psi(c') \text{ and } \theta(c) < \theta(c').$$

The three relations (9) to (11) provide the unit sphere of $\mathbb{R}^3$ with a total ordering representing the chosen poles, and which is easy to compute. In the simpler case of a unique pole α it suffices to take it as north pole, and to take the sum $\theta + \psi$ for Voronoi distance (see below).

5   4-D Segmentations of "Pavie" Image

The image under study, kindly provided by J. Chanussot, represents the university of Pavia. It is composed of 103 bands from 0.43 to 0.86 micrometers. It has already been studied and classified [6]. The first principal component is depicted in Fig.3a, and the three next ones in Fig.2. Their variances are 64.84%, 28.41%, 5.14%, and 0.51% respectively.

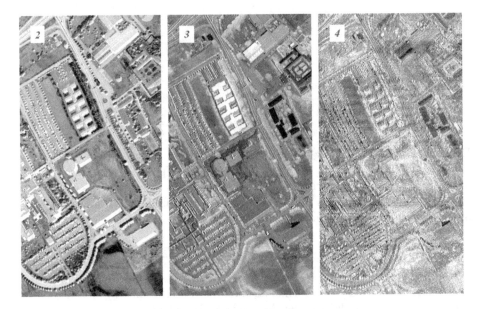

Fig. 2. Principal components n° 2, 3, and 4 of "Pavia" image

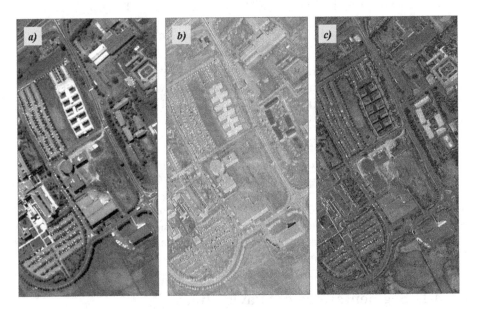

Fig. 3. a) First principal component of "Pavia" image; b) saturation $\mathbb{R}^4$; c) sum of the hues on the unit sphere of $\mathbb{R}^3$

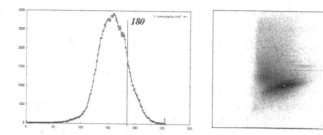

Fig. 4. Histogram of "Pavia" saturation ρ in $\mathbb{R}^4$ (left), and bidimensional histogram of the two hues θ and ψ(right)

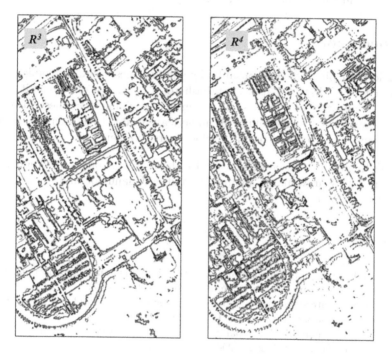

Fig. 5. Composite segmentations from the three ($\mathbb{R}^3$) or four ($\mathbb{R}^4$) first principal components of "Pavie"

The first component is the axis for polar cylindrical coordinates in $\mathbb{R}^4$. In the perpendicular $\mathbb{R}^3$ space, saturation is given by the vector module ρ of Rel.6, and depicted in Fig.3b. Fig.4, left, depicts its histogram. It is unimodal, and its threshold at 180, for separating the zones of more representative hue, versus luminance, was set from the images themselves. The 2-D histogram of the two hues θ and ψ (Fig.4, right) is sufficiently unimodal for extracting the single pole of $\theta = 159$ and $\psi = 162$, indicated in white on the histogram. By taking for distance on the unit sphere the sum of the distances according θ and ψ, with θ

priority, we establish a total ordering $\mathcal{O}_{hue}$ on the unit sphere. It results in a unique hue, represented in Fig.3c. The whole 4-D lattice is given by the product $\mathcal{T}_{lum} \otimes \mathcal{T}_{sat} \otimes \mathcal{T}_{hue}$ where $\mathcal{T}_{hue}$ is the lattice associated with $\mathcal{O}_{hue}$.

The composite segmentation is performed according to the technique already presented in [1]. The luminance of Fig. 3a is segmented by iterated jumps (jump=25) and merging of small particles fusion (area ≤ 5), Similarly the hue of Fig. 3c is segmented by jumps of 35, and merging 5. Saturation ρ of Fig. 3b is used as a local criterion for choosing between the partitions of the luminance and of the hue. The final composite partition is depicted in Fig.5. By comparing with the same technique applied to the first three components only, we observe that the fourth dimension allowed us to segment several supplementary details, in particular in the bottom of the image.

6 Conclusion

A technique for piloting some variables by another ones in multivariable lattices has been proposed. It led us to establish total orderings on the unit sphere that are significative in remote sensing. One can probably free oneself from the discrete assumption, by modelling all numerical variable by Lipschitz functions. Applications of the method to GIS problems are foreseen.

Acknowledgements. The author wish to thank Ch. Ronse, J. Angulo and S. Lefèvre for their valuable comments, and J. Chanussot and Y. Tarabalka for the use of "Pavia" image.

References

1. Angulo, J., Serra, J.: Modeling and segmentation of colour images in polar representations. Image and Vision Computing 25, 475–495 (2007)
2. Angulo, J.: Quaternions colour representation and derived total orderings for morphological operators. Note interne CMM (May 2008)
3. Aptoula, E., Lefèvre, S.: A comparative study on multivariate morphology. Pattern Recognition 40(11), 2914–2929 (2007)
4. Aptoula, E.: Analyse d'images couleur par morphologie mathématique, application à la description, l'annotation et la recherche d'images Thèse d'informatique. Univ. Louis Pasteur, Strasbourg (July 10, 2008)
5. Chanussot, J., Lambert, P.: Total ordering based on space filling curves for multi-valued morphology. In: ISMM 1998, Norwell, MA, USA, pp. 51–58. Kluwer Academic Publishers, Dordrecht (1998)
6. Chanussot, J., Benediktsson, J.A., Fauvel, M.: Classification of Remote Sensing Images from Urban Areas Using a Fuzzy Possibilistic Model. IEEE Trans. Geosci. Remote Sens. 3(1), 40–44 (2006)
7. Ell, T.A., Sangwine, S.J.: Hypercomplex Fourier transform of color images. IEEE Trans. Image Processing 16(1), 22–35 (2007)
8. Evans, A., Gimenez, D.: Extending Connected Operators To Colour Images. In: Proc. Int. Conf. Image Proc. 2008, pp. 2184–2187 (2008)

9. Hanbury, A., Serra, J.: Morphological operators on the unit circle. IEEE Trans. Image Processing 10(12), 1842–1850 (2001)
10. Hanbury, A., Serra, J.: Colour Image Analysis in 3D-polar coordinates. In: Michaelis, B., Krell, G. (eds.) DAGM 2003. LNCS, vol. 2781, pp. 124–131. Springer, Heidelberg (2003)
11. Meyer, F.: Color image segmentation. In: Proc. 4th International Conference on Image Processing and its Applications 1992, pp. 303–306 (1992)
12. Serra, J.: Les treillis pilotes. Rapport Technique CMM-Ecole des Mines de Paris (January 2009)
13. Soille, P.: Constrainted connectivity for hierarchical image partitioning and simplification. IEEE Trans. PAMI 30(7), 1132–1145 (2008)
14. Talbot, H., Evans, C., Jones, R.: Complete ordering and multivariate mathematical morphology: Algorithms and applications. In: ISMM 1998, Norwell, MA, USA, pp. 27–34. Kluwer Academic Publishers, Dordrecht (1998)

Bipolar Fuzzy Mathematical Morphology for Spatial Reasoning

Isabelle Bloch

Télécom ParisTech (ENST), CNRS UMR 5141 LTCI, Paris, France
isabelle.bloch@enst.fr

Abstract. Bipolarity is an important feature of spatial information, involved in the expressions of preferences and constraints about spatial positioning, or in pairs of "opposite" spatial relations such as left and right. Imprecision should also be taken into account, and fuzzy sets is then an appropriate formalism. In this paper, we propose to handle such information based on mathematical morphology operators, extended to the case of bipolar fuzzy sets. The potential of this formalism for spatial reasoning is illustrated on a simple example in brain imaging.

Keywords: bipolar spatial information, fuzzy sets, spatial relations, bipolar fuzzy dilation and erosion, spatial reasoning.

1 Introduction

Spatial reasoning includes two main aspects: knowledge representation, concerning spatial entities and spatial relations, and reasoning on them. In this paper, we consider both imprecision and bipolarity of spatial information. Imprecision should be taken into account to represent vague knowledge about spatial positions or spatial relations (typically directional relations such as left and right) [1]. Bipolarity is important to distinguish between (i) positive information, which represents what is guaranteed to be possible, for instance because it has already been observed or experienced, and (ii) negative information, which represents what is impossible or forbidden, or surely false [2]. The intersection of the positive information and the negative information has to be empty in order to achieve consistency of the representation, and their union does not necessarily cover the whole underlying space, i.e. there is no direct duality between both types of information, leaving room for indifference or indetermination. In this paper, we consider bipolarity of spatial information and propose to handle it as bipolar fuzzy sets (Section 2) using mathematical morphology operators, extended to these representations (Section 3). Some additional properties are included with respect to our previous work [3,4]. We then present some examples of spatial reasoning in Section 4, as the main contribution of this paper.

2 Bipolar Fuzzy Sets

Let $\mathcal{S}$ be the underlying space (the spatial domain for spatial information processing), that is supposed to be bounded and finite here. A bipolar fuzzy set on

M.H.F. Wilkinson and J.B.T.M. Roerdink (Eds.): ISMM 2009, LNCS 5720, pp. 24–34, 2009.

$\mathcal{S}$ is defined by a pair of functions (μ, ν) such that $\forall x \in \mathcal{S}, \mu(x) + \nu(x) \leq 1$. For each point x, $\mu(x)$ defines the membership degree of x (positive information) and $\nu(x)$ the non-membership degree (negative information), while $1 - \mu(x) - \nu(x)$ encodes a degree of neutrality, indifference or indetermination. This formalism allows representing both bipolarity and fuzziness. Concerning semantics, it should be noted that a bipolar fuzzy set does not necessarily represent one physical object or spatial entity, but rather more complex information, potentially issued from different sources.

Let us consider the set $\mathcal{L}$ of pairs of numbers (a, b) in $[0, 1]$ such that $a + b \leq 1$. It is a complete lattice, for the partial order defined as [5]: $(a_1, b_1) \preceq (a_2, b_2)$ iff $a_1 \leq a_2$ and $b_1 \geq b_2$. The greatest element is $(1, 0)$ and the smallest element is $(0, 1)$. The supremum and infimum are respectively defined as: $(a_1, b_1) \vee (a_2, b_2) = (\max(a_1, a_2), \min(b_1, b_2))$, $(a_1, b_1) \wedge (a_2, b_2) = (\min(a_1, a_2), \max(b_1, b_2))$. The partial order $\preceq$ induces a partial order on the set of bipolar fuzzy sets:

$$(\mu_1, \nu_1) \preceq (\mu_2, \nu_2) \text{ iff } \forall x \in \mathcal{S}, \mu_1(x) \leq \mu_2(x) \text{ and } \nu_1(x) \geq \nu_2(x), \qquad (1)$$

and infimum and supremum are defined accordingly. It follows that, if $\mathcal{B}$ denotes the set of bipolar fuzzy sets on $\mathcal{S}$, $(\mathcal{B}, \preceq)$ is a complete lattice.

3   Bipolar Fuzzy Erosion and Dilation

Mathematical morphology on bipolar fuzzy sets has been first introduced in [3]. Once we have a complete lattice, as described in Section 2, it is easy to define algebraic dilations and erosions on this lattice, as operators that commute with the supremum and the infimum, respectively [3]. Their properties are derived from general properties of lattice operators. If we assume that $\mathcal{S}$ is an affine space (or at least a space on which translations can be defined), it is interesting, for dealing with spatial information, to consider morphological operations based on a structuring element. We detail the construction of such morphological operators, extending our preliminary work in [3,4].

Erosion. As for fuzzy sets [6], defining morphological erosions of bipolar fuzzy sets, using bipolar fuzzy structuring elements, requires to define a degree of inclusion between bipolar fuzzy sets. Such inclusion degrees have been proposed in the context of intuitionistic fuzzy sets [7], which are formally (although not semantically) equivalent to bipolar fuzzy sets. With our notations, a degree of inclusion of a bipolar fuzzy set (μ', ν') in another bipolar fuzzy set (μ, ν) is defined as:

$$\inf_{x \in \mathcal{S}} I((\mu'(x), \nu'(x)), (\mu(x), \nu(x))) \qquad (2)$$

where I is an implication operator. Two types of implication can be defined [7], one derived from a bipolar t-conorm $\perp$[1]:

---

[1] A bipolar disjunction is an operator D from $\mathcal{L} \times \mathcal{L}$ into $\mathcal{L}$ such that $D((1, 0), (1, 0)) = D((0, 1), (1, 0)) = D((1, 0), (0, 1)) = (1, 0)$, $D((0, 1), (0, 1)) = (0, 1)$ and that is increasing in both arguments. A bipolar t-conorm is a commutative and associative bipolar disjunction such that the smallest element of $\mathcal{L}$ is the unit element.

$$I_N((a_1, b_1), (a_2, b_2)) = \bot((b_1, a_1), (a_2, b_2)), \tag{3}$$

and one derived from a residuation principle from a bipolar t-norm $\top^2$:

$$I_R((a_1, b_1), (a_2, b_2)) = \sup\{(a_3, b_3) \in \mathcal{L} \mid \top((a_1, b_1), (a_3, b_3)) \preceq (a_2, b_2)\} \tag{4}$$

where $(a_i, b_i) \in \mathcal{L}$ and (b_i, a_i) is the standard negation of (a_i, b_i).

Two types of t-norms and t-conorms are considered in [7] and will be considered here as well:

1. operators called t-representable t-norms and t-conorms, which can be expressed using usual t-norms t and t-conorms T from the fuzzy sets theory [8]:

$$\top((a_1, b_1), (a_2, b_2)) = (t(a_1, a_2), T(b_1, b_2)), \tag{5}$$

$$\bot((a_1, b_1), (a_2, b_2)) = (T(a_1, a_2), t(b_1, b_2)). \tag{6}$$

2. Lukasiewicz operators, which are not t-representable:

$$\top_W((a_1, b_1), (a_2, b_2)) = (\max(0, a_1 + a_2 - 1), \min(1, b_1 + 1 - a_2, b_2 + 1 - a_1)), \tag{7}$$

$$\bot_W((a_1, b_1), (a_2, b_2)) = (\min(1, a_1 + 1 - b_2, a_2 + 1 - b_1), \max(0, b_1 + b_2 - 1)). \tag{8}$$

In these equations, the positive part of $\top_W$ is the usual Lukasiewicz t-norm of a_1 and a_2 (i.e. the positive parts of the input bipolar values). The negative part of $\bot_W$ is the usual Lukasiewicz t-norm of the negative parts (b_1 and b_2) of the input values. The two types of implication coincide for the Lukasiewicz operators [5].

Based on these concepts, we can now propose a definition for morphological erosion.

Definition 1. *Let (μ_B, ν_B) be a bipolar fuzzy structuring element (in $\mathcal{B}$). The erosion of any (μ, ν) in $\mathcal{B}$ by (μ_B, ν_B) is defined from an implication I as:*

$$\forall x \in \mathcal{S}, \varepsilon_{(\mu_B, \nu_B)}((\mu, \nu))(x) = \inf_{y \in \mathcal{S}} I((\mu_B(y - x), \nu_B(y - x)), (\mu(y), \nu(y))), \tag{9}$$

where $\mu_B(y - x)$ denotes the value at point y of μ_B translated at x.

A similar approach has been used for intuitionistic fuzzy sets in [9], but with weaker properties (in particular an important property such as the commutativity of erosion with the conjunction may be lost).

---

[2] A bipolar conjunction is an operator C from $\mathcal{L} \times \mathcal{L}$ into $\mathcal{L}$ such that $C((0, 1), (0, 1)) = C((0, 1), (1, 0)) = C((1, 0), (0, 1)) = (0, 1)$, $C((1, 0), (1, 0)) = (1, 0)$ and that is increasing in both arguments. A bipolar t-norm is a commutative and associative bipolar conjunction such that the largest element of $\mathcal{L}$ is the unit element.

Morphological dilation of bipolar fuzzy sets. Dilation can be defined based on a duality principle or based on the adjunction property. Both approaches have been developed in the case of fuzzy sets, and the links between them and the conditions for their equivalence have been proved in [10,11]. Similarly we consider both approaches to define morphological dilation on $\mathcal{B}$.

Dilation by duality. The duality principle states that the dilation is equal to the complementation of the erosion, by the same structuring element (if it is symmetrical with respect to the origin of $\mathcal{S}$, otherwise its symmetrical is used), applied to the complementation of the original set. Applying this principle to bipolar fuzzy sets using a complementation c (typically the standard negation $c((a, b)) = (b, a)$) leads to the following definition of morphological bipolar dilation.

Definition 2. *Let (μ_B, ν_B) be a bipolar fuzzy structuring element. The dilation of any (μ, ν) in $\mathcal{B}$ by (μ_B, ν_B) is defined from erosion by duality as:*

$$\delta_{(\mu_B,\nu_B)}((\mu,\nu)) = c[\varepsilon_{(\mu_B,\nu_B)}(c((\mu,\nu)))]. \tag{10}$$

Dilation by adjunction. Let us now consider the adjunction principle, as in the general algebraic case. An adjunction property can also be expressed between a bipolar t-norm and the corresponding residual implication as follows:

$$\top((a_1,b_1),(a_3,b_3)) \preceq (a_2,b_2) \Leftrightarrow (a_3,b_3) \preceq I_R((a_1,b_1),(a_2,b_2)). \tag{11}$$

Definition 3. *Using a residual implication for the erosion for a bipolar t-norm $\top$, the bipolar fuzzy dilation, adjoint of the erosion, is defined as:*

$$\delta_{(\mu_B,\nu_B)}((\mu,\nu))(x) = \inf\{(\mu',\nu')(x) \mid (\mu,\nu)(x) \preceq \varepsilon_{(\mu_B,\nu_B)}((\mu',\nu'))(x)\}$$
$$= \sup_{y \in \mathcal{S}} \top((\mu_B(x-y), \nu_B(x-y)), (\mu(y), \nu(y))). \tag{12}$$

Links between both approaches. It is easy to show that the bipolar Lukasiewicz operators are adjoint, according to Equation 11. It has been shown that the adjoint operators are all derived from the Lukasiewicz operators, using a continuous bijective permutation on $[0, 1]$ [7]. Hence equivalence between both approaches can be achieved only for this class of operators. This result is similar to the one obtained for fuzzy mathematical morphology [10,11].

An illustrative example is shown in Figure 1.

Properties.

Proposition 1. *All definitions are consistent: they actually provide bipolar fuzzy sets of $\mathcal{B}$.*

Proposition 2. *In case the bipolar fuzzy sets are usual fuzzy sets (i.e. $\nu = 1 - \mu$ and $\nu_B = 1 - \mu_B$), the definitions lead to the usual definitions of fuzzy dilations and erosions (using classical Lukasiewicz t-norm and t-conorm for the definitions based on the Lukasiewicz operators). Hence they are also compatible with classical morphology in case μ and μ_B are crisp.*

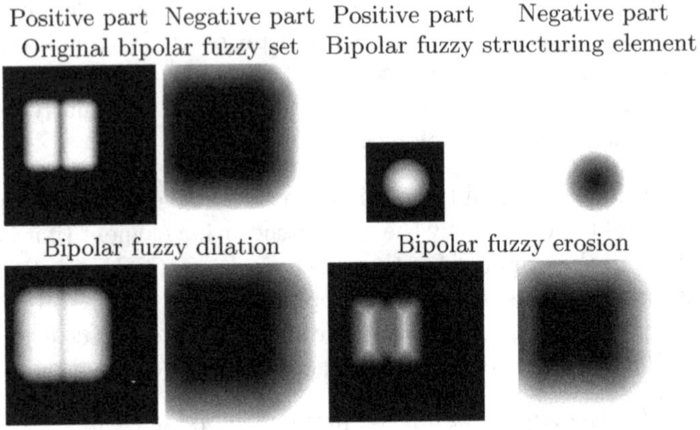

Positive part Negative part Positive part   Negative part
Original bipolar fuzzy set Bipolar fuzzy structuring element

Bipolar fuzzy dilation   Bipolar fuzzy erosion

Fig. 1. Bipolar fuzzy set and structuring element, dilation and erosion

Proposition 3. *The proposed definitions of bipolar fuzzy dilations and erosions commute respectively with the supremum and the infimum of the lattice $(\mathcal{B}, \preceq)$.*

Proposition 4. *The bipolar fuzzy dilation is extensive (i.e. $(\mu, \nu) \preceq \delta_{(\mu_B, \nu_B)}((\mu, \nu)))$ and the bipolar fuzzy erosion is anti-extensive (i.e. $\varepsilon_{(\mu_B, \nu_B)}((\mu, \nu)) \preceq (\mu, \nu))$ if and only if $(\mu_B, \nu_B)(0) = (1, 0)$, where 0 is the origin of the space $\mathcal{S}$ (i.e. the origin completely belongs to the structuring element, without any indetermination).*

Note that this condition is equivalent to the conditions on the structuring element found in classical and fuzzy morphology to have extensive dilations and anti-extensive erosions [12,6].

Proposition 5. *The dilation satisfies the following iterativity property:*

$$\delta_{(\mu_B, \nu_B)}(\delta_{(\mu_B', \nu_B')}((\mu, \nu))) = \delta_{(\delta_{\mu_B}(\mu_B'), 1 - \delta_{(1-\nu_B)}(1-\nu_B'))}((\mu, \nu)). \quad (13)$$

Proposition 6. *Conversely, if we want all classical properties of mathematical morphology to hold true, the bipolar conjunctions and disjunctions used to define intersection and inclusion in $\mathcal{B}$ have be be bipolar t-norms and t-conorms. If both duality and adjunction are required, then the only choice is bipolar Lukasiewicz operators (up to a continuous permutation on $[0, 1]$).*

This new result is very important, since it shows that the proposed definitions are the most general ones to have a satisfactory interpretation in terms of mathematical morphology.

Interpretations. Let us first consider the implication defined from a t-representable bipolar t-conorm. Then the erosion is written as:

$$\varepsilon_{(\mu_B, \nu_B)}((\mu, \nu))(x) = \inf_{y \in \mathcal{S}} \bot((\nu_B(y - x), \mu_B(y - x)), (\mu(y), \nu(y)))$$

$$= (\inf_{y \in \mathcal{S}} T((\nu_B(y - x), \mu(y)), \sup_{y \in \mathcal{S}} t(\mu_B(y - x), \nu(y))). \quad (14)$$

This resulting bipolar fuzzy set has a membership function which is exactly the fuzzy erosion of μ by the fuzzy structuring element $1 - \nu_B$, according to the original definitions in the fuzzy case [6]. The non-membership function is exactly the dilation of the fuzzy set ν by the fuzzy structuring element μ_B.

Let us now consider the derived dilation, based on the duality principle. Using the standard negation, it is written as:

$$\delta_{(\mu_B, \nu_B)}((\mu, \nu))(x) = (\sup_{y \in S} t(\mu_B(x - y), \mu(y)), \inf_{y \in S} T((\nu_B(x - y), \nu(y))). \quad (15)$$

The first term (membership function) is exactly the fuzzy dilation of μ by μ_B, while the second one (non-membership function) is the fuzzy erosion of ν by $1 - \nu_B$, according to the original definitions in the fuzzy case [6].

This observation has a nice interpretation, which well fits with intuition. Let (μ, ν) represent a spatial bipolar fuzzy set, where μ is a positive information for the location of an object for instance, and ν a negative information for this location. A bipolar structuring element can represent additional imprecision on the location, or additional possible locations. Dilating (μ, ν) by this bipolar structuring element amounts to dilate μ by μ_B, i.e. the positive region is extended by an amount represented by the positive information encoded in the structuring element. On the contrary, the negative information is eroded by the complement of the negative information encoded in the structuring element. This corresponds well to what would be intuitively expected in such situations. A similar interpretation can be provided for the bipolar fuzzy erosion.

Similarly, if we now consider the implication derived from the Lukasiewicz bipolar operators (Equations 7 and 8), it is easy to show that the negative part of the erosion is exactly the fuzzy dilation of ν (negative part of the input bipolar fuzzy set) with the structuring element μ_B (positive part of the bipolar fuzzy structuring element), using the Lukasiewicz t-norm. Similarly, the positive part of the dilation is the fuzzy dilation of μ (positive part of the input) by μ_B (positive part of the bipolar fuzzy structuring element), using the Lukasiewicz t-norm. Hence for both operators, the "dilation" part (i.e. negative part for the erosion and positive part for the dilation) has always a direct interpretation and is the same as the one obtained using t-representable operators, for t being the Lukasiewicz t-norm.

In the case the structuring element is non bipolar (i.e. $\forall x \in S, \nu_B(x) = 1 - \mu_B(x)$), then the "erosion" part has also a direct interpretation: the positive part of the erosion is the fuzzy erosion of μ by μ_B for the Lukasiewicz t-conorm; the negative part of the dilation is the erosion of ν by μ_B for the Lukasiewicz t-conorm.

4   Application to Spatial Reasoning

Mathematical morphology provides tools for spatial reasoning at several levels [13]. Its features allow representing objects or object properties, that we do not address here to concentrate rather on tools for representing spatial relations. The notion of structuring element captures the local spatial context, in a fuzzy and bipolar way here, which endows dilation and erosion with a low level spatial reasoning feature, as shown in the interpretation part of Section 3. This is then reinforced by the derived operators (opening, closing, gradient, conditional operations...), as introduced for bipolar fuzzy sets in [14]. At a more global level, several spatial relations between spatial entities can be expressed as morphological operations, in particular using dilations [1,13], leading to large scale spatial reasoning, based for instance on distances [15].

Let us provide a few examples where bipolarity occurs when dealing with spatial information, in image processing or for spatial reasoning applications: when assessing the position of an object in space, we may have positive information expressed as a set of possible places, and negative information expressed as a set of impossible or forbidden places (for instance because they are occupied by other objects). As another example, let us consider spatial relations. Human beings consider "left" and "right" as opposite relations. But this does not mean that one of them is the negation of the other one. The semantics of "opposite" captures a notion of symmetry (with respect to some axis or plane) rather than a strict complementation. In particular, there may be positions which are considered neither to the right nor to the left of some reference object, thus leaving room for some indifference or neutrality. This corresponds to the idea that the union of positive and negative information does not cover all the space. Similar considerations can be provided for other pairs of "opposite" relations, such as "close to" and "far from" for instance.

In this section, we illustrate a typical scenario showing the interest of bipolar representations of spatial relations and of morphological operations on these representations for spatial reasoning.

An example of a brain image is shown in Figure 2, with a few labeled structures of interest.

Let us first consider the right hemisphere (i.e. the non-pathological one). We consider the problem of defining a region of interest for the RPU, based on a known segmentation of RLV and RTH. An anatomical knowledge base or

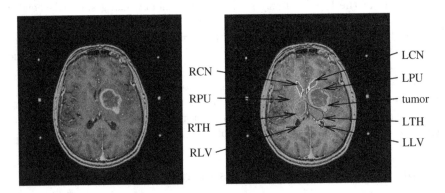

Fig. 2. A slice of a 3D MRI brain image, with a few structures: left and right lateral ventricles (LLV and RLV), caudate nuclei (LCN and RCN), putamen (LPU and RPU) and thalamus (LTH and RTH). A ring-shaped tumor is present in the left hemisphere (the usual "left is right" convention is adopted for the visualization).

ontology provides some information about the relative position of these structures [16,17]:

- directional information: the RPU is exterior (left on the image) of the union of RLV and RTH (positive information) and cannot be interior (negative information);
- distance information: the RPU is quite close to the union of RLV and RTH (positive information) and cannot be very far (negative information).

These pieces of information are represented in the image space based on morphological dilations using appropriate structuring elements [1] (representing the semantics of the relations, as displayed in Figure 3) and are illustrated in Figure 4. A bipolar fuzzy set modeling the direction information is defined as:

$$(\mu_{dir}, \nu_{dir}) = (\delta_{\nu_L}(\text{RLV} \cup \text{RTH}), \delta_{\nu_R}(\text{RLV} \cup \text{RTH})),$$

where ν_L and ν_R define the semantics of left and right, respectively. Similarly a bipolar fuzzy set modeling the distance information is defined as:

$$(\mu_{dist}, \nu_{dist}) = (\delta_{\nu_C}(\text{RLV} \cup \text{RTH}), 1 - \delta_{1-\nu_F}(\text{RLV} \cup \text{RTH})),$$

where ν_C and ν_F define the semantics of close and far, respectively. The neutral area between positive and negative information allows accounting for potential anatomical variability. The conjunctive fusion of the two types of relations is computed as a conjunction of the positive parts and a disjunction of the negative parts:

$$(\mu_{Fusion}, \nu_{Fusion}) = (\min(\mu_{dir}, \mu_{dist}), \max(\nu_{dir}, \nu_{dist})).$$

As shown in the illustrated example, the RPU is well included in the bipolar fuzzy region of interest which is obtained using this procedure. This region can then be efficiently used to drive a segmentation and recognition technique of the RPU.

Let us now consider the left hemisphere, where a ring-shaped tumor is present. The tumor induces a deformation effect which strongly changes the shape of the normal structures, but also their spatial relations, to a less extent. In particular the LPU is pushed away from the inter-hemispheric plane, and the LTH is pushed towards the posterior part of the brain and compressed. Applying the same procedure as for the right hemisphere does not lead to very satisfactory results in this case (see Figure 6). The default relations are here too strict and the resulting region of interest is not adequate: the LPU only satisfies with low

Fig. 3. Fuzzy structuring elements ν_L, ν_R, ν_C and ν_F, defining the semantics of left, right, close and far, respectively

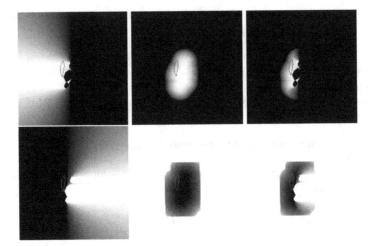

Fig. 4. Bipolar fuzzy representations of spatial relations with respect to RLV and RTH. Top: positive information, bottom: negative information. From left to right: directional relation, distance relation, conjunctive fusion. The contours of the RPU are displayed to show the position of this structure with respect to the region of interest.

Fig. 5. Bipolar fuzzy structuring element (μ_{var}, ν_{var})

degrees the positive part of the information, while it also slightly overlaps the negative part. In such cases, some relations (in particular metric ones) should be considered with care. This means that they should be more permissive, so as to include a larger area in the possible region, accounting for the deformation induced by the tumor. This can be easily modeled by a bipolar fuzzy dilation of the region of interest with a structuring element (μ_{var}, ν_{var}) (Figure 5), as shown in the last column of Figure 6:

$$(\mu'_{dist}, \nu'_{dist}) = \delta_{(\mu_{var}, \nu_{var})}(\mu_{dist}, \nu_{dist}),$$

where (μ_{dist}, ν_{dist}) is defined as for the other hemisphere. Now the obtained region is larger but includes the correct area. This bipolar dilation amounts to dilate the positive part and to erode the negative part, as explained in Section 3.

Let us finally consider another example, where we want to use symmetry information to derive a search region for a structure in one hemisphere, based on the segmentation obtained in the other hemisphere. As an illustrative example, we consider the thalamus, and assume that it has been segmented in the non pathological hemisphere (right). Its symmetrical with respect to the

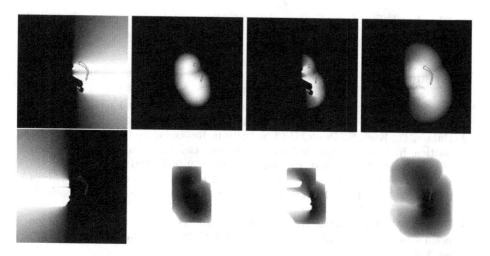

Fig. 6. Bipolar fuzzy representations of spatial relations with respect to LLV and LTH. From left to right: directional relation, distance relation, conjunctive fusion, Bipolar fuzzy dilation. First line: positive parts, second line: negative parts. The contours of the LPU are displayed to show the position of this structure.

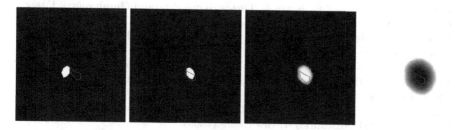

Fig. 7. RTH and its symmetrical, bipolar dilation defining an appropriate search region for the LTH (left: positive part, right: negative part)

inter-hemispheric plane should provide an adequate search region for the LTH in normal cases. Here this is not case, because of the deformation induced by the tumor (see Figure 7). Since the brain symmetry is approximate, a small deviation could be expected, but not as large as the one observed here. Here again a bipolar dilation allows defining a proper region, by taking into account both the deformation induced by the tumor and the imprecision in the symmetry.

5   Conclusion

In this paper, we have shown how a formal extension of mathematical morphology operators to the lattice of bipolar fuzzy sets may be used to represent two important features of spatial information, imprecision on the one hand and bipolarity on the other hand. This formalism can be useful for spatial reasoning, as illustrated on a typical scenario in brain imaging.

References

1. Bloch, I.: Fuzzy Spatial Relationships for Image Processing and Interpretation: A Review. Image and Vision Computing 23(2), 89–110 (2005)
2. Dubois, D., Kaci, S., Prade, H.: Bipolarity in Reasoning and Decision, an Introduction. In: International Conference on Information Processing and Management of Uncertainty, IPMU 2004, Perugia, Italy, pp. 959–966 (2004)
3. Bloch, I.: Dilation and Erosion of Spatial Bipolar Fuzzy Sets. In: Masulli, F., Mitra, S., Pasi, G. (eds.) WILF 2007. LNCS (LNAI), vol. 4578, pp. 385–393. Springer, Heidelberg (2007)
4. Bloch, I.: Mathematical Morphology on Bipolar Fuzzy Sets. In: International Symposium on Mathematical Morphology (ISMM 2007), Rio de Janeiro, Brazil, vol. 2, pp. 3–4 (2007)
5. Cornelis, C., Kerre, E.: Inclusion Measures in Intuitionistic Fuzzy Sets. In: Nielsen, T.D., Zhang, N.L. (eds.) ECSQARU 2003. LNCS (LNAI), vol. 2711, pp. 345–356. Springer, Heidelberg (2003)
6. Bloch, I., Maître, H.: Fuzzy Mathematical Morphologies: A Comparative Study. Pattern Recognition 28(9), 1341–1387 (1995)
7. Deschrijver, G., Cornelis, C., Kerre, E.: On the Representation of Intuitionistic Fuzzy t-Norms and t-Conorms. IEEE Transactions on Fuzzy Systems 12(1), 45–61 (2004)
8. Dubois, D., Prade, H.: Fuzzy Sets and Systems: Theory and Applications. Academic Press, New-York (1980)
9. Nachtegael, M., Sussner, P., Mélange, T., Kerre, E.: Some Aspects of Interval-Valued and Intuitionistic Fuzzy Mathematical Morphology. In: IPCV 2008 (2008)
10. Bloch, I.: Duality vs Adjunction and General Form for Fuzzy Mathematical Morphology. In: Bloch, I., Petrosino, A., Tettamanzi, A.G.B. (eds.) WILF 2005. LNCS, vol. 3849, pp. 354–361. Springer, Heidelberg (2005)
11. Bloch, I.: Duality vs. Adjunction for Fuzzy Mathematical Morphology and General Form of Fuzzy Erosions and Dilations. Fuzzy Sets and Systems 160, 1858–1867 (2009)
12. Serra, J.: Image Analysis and Mathematical Morphology. Academic Press, London (1982)
13. Bloch, I., Heijmans, H., Ronse, C.: Mathematical Morphology. In: Aiello, M., Pratt-Hartman, I., van Benthem, J. (eds.) Handbook of Spatial Logics, ch. 13, pp. 857–947. Springer, Heidelberg (2006)
14. Bloch, I.: A Contribution to the Representation and Manipulation of Fuzzy Bipolar Spatial Information: Geometry and Morphology. In: Workshop on Soft Methods in Statistical and Fuzzy Spatial Information, Toulouse, France, September 2008, pp. 7–25 (2008)
15. Bloch, I.: Geometry of Spatial Bipolar Fuzzy Sets based on Bipolar Fuzzy Numbers and Mathematical Morphology. In: Di Gesù, V., Pal, S.K., Petrosino, A. (eds.) WILF 2009. LNCS (LNAI), vol. 5571, pp. 237–245. Springer, Heidelberg (2009)
16. Waxman, S.G.: Correlative Neuroanatomy, 24th edn. McGraw-Hill, New York (2000)
17. Hudelot, C., Atif, J., Bloch, I.: Fuzzy Spatial Relation Ontology for Image Interpretation. Fuzzy Sets and Systems 159, 1929–1951 (2008)

An Axiomatic Approach to Hyperconnectivity

Michael H.F. Wilkinson

Institute for Mathematics and Computing Science, University of Groningen
P.O. Box 407, 9700 AK Groningen, The Netherlands

Abstract. In this paper the notion of hyperconnectivity, first put forward by Serra as an extension of the notion of connectivity is explored theoretically. Hyperconnectivity operators, which are the hyperconnected equivalents of connectivity openings are defined, which supports both hyperconnected reconstruction and attribute filters. The new axiomatics yield insight into the relationship between hyperconnectivity and structural morphology. The latter turns out to be a special case of the former, which means a continuum of filters between connected and structural exists, all of which falls into the category of hyperconnected filters.

1 Introduction

Connected filters are object-based morphological filters which allow edge preserving filtering based on a range of criteria [1, 2, 3]. Fig. 1 shows the difference between the structural opening and the opening by reconstruction [4]. In some cases, however, such strict edge preservation is not desirable, because thin structures can link up different entities in an image. For example, the thin stripes on the clothes in Fig. 1 link up the face area to other structures nearby. To circumvent some problems with the strictness of the edge preserving nature of these filters, and their inability to handle overlapping objects as separate entities, several solutions have been put forward [5, 6, 7]. One of these is hyperconnectivity, first proposed by Serra [6] and extended in [8]. Recently, hyperconnectivity has moved from a theoretical concept to a practical one, in particular in fuzzy connectivity [9], in fast reconstruction using reconstruction criteria [10], and in hyperconnected attribute filtering using k-flat zones (overlapping connected regions of with grey level total variations no more than k grey levels) [11]. The latter are useful for separation of galaxies from stars in astronomical imaging (see Fig. 2).

In this paper, a new axiomatics for hyperconnectivity is derived. We will first deal with some theoretical preliminaries. After this, connectivity and connectivity openings are treated. Then we replace the hyperconnectivity openings proposed in [8] by operators which return sets of hyperconnected components. It is then shown that hyperconnected counterparts of the connected attribute filters introduced by Breen and Jones [1] can only be constructed using the new framework. Finally, it is shown that any structural morphology can be seen as a special case of hyperconnected filters. This means that a large family of filters exist between the extremes of edge preserving connected filters, and structural filters. A variant of the work in [5] is shown to be part of that family.

M.H.F. Wilkinson and J.B.T.M. Roerdink (Eds.): ISMM 2009, LNCS 5720, pp. 35–46, 2009.
© Springer-Verlag Berlin Heidelberg 2009

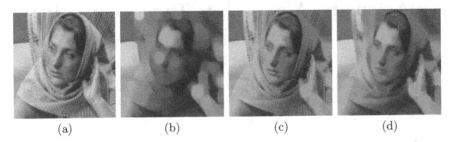

(a)   (b)   (c)   (d)

Fig. 1. Structural, connected, and hyperconnected filters:(a) original image f (b) opening with Euclidean disc of diameter 21 $g = \gamma_{21}f$; (c) connected reconstruction of f by g (d) hyperconnected reconstruction of f by g according to (32)

(a)   (b)   (c)

Fig. 2. Separating galaxies from stars: (a) spiral galaxy M81, original image, courtesy Giovanni Benintende; (b) stars suppressed by an area attribute filter with $2000 \leq$ area ≤ 240000; (c) k-flat hyperconnected variant of (b), showing improved suppression of stellar, and better retention of galactic detail

2   Theory

Let E denote some *finite*, universal, non-empty set, and $\mathcal{P}(E)$ the set of all subsets of E. $\mathcal{P}(E)$is also finite. A cover $\mathcal{A} = \{A_i\}$ of E is a subset of $\mathcal{P}(E)$ such that $\cup_i A_i = E$. A partition $\mathcal{A} = \{A_i\}$ of E is a cover such that $A_i \cap A_j = \emptyset$ for all $i \neq j$, and all A_i are non-empty. Covers of any $X \subseteq E$ can be defined likewise. Because covers and partitions are sets of subsets of E they are elements of $\mathcal{P}(\mathcal{P}(E))$. To avoid confusion, $\emptyset$ denotes the least element of $\mathcal{P}(E)$, and $\emptyset_{\mathcal{P}(E)}$ denotes least element of $\mathcal{P}(\mathcal{P}(E))$.

A cover $\mathcal{A}$, or indeed any element of $\mathcal{P}(\mathcal{P}(E))$ will be called *redundant* if there exists at least one pair of elements $A_i, A_j \in \mathcal{A}$ such that $A_i \subset A_j$. Obviously, partitions are non-redundant covers. We denote the set of all non-redundant subsets of $\mathcal{P}(E)$ as $\mathcal{N}(\mathcal{P}(E))$.

Any redundant cover can be reduced to a non-redundant cover by means of a *binary reduction operator* $\Phi_\subset$. This reduces any redundant subset $\mathcal{A} \subseteq \mathcal{P}(E)$ to the largest, non-redundant subset of $\mathcal{A}$.

Definition 1. *The binary reduction operator* $\Phi_C : \mathcal{P}(\mathcal{P}(E)) \rightarrow \mathcal{N}(\mathcal{P}(E))$ *is defined as*

$$\Phi_C(\mathcal{A}) = \mathcal{A} \setminus \{A_i \in \mathcal{A} \mid \exists A_j \in \mathcal{A} : A_i \subset A_j\}. \tag{1}$$

It is important to observe that if E is not finite, $\Phi_C(\mathcal{A})$ might be empty. Let $E = [0,1]$ and $\mathcal{A} = \{[0, 1-\frac{1}{n}] : n \in \mathbb{N}\}$. It can easily be verified that $\Phi_C(\mathcal{A}) = \emptyset$ in this case. This problem does not arise in finite, discrete images used in practice. Obviously, Φ_C has the following property

Proposition 1. *For any* $\mathcal{A} \in \mathcal{P}(\mathcal{P}(E))$

$$\bigcup \mathcal{A} = \bigcup \Phi_C(\mathcal{A}) \tag{2}$$

Proof. Because $\Phi_C(\mathcal{A}) \subseteq \mathcal{A}$ by definition, we only need to show that all elements of $\bigcup \mathcal{A}$ are contained in $\bigcup \Phi_C(\mathcal{A})$. Consider a point $x \in \bigcup \mathcal{A}$. This means that there is some $A_i \in \mathcal{A}$ such that $x \in A_i$. If $A_i \in \Phi_C(\mathcal{A})$, x is obviously contained in $\bigcup \Phi_C(\mathcal{A})$. If $A_i \notin \Phi_C(\mathcal{A})$, there must exist an $A_j \in \Phi_C(\mathcal{A})$ such that $A_i \subset A_j$, and x is also contained in $\bigcup \Phi_C(\mathcal{A})$.

We can define a partial order on $\mathcal{N}(\mathcal{P}(E))$ as

$$\mathcal{A} \preccurlyeq \mathcal{B} \equiv \forall A_i \in \mathcal{A} \, \exists \, B_j \in \mathcal{B} : A_i \subseteq B_j. \tag{3}$$

This is the same partial order as used for partitions in [12]. Suppose we have some elements $\mathcal{C}_i$ of $\mathcal{N}(\mathcal{P}(E))$, with $i \in I$, and I some index set, under $\preccurlyeq$ the infimum is

$$\bigwedge_{i \in I} \mathcal{C}_i = \Phi_C\left(\left\{\bigcap_{i \in I} D_i \mid D_i \in \mathcal{C}_i\right\}\right), \tag{4}$$

i.e. we first compute all sets which are intersections of one element from each of the sets $\mathcal{C}_i$. These are the maximal sets which are subset of some set in *each* of the $\mathcal{C}_i$. In general this set is redundant, so we map it back to $\mathcal{N}(\mathcal{P}(E))$ using Φ_C. If the $\mathcal{C}_i$ are partitions, (4) is equal to the infimum of partitions in [12]. The supremum is given by

$$\bigvee_{i \in I} \mathcal{C}_i = \Phi_C\left(\bigcup_{i \in I} \mathcal{C}_i\right), \tag{5}$$

i.e. we create a new cover by first combining all elements of all $\mathcal{C}_i$, and then removing any redundant ones. For any $D \in \mathcal{C}_i$ there exist an element $S \in \bigvee_{i \in I} \mathcal{C}_i$ such that $D \subseteq S$. Conversely, because for any $S \in \bigvee_{i \in I} \mathcal{C}_i$ there exists a $\mathcal{C}_i$ such that $S \in \mathcal{C}_i$. Therefore, we cannot replace any $S \in \bigvee_{i \in I} \mathcal{C}_i$ by some smaller set, without violating $\mathcal{C}_i \preccurlyeq \bigvee_{i \in I} \mathcal{C}_i$. Therefore (5) defines a supremum under $\preccurlyeq$. Within $\mathcal{N}(\mathcal{P}(E))$ the least element under $\preccurlyeq$ is $\emptyset_{\mathcal{P}(E)}$ and the maximal element is $\{E\}$. If $\mathcal{A}_1 \preccurlyeq \mathcal{A}_2$ for two partitions or covers we state that $\mathcal{A}_1$ is finer than $\mathcal{A}_2$, or, equivalently, $\mathcal{A}_2$ is coarser than $\mathcal{A}_1$.

Note that $\preccurlyeq$ is not a partial order on $\mathcal{P}(\mathcal{P}(E))$. Suppose I have some redundant $\mathcal{A} \in \mathcal{P}(\mathcal{P}(E))$, i.e., $A_i \subset A_j$ for some $A_i, A_j \in \mathcal{A}$. We then have

$$\mathcal{A} \preccurlyeq \mathcal{A} \setminus \{A_i\} \ \wedge \ \mathcal{A} \setminus \{A_i\} \preccurlyeq \mathcal{A} \tag{6}$$

but

$$\mathcal{A} \setminus \{A_i\} \neq \mathcal{A}. \tag{7}$$

2.1 Connectivity

Connectivity such as is used in morphological filtering is defined through the notion of connectivity classes or *connections* [13, 14, 6].

Definition 2. *A connection* $\mathcal{C} \subseteq \mathcal{P}(E)$ *is a set of sets with the following two properties:*

1. $\emptyset \in \mathcal{C}$ *and* $\{x\} \in \mathcal{C}$ *for all* $x \in E$
2. *for each family* $\{C_i\} \subset \mathcal{C}$, $\cap C_i \neq \emptyset$ *implies* $\cup C_i \in \mathcal{C}$.

Any set $C \in \mathcal{C}$ is said to be connected. Using such a notion of connectivity, any set $X \in \mathcal{P}(E)$ can be partitioned into connected components. These are the connected subsets of X of maximal extent, i.e. if $C \subseteq X$ and $C \in \mathcal{C}$ and there exists no set $D \in \mathcal{C}$ such that $C \subset D \subseteq X$, then C is a connected component of X. Let $\mathcal{C}_X$ be defined as

$$\mathcal{C}_X = \{C \in \mathcal{C} \mid C \subseteq X\}, \tag{8}$$

in other words $\mathcal{C}_X$ is the set of all connected subsets of X. $\mathcal{C}_X$ is obviously a cover of X because for every $x \in X$ $\{x\} \in \mathcal{C}_X$. Therefore every $x \in X$ is represented in the union of all elements of $\mathcal{C}_X$. The set of all connected components $\mathcal{C}_X^*$ of X is simply

$$\mathcal{C}_X^* = \Phi_\subset(\mathcal{C}_X). \tag{9}$$

It is well known that this constitutes a partition of X because any $C, D \in \mathcal{C}_X^*$ are either disjoint or equal.

Connected components can be accessed through *connectivity openings* [6]:

Definition 3. *The binary connectivity opening* Γ_x *of X at a point $x \in E$ is given by*

$$\Gamma_x(X) = \begin{cases} \bigcup\{C_i \in \mathcal{C} \mid x \in C_i \wedge C_i \subseteq X\} & \text{if } x \in X \\ \emptyset & \text{otherwise.} \end{cases} \tag{10}$$

In this definition the notion of maximum extent is derived by taking the union of all connected subsets of X containing x. It can readily be shown that this is equivalent to

$$\Gamma_x(X) = \begin{cases} C_i \in \mathcal{C}_X^* : x \in C_i & \text{if } x \in X \\ \emptyset & \text{otherwise.} \end{cases} \tag{11}$$

This equivalence stems from the fact that connected subsets of X which contain x have a non-empty intersection, and that their union is therefore connected.

An important theorem links connectivity openings to connections [6].

Theorem 1. *The datum of a connection $\mathcal{C}$ in $\mathcal{P}(E)$ is equivalent to the family* $\{\Gamma_x, x \in E\}$ *of openings on x such that:*

1. Γ_x *is an algebraic opening marked by* $x \in E$
2. *for all* $x \in E$, *we have* $\Gamma_x(\{x\}) = \{x\}$
3. *for all* $X \in \mathcal{P}(E)$ *and all* $x \in E$, *we have that* $x \notin X \Rightarrow \Gamma_x(X) = \emptyset$.
4. *for all* $X \in \mathcal{P}(E)$, $x, y \in E$, *if* $\Gamma_x(X) \cap \Gamma_y(X) \neq \emptyset \Rightarrow \Gamma_x(X) = \Gamma_y(X)$, *i.e.* $\Gamma_x(X)$ *and* $\Gamma_y(X)$ *are equal or disjoint.*

2.2   Hyperconnectivity

Hyperconnectivity is a generalization of connectivity, which generalizes the second condition of Definition 2 [6]. Instead of using a non-empty intersection, we can use any *overlap criterion* $\perp$ which is defined as follows.

Definition 4. *An overlap criterion in* $\mathcal{P}(E)$ *is a mapping* $\perp : \mathcal{P}(\mathcal{P}(E)) \rightarrow \{0, 1\}$ *such that* $\perp$ *is decreasing, i.e., for any* $\mathcal{A}, \mathcal{B} \subseteq \mathcal{P}(E)$

$$\mathcal{A} \subseteq \mathcal{B} \quad \Rightarrow \quad \perp(\mathcal{B}) \leq \perp(\mathcal{A}). \tag{12}$$

Any $\mathcal{A} \subseteq \mathcal{P}(E)$ for which $\perp(\mathcal{A}) = 1$ is said to be *overlapping*, otherwise $\mathcal{A}$ is non-overlapping. We can now define a *hyperconnectivity class* or *hyperconnection* as follows.

Definition 5. *A hyperconnection* $\mathcal{H} \subseteq \mathcal{P}(E)$ *is a set of sets with the following two properties:*

1. $\emptyset \in \mathcal{H}$ *and* $\{x\} \in \mathcal{H}$ *for all* $x \in E$
2. *for each family* $\{H_i\} \subset \mathcal{H}$, $\perp(\{H_i\}) = 1$ *implies* $\bigcup_i H_i \in \mathcal{H}$,

with $\perp$ *an overlap criterion such that* $\perp(\{H_i\}) \Rightarrow \cap_i H_i \neq \emptyset$.

Any set $H \in \mathcal{H}$ is said to be hyperconnected. Note that inserting the overlap criterion

$$\perp_\cap(\{H_i\}) = \begin{cases} 1 & \text{if } \bigcap_i H_i \neq \emptyset \\ 0 & \text{otherwise,} \end{cases} \tag{13}$$

into Definition 5 just yields a connection, showing that a connection is a special case of hyperconnection [6].

As can be seen from Definition 5, $\perp_\cap$ is the least strict overlap criterion to be used in a hyperconnection, i.e., $\perp(\{H_i\}) \leq \perp_\cap(\{H_i\})$ in general. For example we might require that the intersection contains a ball B_r of some diameter r for which $B_r \subseteq \bigcap_i H_i$. This leads to a "viscous" hyperconnectivity [10], which has been used to implement hyperconnected reconstruction shown in Fig. 1(d).

Like the notion of *connected* components for connection, we need to define the notion of *hyperconnected component*, which are hyperconnected subsets of X of maximal extent. In complete analogy with connected components we can first define the set $\mathcal{H}_X$ of all hyperconnected subsets of $X \in \mathcal{P}(E)$:

$$\mathcal{H}_X = \{H \in \mathcal{H} \mid H \subseteq X\}, \tag{14}$$

which is a cover of X for the same reasons as for $\mathcal{C}_X$. The set of hyperconnected components $\mathcal{H}_X^*$ is defined equivalently

$$\mathcal{H}_X^* = \Phi_\subset(\mathcal{H}_X). \tag{15}$$

Note that $\mathcal{H}_X^*$ is not necessarily a partition of X, because two hyperconnected components H_j, H_k may have a non-zero intersection, but $H_j \cup H_k$ need not be a member of $\mathcal{H}_X$ if $\perp(\{H_j, H_k\}) = 0$.

Braga-Neto and Goutsias [8] define a hyperconnectivity opening H_x as follows

Definition 6. *The binary hyperconnectivity opening H_x of X at point $x \in E$ is given by*

$$H_x(X) = \begin{cases} \bigcup \{H_i \in \mathcal{H}_X \mid x \in H_i\} & \text{if } x \in X \\ \emptyset & \text{otherwise.} \end{cases} \tag{16}$$

Unlike the connectivity opening Γ_x, which always returns a connected set, the hyperconnectivity opening H_x does not necessarily return a hyperconnected set, as pointed out by Braga-Neto and Goutsias in [8]. In this paper I propose a different approach.

Instead of the hyperconnectivity opening, we introduce the *hyperconnectivity operator* $\Upsilon_x : \mathcal{P}(E) \to \mathcal{P}(\mathcal{P}(E))$ which returns a set of hyperconnected sets. In the case of the connectivity opening in definition 3, we capture the notion of maximal extent by taking the union of all connected sets within X which contain the point x. This is not possible in the hyperconnected case, where we use the more explicit formulation using set inclusion used in the definition of hyperconnected components.

Definition 7. *The hyperconnectivity operator $\Upsilon_x : \mathcal{P}(E) \to \mathcal{P}(\mathcal{P}(E))$ associated with hyperconnection $\mathcal{H}$ is defined as*

$$\Upsilon_x(X) = \begin{cases} \Phi_{\subset}(\{H_i \in \mathcal{H}_X \mid x \in H_i\}), & \text{if } x \in X \\ \{\emptyset\} & \text{otherwise,} \end{cases} \tag{17}$$

In other words, Υ_x extracts the set of hyperconnected components of X containing x. It is obvious that the relationship between Υ_x and H_x is a simple one:

$$H_x(X) = \bigcup_{H_i \in \Upsilon_x(X)} H_i. \tag{18}$$

Fig. 3 illustrates the difference between the two operators.

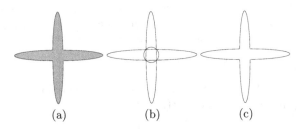

(a)   (b)   (c)

Fig. 3. Hyperconnectivity opening vs. hyperconnectivity operator: (a) binary image X; (b) outlines of hyperconnected components H_1, H_2, H_3 for some hypothetical hyperconnection $\mathcal{H}$; (c) outline of union of these hyperconnected components. Hyperconnectivity opening $H_x(X)$ returns the set outlined in (c) for any x in the intersection $\bigcap_{i=1}^{3} H_i$, whereas hyperconnectivity operator $\Upsilon_x(X)$ returns one or more of the sets outlined in (b).

We now define the properties a family of mappings $\Upsilon_x : \mathcal{P}(E) \rightarrow \mathcal{P}(\mathcal{P}(E))$ requires to define a hyperconnection. A few properties are "inherited" from connectivity openings:

1. $\Upsilon_x(H_i) = \{H_i\}$ for all $H_i \in \Upsilon_x(X)$ for all $X \in \mathcal{P}(E)$ and all $x \in E$;
2. $H_i \subseteq X$ for all $H_i \in \Upsilon_x(X)$ for all $X \in \mathcal{P}(E)$ and all $x \in X$;
3. for any $X, Y \in \mathcal{P}(E)$ we have $X \subseteq Y \Rightarrow \Upsilon_x(X) \preccurlyeq \Upsilon_x(Y)$ for all $x \in X$;
4. for all $x \in E$ we have $\Upsilon_x(\{x\}) = \{\{x\}\}$
5. for all $X \in \mathcal{P}(E)$, and all $x \in E$ we have $x \notin X \Rightarrow \Upsilon_x(X) = \{\emptyset\}$;
6. for any $H_i \in \Upsilon_x(X)$, $y \in H_i$ implies $H_i \in \Upsilon_y(X)$;
7. for all $x \in E$ and all $X \in \mathcal{P}(E)$, and any $H_i, H_j \in \Upsilon_x(X)$ we have $H_i \neq H_j \Rightarrow \perp(\{H_i, H_j\}) = 0$.

The first property ensures each $H_i \in \Upsilon_x(X)$ is hyperconnected according to the associated hyperconnection $\mathcal{H}$, and it contains x, it is the largest hyperconnected set contained in itself. Therefore, by definition 7, it is the only set $\Upsilon_x(H_i)$ should return. The second property ensures any hyperconnected component of X is a subset of X.

The third property is increasingness in the sense of (3), which can be shown as follows. Let $X \subseteq Y$. In this case any $H_i \in \Upsilon_x(X)$ is a subset of Y, through property 2. This means that either $H_i \in \Upsilon_x(Y)$, or there exists an $H_j \in \Upsilon_x(Y)$ such that $H_i \subset H_j$. Because all sets in $\Upsilon_x(X)$ have a set in $\Upsilon_x(Y)$ which is a superset or equal, the union of all members of $\Upsilon_x(X)$ is a subset of the union of all sets in $\Upsilon_x(Y)$.

The fourth property ensures that all singletons are members of $\mathcal{H}$, and the fifth that each hyperconnected component is marked *only* by its members.

The sixth property can be derived as follows. Because $H_i \in \Upsilon_x(X)$, there exists no hyperconnected set $H_j \subseteq X$, such that $H_i \subset H_j$. If $y \in H_i$ but $H_i \notin \Upsilon_y(X)$, this would imply that there is some $H_j \subseteq X$, such that $H_i \subset H_j$, leading to contradiction. This also ensures that each hyperconnected component is marked by *all* its members.

The seventh property is related, and states that no two different sets $H_i, H_j \in \Upsilon_x(X)$ can overlap in the sense of $\perp$. If they did, $H_i \cup H_j \in \mathcal{H}$ and $x \in H_i \cup H_j$. This means there exists a hyperconnected superset of both H_i and H_j containing x, and they should therefore not be members of $\Upsilon_x(X)$.

2.3   Relationship with Connectivity Openings

We will now investigate how the properties of hyperconnectivity operators relate to those of connectivity openings. Let $\#\Upsilon_x(X)$ denote the cardinality of $\Upsilon_x(X)$.

Proposition 2. *A hyperconnection $\mathcal{H}$ is a connection if and only if*

$$\#\Upsilon_x(X) = 1 \quad \text{for all } x \in E \text{ and all } X \in \mathcal{P}(E), \tag{19}$$

with Υ_x the hyperconnectivity operator associated with $\mathcal{H}$. In this case $H_x(X) = \bigcup_{H_i \in \Upsilon_x(X)} H_i$ is a connectivity opening.

Proof. If $\#\Upsilon_x(X) > 1$ for some $x \in E$ and some $X \in \mathcal{P}(E)$, $\mathcal{H}$ cannot be a connection because there are at least two hyperconnected components of X to which x belongs. Therefore, there are at least two sets $H_1, H_2 \in \mathcal{H}$ with non-empty intersection, but for which $H_1 \cup H_2 \notin \mathcal{H}$. This violates property 3 of Definition 2, and $\mathcal{H}$ is not a connection.

If $\#\Upsilon_x(X) = 1$ for all $x \in E$ and all $X \in \mathcal{P}(E)$ then the hyperconnected opening H_x is just a way of extracting the single element from $\Upsilon_x(X)$, i.e. $H_x(X) \in \Upsilon_x(X)$, implying $H_x(X) \in \mathcal{H}$ for all $x \in E$ and all $X \in \mathcal{P}(E)$. It has been shown that H_x is an algebraic opening [8], proving the first requirement of Theorem 1.

The second requirement of Theorem 1 follows from property 4, which states that $\Upsilon_x(\{x\}) = \{\{x\}\}$ for all hyperconnectivity operators, and therefore $H_x(\{x\}) = \{x\}$ for all $X \in \mathcal{P}(E)$. The third requirement derives from property 5, i.e. $\Upsilon_x(X) = \{\emptyset\}$ if $x \notin X$, which implies $H_x(X) = \emptyset$ for all $x \notin X$, for all $X \in \mathcal{P}(E)$.

The fourth requirement of Theorem 1 derives from property 6 above. If $y \in H_x(X)$ it follows from property 6 that $H_x(X) \in \Upsilon_y(X)$, and because $\#\Upsilon_y(X) = 1$, it follows that $H_x(X) = H_y(X)$. If $y \notin H_x(X)$, suppose that there exists some $z \in H_x(X) \cap H_y(X)$. For the previously given reasons, this implies $H_z(X) = H_x(X) = H_y(X)$, and therefore $y \in H_x(X)$, leading to contradiction. Therefore $y \notin H_x(X)$ implies $H_x(X) \cap H_y(X) = \emptyset$. Thus H_x is a connectivity opening.

Because $H_x \in \mathcal{H}$ for all $x \in E$ and $X \in \mathcal{P}(E)$, $\mathcal{H}$ is a connectivity class associated with the family of connectivity openings $\{H_x, x \in E\}$, proving Proposition 2.

2.4 Hyperconnected Filters

We will now turn to hyperconnected attribute filters, which were not considered by either Serra or Braga-Neto and Goutsias. Hyperconnected attribute filters can be defined in much the same way as connected attribute filters. We do this using a *trivial filter* $\Psi_\Lambda(H)$ which returns H if the criterion $\Lambda(H) = 1$ and $\emptyset$ otherwise. Let $\Psi_\Lambda(\mathcal{H}_X^*)$ be shorthand for the subset of all $H_i \in \mathcal{H}_X^*$ for which $\Lambda(H_j) = 1$.

Definition 8. *A hyperconnected attribute filter* $\Psi^\Lambda : \mathcal{P}(E) \to \mathcal{P}(E)$ *based on criterion* $\Lambda : \mathcal{H} \to \{0,1\}$ *is defined as*

$$\Psi^\Lambda(X) = \bigcup_{x \in X} \bigcup_{H_i \in \Upsilon_x(X)} \Psi_\Lambda(H_i) = \bigcup_{H_j \in \mathcal{H}_X^*} \Psi_\Lambda(H_j) = \bigcup_{H_k \in \Psi_\Lambda(\mathcal{H}_X^*)} H_k, \qquad (20)$$

We can define an alternative attribute filter Ψ_H^Λ using hyperconnectivity openings H_x as

$$\Psi_H^\Lambda = \bigcup_{x \in X} \Psi_\Lambda(H_x(X)) = \bigcup_{x \in X} \Psi_\Lambda\left(\bigcup \Upsilon_x(X)\right) \neq \bigcup_{x \in X} \bigcup_{H_i \in \Upsilon_x(X)} \Psi_\Lambda(H_i). \qquad (21)$$

Here we see a clear distinction between the framework using hyperconnected openings H_x versus the proposed framework using operators Υ_x, because Ψ_Λ does

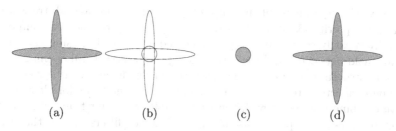

Fig. 4. Hyperconnected attribute filter with criterion Λ according to (22): (a) original images; (b) outlines of hyperconnected components; (c) union of trivial thinnings applied to hyperconnected components; (d) trivial thinning applied to union of hyperconnected components

not necessarily commute with set union. Consider the non-increasing criterion for 2-D images

$$\Lambda(H) = \begin{cases} 1 & \text{if } \Delta_x(H) = \Delta_y(H) \\ 0 & \text{otherwise,} \end{cases} \tag{22}$$

in which $\Delta_x(H)$ and $\Delta_y(H)$ are the maximal extents in x and y direction respectively. This requires that the minimum enclosing, axis-aligned rectangle is a square. Fig. 4 demonstrates the different outcomes of attribute filtering using hyperconnectivity openings and hyperconnectivity operators. The small circle in the centre is not seen as a separate entity by the H_x, whereas the "cross" preserved by Ψ_H^Λ is not hyperconnected.

3 Relationship to Structural Filters

In this section we will show the relationship with structural morphology. Let $S \subseteq E$ be an arbitrary structuring element centred at the origin $\mathbf{0}$, and $\mathcal{S}$ be the set of singletons in E, i.e.

$$\mathcal{S} = \{\{x\}|x \in E\}. \tag{23}$$

Furthermore, consider a finite chain $\mathcal{A} \subseteq \mathcal{P}(E)$, i.e. a totally ordered ordered family of sets under $\subseteq$ such that for an appropriate index set I, $A_i \subseteq A_j$ for any $i \leq j$. Obviously, if $\mathcal{A}$ is a chain, so is any subset of $\mathcal{A}$. Furthermore,

$$\bigcup_{i \in I} A_i = A_{\max I}, \tag{24}$$

provided E is finite. We can now show that the following set

$$\mathcal{H}_S = \{\emptyset\} \cup \mathcal{S}, \cup\{\{x\} \oplus S, x \in E\}, \tag{25}$$

is a hyperconnection, if provided with the overlap criterion

$$\perp_0(\mathcal{A}) = \begin{cases} 1 & \text{if } \mathcal{A} \text{ is a finite chain} \\ 0 & \text{otherwise} \end{cases} \tag{26}$$

In other words, $\mathcal{H}_S$ consist of the empty set, all singletons, and all translates of S. The overlap criterion states that only chains of hyperconnected components overlap. It is easily seen that a hyperconnected area opening using $\mathcal{H}_S$ with an area threshold between 1 and the area of S is just the structural opening with S. Thus, *any* structural opening using any structuring element can be represented as a hyperconnected area opening. By duality, the same holds for closings.

If we combine this result with the well-established result from Serra [6] that connected filters are a special case of hyperconnected filters, we see that hyperconnected filters form a family of filters in between the two extremes. An example of such a filter is inspired by [5, 15], but now based on hyperconnected filters. Let B be a ball centred on the origin, and $\mathcal{C}$ some connection on E. Consider

$$\mathcal{H}_B = \{\emptyset\} \cup \mathcal{S} \cup \{H \in \mathcal{P}(E) \mid \exists C \in \mathcal{C} : H = \delta_B C\}, \tag{27}$$

which is just the set of all dilates by B of all connected sets, augmented with the empty set and all singletons. This set is a hyperconnection with overlap criterion

$$\perp_B(\{A_i\}) = \bigcup_i (\epsilon_B A_i) \neq \emptyset. \tag{28}$$

This overlap criterion is true if and only if the intersection of all sets A_i eroded by B is non-empty. Equivalently, the intersection of A_i must contain at least one translate of B. In this hyperconnectivity, any image is constructed from a series of hyperconnected components which all lie within $\gamma_B f$ and a series of singletons which lie in $f - \gamma_B f$. Reconstruction from markers becomes

$$\rho_{\mathcal{H}_B}(f|g) = \delta_B \rho(\epsilon_B f | \epsilon_B g). \tag{29}$$

Thus, we erode the image and the marker, and then reconstruct all those parts of the eroded image which are marked by the eroded marker. This means those parts of f which overlap with g in the sense of $\perp_B$. After this, we dilate the result to reconstitute the hyperconnected components retained in the reconstruction. If marker g is obtained by an opening with some ball B_r, we can move (more-or-less) continuously from a structural opening, when $B_r \subseteq B$, through a "viscous" hyperconnected reconstruction ($B_0 \subset B \subset B_r$) to connected reconstruction when $B = B_0$, as in [5].

A drawback of this approach is that the end result of this is a subset of $\gamma_B f$ except when singletons are included in the result. This could seriously reduce the edge-preserving qualities of this filter. We can partly amend this by performing a geodesic dilation within f, similar to [5]. The geodesic dilation by a unit ball $\bar{\delta}_X^1$ within X is defined as

$$\bar{\delta}_X^1 Y = X \cap \delta^1 Y. \tag{30}$$

with δ^1 the dilation by a unit ball.

$$\mathcal{H}_B^X = \{\emptyset\} \cup \mathcal{S} \cup \{H \in \mathcal{P}(E) \mid \exists C \in \mathcal{C} : H = \bar{\delta}_X^1 \delta_B C\}. \tag{31}$$

This is a hyperconnection under the the overlap criterion from (28). In this case we simply perform a geodesic dilation by a unit ball after the processing, i.e.:

$$\rho_{\mathcal{H}_B^f}(f|g) = \bar{\delta}_f^1 \delta_B \rho(\epsilon_B f | \epsilon_B g), \tag{32}$$

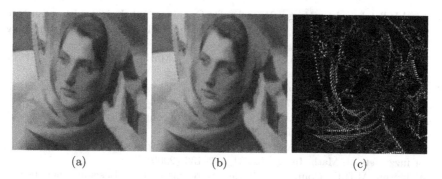

Fig. 5. Viscous hyperconnections: (a) reconstruction of Fig 4(a) by Fig 4(b) according to (29); (b) same according to (32); (c) difference (contrast stretched)

as put forward in [10]. The difference between reconstruction according to (29) and (32) is quite small, as shown in Fig. 5.

4   Conclusion

In this paper new axiomatics for hyperconnected filters have been introduced. It has been shown that this is needed to define hyperconnected attribute filters. Before these are of any practical use, efficient algorithms for these filters must be devised. Currently work is in progress to extend the work in [10] to attribute filters in general. A drawback of the formulation chosen is that it applies to finite images, and work is in progress to obtain a more general result. An important conclusion is the relationship to structural filters. This means that there is a (semi-)continuum of operators stretching from the edge-preserving connected filters to structural filters, all of which are hyperconnected. The relationship to path openings [16] and attribute-space connectivity [7] is explored in the next paper in this volume [17].

Acknowledgements. I would like to thank the anonymous reviewers for many valuable comments which have materially improved the content of this paper.

References

1. Breen, E.J., Jones, R.: Attribute openings, thinnings and granulometries. Comp. Vis. Image Understand. 64(3), 377–389 (1996)
2. Salembier, P., Oliveras, A., Garrido, L.: Anti-extensive connected operators for image and sequence processing. IEEE Trans. Image Proc. 7, 555–570 (1998)
3. Heijmans, H.J.A.M.: Connected morphological operators for binary images. Comp. Vis. Image Understand. 73, 99–120 (1999)
4. Klein, J.C.: Conception et réalisation d'une unité logique pour l'analyse quantitative d'images. PhD thesis, Nancy University, France (1976)

5. Terol-Villalobos, I.R., Vargas-Vázquez, D.: Openings and closings with reconstruction criteria: a study of a class of lower and upper levelings. J. Electron. Imaging 14(1), 013006 (2005)
6. Serra, J.: Connectivity on complete lattices. J. Math. Imag. Vis. 9(3), 231–251 (1998)
7. Wilkinson, M.H.F.: Attribute-space connectivity and connected filters. Image Vis. Comput. 25, 426–435 (2007)
8. Braga-Neto, U., Goutsias, J.: A theoretical tour of connectivity in image processing and analysis. J. Math. Imag. Vis. 19, 5–31 (2003)
9. Nempont, O., Atif, J., Angelini, E., Bloch, I.: A new fuzzy connectivity measure for fuzzy sets. J. Math. Imag. Vis. 34, 107–136 (2009)
10. Wilkinson, M.H.F.: Connected filtering by reconstruction: Basis and new results. In: Proc. Int. Conf. Image Proc. 2008, pp. 2180–2183 (2008)
11. Ouzounis, G.K.: Generalized Connected Morphological Operators for Robust Shape Extraction. PhD thesis, University of Groningen (2009) ISBN: 978-90-367-3698-5
12. Serra, J.: A lattice approach to image segmentation. J. Math. Imag. Vis. 24(1), 83–130 (2006)
13. Serra, J.: Mathematical morphology for Boolean lattices. In: Serra, J. (ed.) Image Analysis and Mathematical Morphology, II: Theoretical Advances, pp. 37–58. Academic Press, London (1988)
14. Ronse, C.: Set-theoretical algebraic approaches to connectivity in continuous or digital spaces. J. Math. Imag. Vis. 8, 41–58 (1998)
15. Serra, J.: Viscous lattices. J. Math. Imag. Vis. 22(2-3), 269–282 (2005)
16. Heijmans, H., Buckley, M., Talbot, H.: Path openings and closings. J. Math. Imag. Vis. 22, 107–119 (2005)
17. Wilkinson, M.H.F.: Hyperconnectivity, attribute-space connectivity and path-openings: Theoretical relationships. In: Wilkinson, M.H.F., Roerdink, J.B.T.M. (eds.) ISMM 2009. LNCS, vol. 5720, pp. 47–58. Springer, Heidelberg (2009)

Hyperconnectivity, Attribute-Space Connectivity and Path Openings: Theoretical Relationships

Michael H.F. Wilkinson

Institute for Mathematics and Computing Science, University of Groningen,
P.O. Box 407, 9700 AK Groningen, The Netherlands

Abstract. In this paper the relationship of hyperconnected filters with path openings and attribute-space connected filters is studied. Using a recently developed axiomatic framework based on hyperconnectivity operators, which are the hyperconnected equivalents of connectivity openings, it is shown that path openings are a special case of hyperconnected area openings. The new axiomatics also yield insight into the relationship between hyperconnectivity and attribute-space connectivity. It is shown any hyperconnectivity is an attribute-space connectivity, but that the reverse is not true.

1   Introduction

Connected filters are edge-preserving morphological filters [1,2,3]. All connected filters are based on a notion of connectivity [4,5], which cannot deal with overlapping of image components. To circumvent this inability to handle overlapping objects as separate entities, two solutions have been put forward: (i) hyperconnectivity [5,6], and attribute-space connectivity [7].

In the previous paper in these proceedings [8] a new axiomatic approach to hyperconnectivity has been presented. This axiomatics first of all leads to the definition of hyperconnected attribute filters. Furthermore, it is shown that structural filters can be seen as a special case of hyperconnected attribute filters. Given this, and the fact that Serra [5] already showed that connected filters are a special case of hyperconnected filters, the whole family of hyperconnected filters must bridge a gap between the two. As we move towards connected filters, they become more edge preserving, as we approach the "structural side" they become less so. In [8], it was also shown that reconstruction using reconstruction criteria [9] can also be interpreted as a hyperconnected filter. These were obvious candidates to study, because they are designed to be "tunably edge preserving" depending on the reconstruction criteria used. Another obvious candidate for inclusion in this bridge region are path openings [10]. In this paper it will be shown that they too can be seen as hyperconnected filters.

Another open question which can be answered in the new axiomatic framework is that of the relationship between hyperconnectivity and attribute-space

M.H.F. Wilkinson and J.B.T.M. Roerdink (Eds.): ISMM 2009, LNCS 5720, pp. 47–58, 2009.

connectivity, which also allows overlap between structures [7]. After some theoretical preliminaries, the axiomatic framework for hyperconnectivity will be presented briefly, without the proofs, for which the reader is referred to [8]. Then the link with path openings is studied. Finally, it will be shown that any hyperconnectivity can be defined as an attribute-space connectivity, but not the reverse.

2   Theory

The axiomatics presented in [8] are restricted to finite images, therefore in the rest of this paper E denotes some *finite*, universal, non-empty set, and $\mathcal{P}(E)$ the set of all subsets of E. A cover $\mathcal{A} = \{A_i\}$ of E is any subset of $\mathcal{P}(E)$ such that $\cup_i A_i = E$. A partition $\mathcal{A} = \{A_i\}$ of E is a cover such that $A_i \cap A_j = \emptyset$ for all $i \neq j$, and all A_i are non-empty. Covers of any $X \subseteq E$ can be defined in the same way. Because covers and partitions are sets of subsets of E they are elements of $\mathcal{P}(\mathcal{P}(E))$. To avoid confusion, $\emptyset$ denotes the least element of $\mathcal{P}(E)$, and $\emptyset_{\mathcal{P}(E)}$ denotes least element of $\mathcal{P}(\mathcal{P}(E))$. Any $\mathcal{A} \in \mathcal{P}(\mathcal{P}(E))$ will be called *redundant* if there exists at least one pair of elements $A_i, A_j \in \mathcal{A}$ such that $A_i \subset A_j$. We denote the set of all non-redundant subsets of $\mathcal{P}(E)$ as $\mathcal{N}(\mathcal{P}(E))$.

In [8], it was shown that $\mathcal{N}(\mathcal{P}(E))$ combined with the partial order $\preccurlyeq$

$$\mathcal{A} \preccurlyeq \mathcal{B} \equiv \forall A_i \in \mathcal{A} \, \exists \, B_j \in \mathcal{B} : A_i \subseteq B_j, \tag{1}$$

is a complete lattice with $\emptyset_{\mathcal{P}(E)}$ as the least element and $\{E\}$ as the greatest in $\mathcal{N}(\mathcal{P}(E))$.

To define hyperconnectivity operators we first must define *binary reduction operator* $\Phi_\subset$. This reduces any redundant subset $\mathcal{A} \subseteq \mathcal{P}(E)$ to the largest, non-redundant subset of $\mathcal{A}$ (in the finite case!).

Definition 1. *The binary reduction operator* $\Phi_\subset : \mathcal{P}(\mathcal{P}(E)) \to \mathcal{N}(\mathcal{P}(E))$ *is defined as*

$$\Phi_\subset(\mathcal{A}) = \mathcal{A} \setminus \{A_i \in \mathcal{A} \mid \exists A_j \in \mathcal{A} : A_i \subset A_j\}. \tag{2}$$

Thus, $\Phi_\subset(\mathcal{A})$ extracts the maximal elements of $\mathcal{A}$.

Both hyperconnectivity and attribute-space connectivity are based on connectivity. Central to connectivity is the notion of a connection [11, 4, 5, 12]:

Definition 2. *A connection $\mathcal{C} \subseteq \mathcal{P}(E)$ is a set of sets with the following two properties:*

1. *$\emptyset \in \mathcal{C}$ and $\{x\} \in \mathcal{C}$ for all $x \in E$*
2. *for each family $\{C_i\} \subset \mathcal{C}$, $\cap C_i \neq \emptyset$ implies $\cup C_i \in \mathcal{C}$.*

Any set $C \in \mathcal{C}$ is said to be connected. Using connections, connected components (connected subsets of maximal extent) of images can be defined.

2.1 Hyperconnectivity

Hyperconnectivity generalizes the second condition of Definition 2 [5]. It replaces non-empty intersection by any *overlap criterion* $\perp$ defined as follows.

Definition 3. *An overlap criterion in $\mathcal{P}(E)$ is a mapping $\perp : \mathcal{P}(\mathcal{P}(E)) \rightarrow \{0,1\}$ such that $\perp$ is decreasing, i.e., for any $\mathcal{A}, \mathcal{B} \subseteq \mathcal{P}(E)$*

$$\mathcal{A} \subseteq \mathcal{B} \quad \Rightarrow \quad \perp(\mathcal{B}) \leq \perp(\mathcal{A}). \tag{3}$$

Any $\mathcal{A} \subseteq \mathcal{P}(E)$ for which $\perp(\mathcal{A}) = 1$ is *overlapping*, otherwise $\mathcal{A}$ is non-overlapping. A *hyperconnection* is defined as:

Definition 4. *A hyperconnection $\mathcal{H} \subseteq \mathcal{P}(E)$ is a set of sets with the following two properties:*

1. *$\emptyset \in \mathcal{H}$ and $\{x\} \in \mathcal{H}$ for all $x \in E$*
2. *for each family $\{H_i\} \subset \mathcal{H}$, $\perp(\{H_i\}) = 1$ implies $\bigcup_i H_i \in \mathcal{H}$,*

with $\perp$ an overlap criterion such that $\perp(\{H_i\}) \Rightarrow \cap_i H_i \neq \emptyset$.

As in the case of connections, any set $H \in \mathcal{H}$ is said to be hyperconnected. Serra [5] already showed that connectivity is a special case of hyperconnectivity. For hyperconnectivity to be a useful concept, we need useful overlap criteria. An example explored in [13, 8] required that the intersection contains a ball B_r of some diameter r for which $B_r \subseteq \bigcap_i H_i$, yielding a hyperconnected reconstruction similar to viscous lattices [14].

We can now define *hyperconnected component*, which are hyperconnected subsets of X of maximal extent. As before, we define the set $\mathcal{H}_X$ of all hyperconnected subsets of $X \in \mathcal{P}(E)$:

$$\mathcal{H}_X = \{H \in \mathcal{H} \mid H \subseteq X\}, \tag{4}$$

which is a cover of X for the same reasons as for $\mathcal{C}_X$. The set of hyperconnected components $\mathcal{H}_X^*$ is defined equivalently

$$\mathcal{H}_X^* = \Phi_\subset(\mathcal{H}_X). \tag{5}$$

Note that $\mathcal{H}_X^*$ is a non-redundant cover of X, but not necessarily a partition. This is because two hyperconnected components H_j, H_k may have a non-zero intersection, but $H_j \cup H_k$ need not be a member of $\mathcal{H}_X$ if $\perp(\{H_j, H_k\}) = 0$.

In [8] *hyperconnectivity operators* $\Upsilon_x : \mathcal{P}(E) \rightarrow \mathcal{P}(\mathcal{P}(E))$ which return a set of hyperconnected sets were introduced. Instead of using set-union to capture maximal extent, as in [11], we use the reduction opening to retrieve the hyperconnected sets of maximal extent:

Definition 5. *The hyperconnectivity operator $\Upsilon_x : \mathcal{P}(E) \rightarrow \mathcal{P}(\mathcal{P}(E))$ associated with hyperconnection $\mathcal{H}$ is defined as*

$$\Upsilon_x(X) = \begin{cases} \Phi_\subset(\{H_i \in \mathcal{H}_X \mid x \in H_i\}), & \text{if } x \in X \\ \{\emptyset\} & \text{otherwise,} \end{cases} \tag{6}$$

Thus, Υ_x extracts the set of hyperconnected components of X containing x. In [8] the following properties required of a family of mappings $\Upsilon_x : \mathcal{P}(E) \to \mathcal{P}(\mathcal{P}(E))$ to define a hyperconnection were derived:

1. $\Upsilon_x(H_i) = \{H_i\}$ for all $H_i \in \Upsilon_x(X)$ for all $X \in \mathcal{P}(E)$ and all $x \in E$;
2. $H_i \subseteq X$ for all $H_i \in \Upsilon_x(X)$ for all $X \in \mathcal{P}(E)$ and all $x \in X$;
3. for any $X, Y \in \mathcal{P}(E)$ we have $X \subseteq Y \Rightarrow \Upsilon_x(X) \preccurlyeq \Upsilon_x(Y)$ for all $x \in X$;
4. for all $x \in E$ we have $\Upsilon_x(\{x\}) = \{\{x\}\}$
5. for all $X \in \mathcal{P}(E)$, and all $x \in E$ we have $x \notin X \Rightarrow \Upsilon_x(X) = \{\emptyset\}$;
6. for any $H_i \in \Upsilon_x(X)$, $y \in H_i$ implies $H_i \in \Upsilon_y(X)$;
7. for all $x \in E$ and all $X \in \mathcal{P}(E)$, and any $H_i, H_j \in \Upsilon_x(X)$ we have $H_i \neq H_j \Rightarrow \perp(\{H_i, H_j\}) = 0$.

For a full discussion of the meaning of these properties, the reader is referred to [8]. The most important property in this context is property 3, which states that Υ_x is increasing in the sense of (1).

3   Relationship to Path-Openings

Path openings [10,15] can be defined as unions of openings with a range of different structuring element, each consisting of a path of a given length. Their edge preserving properties lie in between classical structural morphology and connected filters. As such they are ideal candidates for the status of hyperconnected filter.

Following Heijmans et al. [10], E denotes the discrete grid on which a directed graph is defined by means of a binary adjacency relation '$\mapsto$', and $x \mapsto y$ means an *oriented* edge between x and y exists. Using this adjacency relation, which is neither reflexive nor symmetric in general, oriented paths can be defined. A path of length L is an $L + 1$-tuple $\mathbf{a} = (a_0, a_1, \ldots, a_L)$ such that $a_k \mapsto a_{k+1}$ for all $k = 0, 1, \ldots, L - 1$. The set of points contained in path $\mathbf{a}$ is denoted as $\mathcal{V}(\mathbf{a})$, i.e., $\mathcal{V}(\mathbf{a}) = \{a_0, a_1, \ldots, a_L\}$. Note that the adjacency relation must be chosen such that loop-backs and self-intersections are impossible. The set of all paths of length L is denoted Π_L. The set of all paths of all lengths of 0 (singletons) and higher will be denoted Π_*.

Any oriented path $\mathbf{a}$ can also be considered as a directed graph $(\mathcal{V}, \mathcal{E})$ with vertices given by $\mathcal{V}(\mathbf{a})$ and edges $\mathcal{E}(\mathbf{a}) = \{(a_k, a_{k+1})\}$ with $k = 0, 1, \ldots, L - 1$. If we define a *source* as a vertex with in-order of 0, and a *sink* as a vertex with an out-order of 0, we can define paths as connected, directed graphs which have exactly one source, exactly one sink, and for which all non-source vertices have an in-order of exactly one, and all non-sink vertices have an out-order of exactly one. For any path $\mathbf{a} \in \Pi_L$ we will have the convention that a_0 denotes the source, and a_L the sink.

The union of two graphs is simply the union of their vertices, and the union of the edges:

$$\mathbf{a} \cup \mathbf{b} = \Big((\mathcal{V}(\mathbf{a}) \cup \mathcal{V}(\mathbf{b}), (\mathcal{E}(\mathbf{a}) \cup \mathcal{E}(\mathbf{b}) \Big), \tag{7}$$

and likewise for intersection. Of course, the intersections or unions of two paths are not necessarily paths (though they remain directed graphs).

Now, let a *continuous sub-path* of length K of $\mathbf{a} \in \Pi_L$ be any K-tuple

$$\mathbf{b} = (b_0, b_1, \ldots, b_K)$$
$$= (a_k, a_{1+k}, \ldots, a_{K+k}) \tag{8}$$

for some value of $k \in \{0, 1, \ldots, L - K\}$. Obviously, any such $\mathbf{b} \in \Pi_K$. This equivalent to saying that a continuous sub-path is a connected subgraph of $\mathbf{a}$. This can be used to define the following relation between paths.

Definition 6. *The binary head-tail overlap* $\perp_{ht} : \Pi_* \times \Pi_* \to \{0, 1\}$ *returns 1 for any* $\mathbf{a} = (a_0, a_1, \ldots, a_L)$ *and* $\mathbf{b} \in (b_0, b_1, \ldots, a_K)$, *and any choice of* $K, L \in \mathbb{Z}^+$, *if* $\mathbf{a} \cap \mathbf{b}$ *is a non-empty, continuous sub-path of both* $\mathbf{a}$ *and* $\mathbf{b}$ *such that for some natural* $i \geq 0$

1. *for every* t *with* $0 \leq t \leq \min(L - i, K)$ *we have* $a_{t+i} = b_t$ *or*
2. *for every* t *with* $0 \leq t \leq \min(K - i, L)$ *we have* $b_{t+i} = a_t$

and 0 otherwise.

This means that either the tail of $\mathbf{a}$ overlaps the head of $\mathbf{b}$, or the head of $\mathbf{a}$ overlaps the tail of $\mathbf{b}$, or they are nested within each other, preserving the order.

Let $\mathbf{c} = \mathbf{a} \cup \mathbf{b}$ and $\mathbf{d} = \mathbf{a} \cap \mathbf{b}$. Consider the graph $\mathbf{c}$ in the case that $\perp_{ht}(\mathbf{a}, \mathbf{b}) = 1$. Note that only the in-order or out-order of the vertices in the intersection $\mathbf{d}$ can change. The condition that $\mathbf{a}$ and $\mathbf{b}$ is a non-empty, continuous sub-path of both $\mathbf{a}$ and $\mathbf{b}$ in Definition 6 means $\mathbf{d} \in \Pi_M$, for some $M \in \{0, 1, \ldots, \min(K, L)\}$. It also means that for all of the vertices $d_i \in \mathcal{V}(\mathbf{d})$, except for d_M, the outgoing edges are the same in both $\mathbf{a}$ and $\mathbf{b}$, and likewise for the incoming edges of all nodes $d_i \in \mathcal{V}(\mathbf{d})$, except for d_0. Thus, in the union $\mathbf{c}$, the out-order of the nodes contained in $\mathcal{V}(\mathbf{d}) \setminus \{d_M\}$ is exactly one. Likewise, in $\mathbf{c}$, all the nodes contained in $\mathcal{V}(\mathbf{d}) \setminus \{d_0\}$ must have an in-order of exactly one.

If condition 1 in Definition 6 holds, the source b_0 of $\mathbf{b}$ is in the intersection $\mathbf{d}$. If $i = 0$ both sources a_0 and b_0 are in $\mathcal{V}(\mathbf{d})$, thus $a_0 = b_0 = d_0$, and the in-order of this vertex in $\mathbf{c}$ is 0, i.e. it is a source. If $i > 0$ only $b_0 \in \mathcal{V}(\mathbf{d})$, i.e. $b_0 = d_0$, and element $a_{i-1} \in \mathcal{V}(\mathbf{a})$ exists such that $a_{i-1} \mapsto a_i = b_0$, and the in-order of the vertex corresponding to d_0 in the union $\mathbf{c}$ is 1. It also means that $\mathbf{c}$ has a single source, i.e., a_0. If $L - i < K$, it follows that $a_L = d_M$. Therefore, element b_{L-i+1} exists such that $a_L = b_{L-i} \mapsto b_{L-i+1}$, and the out-order of the vertex corresponding to d_M in the union $\mathbf{c}$ is 1, and b_K is the single sink for the union $\mathbf{c}$. If $L - i = K$, both a_L and b_K are members of $\mathcal{V}(\mathbf{d})$, so that $a_L = b_K = d_M$ and the out-order of this vertex in $\mathbf{c}$ is 0, i.e. it is a sink. Finally, if $L - i > K$, $\mathbf{b}$ is a continuous sub-path of $\mathbf{a}$ and their union $\mathbf{c} = \mathbf{a}$. In all cases, the union is an oriented path.

If the second condition of Definition 6 is true, the same reasoning holds with the roles of $\mathbf{a}$ and $\mathbf{b}$ interchanged. Thus, if $\perp_{ht}(\mathbf{a}, \mathbf{b}) = 1$, then $\mathbf{c} = \mathbf{a} \cup \mathbf{b}$ is an oriented path, because the union contains one source, one sink, and the in-order of all non-source vertices is one, and the out-order of all non-sink vertices is also one. Now consider the following overlap criterion for multiple paths $\mathcal{A} = \{\mathbf{a}_i\}$.

Definition 7. *The* head-tail overlap $\perp_{HT} : \mathcal{P}(\Pi_*) \to \{0,1\}$ *is defined as*

$$\perp_{HT}(\mathcal{A}) = \begin{cases} 1 & \text{if } \perp_{ht}(\mathbf{a}_i, \mathbf{a}_j) = 1 \text{ for all } \mathbf{a}_i, \mathbf{a}_j \in \mathcal{A}, \\ 0 & \text{otherwise}. \end{cases} \tag{9}$$

It is possible to show by induction that for any number of paths $\perp_{HT}(\{\mathbf{a}_i\}) = 1$ implies $\bigcup_i \mathbf{a}_i$ is a path. This means that given $\perp_{HT}$ as overlap criterion,

$$\mathcal{H}_\Pi = \{\emptyset\} \cup \Pi_* \tag{10}$$

is a hyperconnection. In the binary case a path opening of length L preserves all pixels $x \in X$ which belong to a path $\mathbf{a}$ of at least length L, for which $\mathcal{V}(\mathbf{a}) \subseteq X$. This is equivalent to performing a hyperconnected area opening using $\mathcal{H}_\Pi$ as hyperconnection.

4   Attribute-Space Connectivity

In [7] it was proposed to solve various problems in connectivity by transforming the binary image $X \subset E$ into a higher-dimensional *attribute space* $E \times A$ in which A is some space encoding the local properties or attributes of pixels in any image. To embed the image in $E \times A$ an operator $\Omega : \mathcal{P}(E) \to \mathcal{P}(E \times A)$ is used, i.e., $\Omega(X)$ is a binary image in $E \times A$. Typically $A \subseteq \mathbb{R}$ or $\mathbb{Z}$. The reverse operator $\omega : \mathcal{P}(E \times A) \to \mathcal{P}(E)$, projects $\Omega(X)$ back onto X. The requirements according to [7] are:

Definition 8. *An* attribute-space transform pair (Ω, ω) *from* $E \leftrightarrow E \times A$, *is a pair of operators such that:*

1. $\Omega : \mathcal{P}(E) \to \mathcal{P}(E \times A)$ *is a mapping such that for any* $X \in \mathcal{P}(E)$, *each point* $x \in X$ *has at least one corresponding point* $(x, a) \in \Omega(X)$, *with* $a \in A$,
2. $\Omega(\emptyset) = \emptyset$,
3. $\Omega(\{x\}) \in \mathcal{C}_{E \times A}$ *for all* $x \in E$,
4. $\omega : \mathcal{P}(E \times A) \to \mathcal{P}(E)$ *is a mapping such that for any* $Y \in \mathcal{P}(E \times A)$, *every* $(x, a) \in Y$ *is projected to* $x \in \omega(Y)$,
5. $\omega(\Omega(X)) = X$ *for all* $X \in \mathcal{P}(E)$,
6. ω *is increasing.*

Note that $\mathcal{C}_{E \times A}$ is the connection used in $E \times A$. Axiom four defines ω to be a projection, which is increasing anyway, so axiom 6 is redundant. Furthermore, even though $\omega(\Omega(X)) = X$ for all $X \in \mathcal{P}(E)$, $\Omega(\omega(Y)) = Y$ will not in general hold for all $Y \in \mathcal{P}(E \times A)$. Using the above we can define the notion of *attribute-space connection*.

Definition 9. *An* attribute-space connection $\mathcal{A} \subseteq \mathcal{P}(E)$ *on universal set* E *generated by an attribute-space transform pair* (Ω, ω) *and connection* $\mathcal{C}_{E \times A}$ *on* $E \times A$, *is defined as*

$$\mathcal{A}_\Omega = \{C \in \mathcal{P}(E) | \Omega(C) \in \mathcal{C}_{E \times A}\} \tag{11}$$

Properties 2 and 3 in Definition 8 mean that singletons and the empty set are members of $\mathcal{A}_\Omega$, as in the case of (hyper)connections. Note that only those elements of $\mathcal{C}_{E \times A}$ which are in the range $\mathcal{R}_\Omega$ of Ω correspond to an element of $\mathcal{A}_\Omega$, with

$$\mathcal{R}_\Omega = \{X \in \mathcal{P}(E \times A) | \exists Y \in \mathcal{P}(E) : \Omega(Y) = X\}. \tag{12}$$

For Definition 9 to be of practical use, we might want to replace axiom 6 in Definition 8 by a stronger one

6. $\omega(\Gamma_x(\Omega(X))) \in \mathcal{A}_\Omega$ for all $X \in \mathcal{P}(E)$ and all $x \in \mathcal{P}(E \times A)$.

This guarantees that the cover of X generated by the attribute-space connection consists only of members of the attribute-space connection. Not all attribute-space connections in [7] adhere to this. Attribute-space transform pairs adhering to this new axiom will be called *strong* attribute-space transform pairs. Using the above concepts Attribute-space connected filters can also be defined [7].

Definition 10. *An attribute-space connected filter* $\Psi^A : \mathcal{P}(E) \to \mathcal{P}(E)$ *is defined as*

$$\Psi^A(X) = \omega(\Psi(\Omega(X))) \tag{13}$$

with $X \in \mathcal{P}(E)$ *and* $\Psi : \mathcal{P}(E \times A) \to \mathcal{P}(E \times A)$ *a connected filter, and* (Ω, ω) *an attribute-space transform pair.*

Therefore attribute-space connected filters first map the image to a higher dimensional space, then apply a connected filter and project the result back. If Ψ is anti-extensive (or extensive), so is Ψ^A due to the increasingness of ω.

We will now develop a new, *strong* width space, i.e., one which yields an idempotent *cover* of the image domain, unlike the width space in [7]. The morphological skeleton $SK(X)$ of X with S.E. B is defined as

$$SK(X) = \bigcup_{n=0}^{N} S_n(X) \tag{14}$$

with

$$S_n(X) = X \ominus B_n - (X \ominus B_n) \circ B \tag{15}$$

with B_n the n-fold dilation of the origin $\{0\}$ with itself (i.e. $X \ominus B_0 = X$), and N denotes the largest integer such that $S_N(X) \neq \emptyset$. As is well known, any set X can be exactly reconstructed from its skeleton sets $S_n(X)$ by:

$$X = \bigcup_{n=0}^{N} S_n(X) \oplus B_n \tag{16}$$

The attribute-space transform Ω_{SK} is defined as

$$\Omega_{SK}(X) = \bigcup_{n=0}^{N} \left(S_n(X) \oplus B_n \right) \times \{n\} \tag{17}$$

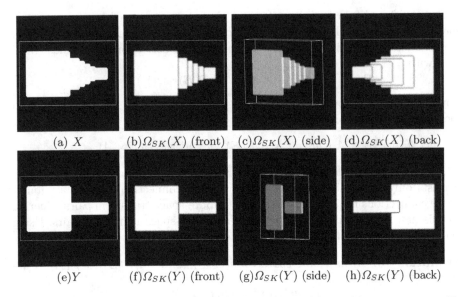

(a) X (b)$\Omega_{SK}(X)$ (front)   (c)$\Omega_{SK}(X)$ (side)   (d)$\Omega_{SK}(X)$ (back)

(e)Y (f)$\Omega_{SK}(Y)$ (front)   (g)$\Omega_{SK}(Y)$ (side)   (h)$\Omega_{SK}(Y)$ (back)

Fig. 1. Attribute-space transforms of two binary images: (a) and (e) binary images X and Y each containing a single (classical) connected component (b), (c), and (d) three iso-surface views of $\Omega_{SK}(X)$ showing a single connected component in $E \times A$; (f), (g) and (h) same for $\Omega_{SK}(Y)$, showing two connected components in $E \times A$

in which $\times$ represents the Cartesian product of the sets, i.e.

$$X \times \{a\} = \{(x,a) \in E \times A \mid x \in X \wedge a \in A\}. \tag{18}$$

It can immediately be seen that projecting $\Omega_{SK}(X)$ back onto E obtains X for all $X \in \mathcal{P}(E)$. An example is shown in Fig. 1. The top image X is considered as one component due to the slow change in width, whereas the abrupt change in width causes a split in Y. Because (14) and (16) show that any X is decomposed into balls of maximal diameter, it can readily be shown that any attribute-space connected component $C = \Gamma_x(\Omega_{SK}(X))$, is a set of maximal balls itself, when projected back onto E. Therefore, for any such $C \in \mathcal{C}_{E \times A}$, $\omega(C)$ will be decomposed into the same set of balls by Ω_{SK}, which will again be element of $\mathcal{C}_{E \times A}$.

In the case of a strong attribute-space transform pair, we can define a family of operators $\{\Theta_x, x \in E\}$, $\Theta_x : \mathcal{P}(E) \to \mathcal{P}(\mathcal{P}(E))$, which return the attribute-space connected components of X at any point $x \in E$.

Definition 11. *The attribute-space connectivity operator* $\Theta_x : \mathcal{P}(E) \to \mathcal{P}(\mathcal{P}(E))$ *associated with strong attribute-space transform pair* (Ω, ω) *is defined as*

$$\Theta_x(X) = \begin{cases} \{\omega(\Gamma_{x,a}(\Omega(X)))|a \in A : (x,a) \in \Omega(X)\} & \textit{if } x \in X \\ \{\emptyset\} & \textit{otherwise.} \end{cases} \tag{19}$$

This operator is the attribute-space connected equivalent of the hyperconnectivity operator Υ_x. In this case Θ_x extracts all connected components in the attribute space which contain at least one point (x, a).

4.1   Hyperconnectivity – Attribute-Space Connectivity Relationship

In [7] it was conjectured that. although hyperconnectivity and attribute-space connectivity share certain properties, they were different. However, no proof of this conjecture has been given since. Using the new axiomatics of hyperconnectivity, we can now prove the following theorem.

Theorem 1. *For every hyperconnection $\mathcal{H}$ on any finite, non-empty, universal set E there exists an attribute-space-transform pair $(\Omega_{\mathcal{H}}, \omega_{\mathcal{H}})$ and connection $\mathcal{C}_{E \times A}$ on attribute space $E \times A$ with associated attribute-space connection $\mathcal{A}_{\mathcal{H}}$ such that*

$$\mathcal{A}_{\mathcal{H}} = \mathcal{H}. \tag{20}$$

In other words, the set of all hyperconnections form a subset of the set of all attribute-space connections.

Proof. Let $\Lambda_{\mathcal{H}}$ be an index set over $\mathcal{H}$, and $\lambda_{\mathcal{H}} : \mathcal{H} \to \Lambda_{\mathcal{H}}$ be a function returning the index of any member of $\mathcal{H}$. We can now construct an attribute-space-transform pair $(\Omega_{\mathcal{H}}, \omega_{\mathcal{H}})$ between E and $E \times \Lambda_{\mathcal{H}}$ as follows.

$$\Omega_{\mathcal{H}}(X) = \bigcup_{x \in X} \bigcup_{H_i \in \Upsilon_x(X)} \left(H_i \times \{\lambda_{\mathcal{H}}(H_i)\} \right), \tag{21}$$

Thus $\Omega_{\mathcal{H}}$ computes each hyperconnected component H_i of X, and shifts it along the $\Lambda_{\mathcal{H}}$ dimension in $E \times \Lambda_{\mathcal{H}}$ to he location given by $\lambda_{\mathcal{H}}(H_i)$. The reverse operator $\omega_{\mathcal{H}} : \mathcal{P}(E \times \Lambda_{\mathcal{H}}) \to \mathcal{P}(E)$ is simply

$$\omega_{\mathcal{H}}(Y) = \{x \in E : \exists (x, a) \in Y\}. \tag{22}$$

It is easily verified that this is a attribute-space-transform pair according to Definition 8. We must also define an appropriate connection on $E \times \Lambda_{\mathcal{H}}$. First we define operator $\Omega_{\Lambda} : \mathcal{P}(E \times \Lambda_{\mathcal{H}}) \to \mathcal{P}(\Lambda_{\mathcal{H}})$

$$\Omega_{\Lambda}(Y) = \{a \in \Lambda_{\mathcal{H}} : \exists (x, a) \in Y\}. \tag{23}$$

This simply projects every point in Y to its location on the $\Lambda_{\mathcal{H}}$-dimension. Let $\mathcal{C}_{E \times \Lambda_{\mathcal{H}}}$ be defined as follows

$$\mathcal{C}_{E \times \Lambda_{\mathcal{H}}} = \{\emptyset\} \cup \{Y \in E \times \Lambda_{\mathcal{H}} \mid \exists a \in \Lambda_{\mathcal{H}} : \Omega_{\Lambda}(Y) = \{a\}\}. \tag{24}$$

Thus, all Y which are confined to a single plane in $E \times \Lambda_{\mathcal{H}}$ are considered connected. It is easy to show this is a connection: (i) $\emptyset \in \mathcal{C}_{E \times \Lambda_{\mathcal{H}}}$, and singletons $\{(x, a)\} \in \mathcal{C}_{E \times \Lambda_{\mathcal{H}}}$ for every $(x, a) \in E \times \Lambda_{\mathcal{H}}$, and (ii) if we have any set $\{C_i\} \subset$

$\mathcal{C}_{E \times \Lambda_{\mathcal{H}}}$, the intersection $\bigcap_i C_i \neq \emptyset$ if and only if there exists a single $a \in \Lambda_{\mathcal{H}}$ such that

$$\Omega_\Lambda(C_i) = \{a\}, \quad \text{for all } C_i. \quad \Rightarrow \quad \Omega_\Lambda\left(\bigcup_i C_i\right) = \{a\}, \tag{25}$$

which means that the second property in Definition 2 holds as well. Thus attribute-space connection $\mathcal{A}_{\mathcal{H}}$ is given by

$$\mathcal{A}_{\mathcal{H}} = \{C \in \mathcal{P}(E) | \Omega_{\mathcal{H}}(C) \in \mathcal{C}_{E \times \Lambda_{\mathcal{H}}}\} \tag{26}$$

For any element $H_i \in \mathcal{H}$ it is obvious that

$$\Omega_{\mathcal{H}}(H_i) = H_i \times \{\lambda_{\mathcal{H}}(H_i)\}, \tag{27}$$

and therefore,

$$\Omega_\Lambda(\Omega_{\mathcal{H}}(H_i)) = \{\lambda_{\mathcal{H}}(H_i)\}. \tag{28}$$

Thus $\Omega_{\mathcal{H}}(H_i) \in \mathcal{C}_{E \times \Lambda_{\mathcal{H}}}$, and

$$X \in \mathcal{H} \Rightarrow X \in \mathcal{A}_{\mathcal{H}}. \tag{29}$$

For any set $X \notin \mathcal{H}$ the cardinality of the set $\mathcal{H}_X^*$ of its hyperconnected components must be larger than 1. Therefore, the cardinality of $\Omega_\Lambda(\Omega_{\mathcal{H}}(X)) > 1$, and $\Omega_{\mathcal{H}}(X) \notin \mathcal{C}_{E \times \Lambda_{\mathcal{H}}}$, proving

$$X \notin \mathcal{H} \Rightarrow X \notin \mathcal{A}_{\mathcal{H}}, \tag{30}$$

which together with (29) proves Theorem 1.

This proves hyperconnectivity to be a special case of attribute-space connectivity. The difference between hyperconnectivity and attribute-space connectivity is formulated in the following theorem.

Theorem 2. *There exist strong attribute-space connections which are not hyperconnections.*

Proof. We need only prove the existence of one case. It can readily be shown that Ω_{SK} yields a strong, width-based attribute-space connection which allows non-increasing decomposition of a binary image into attribute-space connected components, i.e., the corresponding attribute-space connectivity operators Θ_x^{SK} are non-increasing in the sense of (1). This is shown in Fig. 2. In this figure there are two binary images X and Y, with $X \subset Y$. Each is split into two attribute-space connected components (X_1, X_2) and (Y_1, Y_2) by Ω_{SK}. Let X_1 and Y_1 denote the narrower of the two components in each attribute space transform respectively. Obviously $\omega(Y_1) \subset \omega(X_1)$, and Θ_x^{SK} must be non-increasing. This attribute-space connection is therefore *not* a hyperconnection due to the requirement of increasingness of all Υ_x, proving Theorem 2.

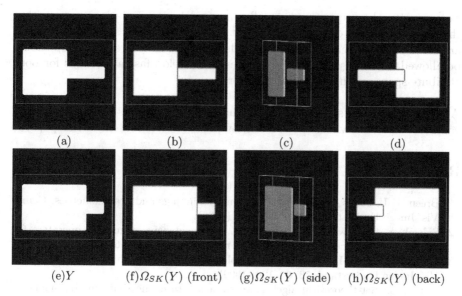

(a)   (b)   (c)   (d)

(e)Y (f)$\Omega_{SK}(Y)$ (front)   (g)$\Omega_{SK}(Y)$ (side)   (h)$\Omega_{SK}(Y)$ (back)

Fig. 2. Non-increasingness of attribute-space connected components: (a) and (e) binary images X and Y, with $X \subset Y$; (b), (c) and (d) three iso-surface views of $\Omega_{SK}(X)$; (f), (g) and (h) three iso-surface views of $\Omega_{SK}(Y)$. Both X and Y have two attribute-space connected components Clearly the narrower component Y_1 in $\Omega_{SK}(Y)$ is smaller than the narrower component X_1 in $\Omega_{SK}(X)$, demonstrating non-increasingness.

5   Conclusion

This paper extends the theoretical work in [8], and shows the utility of the new axiomatic framework in two important ways. First, it has been shown that the path openings fall into the general class of hyperconnected filters. This complements the work in [8], in which it was already shown that structural filters and reconstruction using reconstruction criteria were special cases of hyperconnectivity. The question that this raises is whether all increasing filters can be captured by hyperconnectivity in some way. Furthermore, which other, as yet unknown filters lie within the domain of hyperconnected filters.

The second important conclusion is that hyperconnectivity is a special case of attribute-space connectivity. This almost seems to immediately undermine the utility of hyperconnectivity. However, this result does not mean that hyperconnectivity is not important in its own right. A difference between the two approaches is that hyperconnectivity focuses on the type of overlap, whereas attribute-space connectivity focuses on the type of transform used to decompose the image into its constituent components. These complementary viewpoints may help to solve very different problems.

The work presented here is very much related to other work on connectivity theory, in particular partial connections and connective segmentation [12, 16]. Attribute-space connections and hyperconnections could open up a way in which we can relax the requirement usually imposed in segmentation, namely that the resulting image components (object and background) are disjoint. There are cases when no sensible, strict boundary can be drawn, and overlap should be allowed. Currently, work is in progress to develop fast algorithms for both attribute-space and hyperconnected filters.

Acknowledgements. I would like to thank the anonymous reviewers for their valuable comments which have materially improved the content of this paper.

References

1. Breen, E.J., Jones, R.: Attribute openings, thinnings and granulometries. Comp. Vis. Image Understand. 64(3), 377–389 (1996)
2. Salembier, P., Oliveras, A., Garrido, L.: Anti-extensive connected operators for image and sequence processing. IEEE Trans. Image Proc. 7, 555–570 (1998)
3. Heijmans, H.J.A.M.: Connected morphological operators for binary images. Comp. Vis. Image Understand. 73, 99–120 (1999)
4. Ronse, C.: Set-theoretical algebraic approaches to connectivity in continuous or digital spaces. J. Math. Imag. Vis. 8, 41–58 (1998)
5. Serra, J.: Connectivity on complete lattices. J. Math. Imag. Vis. 9(3), 231–251 (1998)
6. Braga-Neto, U., Goutsias, J.: A theoretical tour of connectivity in image processing and analysis. J. Math. Imag. Vis. 19, 5–31 (2003)
7. Wilkinson, M.H.F.: Attribute-space connectivity and connected filters. Image Vis. Comput. 25, 426–435 (2007)
8. Wilkinson, M.H.F.: An axiomatic approach to hyperconnectivity. In: Wilkinson, M.H.F., Roerdink, J.B.T.M. (eds.) ISMM 2009. LNCS, vol. 5720, pp. 35–46. Springer, Heidelberg (2009)
9. Terol-Villalobos, I.R., Vargas-Vázquez, D.: Openings and closings with reconstruction criteria: a study of a class of lower and upper levelings. J. Electron. Imaging 14(1), 013006 (2005)
10. Heijmans, H., Buckley, M., Talbot, H.: Path openings and closings. J. Math. Imag. Vis. 22, 107–119 (2005)
11. Serra, J.: Mathematical morphology for Boolean lattices. In: Serra, J. (ed.) Image Analysis and Mathematical Morphology, II: Theoretical Advances, pp. 37–58. Academic Press, London (1988)
12. Ronse, C.: Partial partitions, partial connections and connective segmentation. J. Math. Imag. Vis. 32(2), 97–125 (2008)
13. Wilkinson, M.H.F.: Connected filtering by reconstruction: Basis and new results. In: Proc. Int. Conf. Image Proc. 2008, pp. 2180–2183 (2008)
14. Serra, J.: Viscous lattices. J. Math. Imag. Vis. 22(2-3), 269–282 (2005)
15. Talbot, H., Appleton, B.: Efficient complete and incomplete path openings and closings. Image Vis. Comput. 25(4), 416–425 (2007)
16. Serra, J.: A lattice approach to image segmentation. J. Math. Imag. Vis. 24(1), 83–130 (2006)

Constrained Connectivity
and Transition Regions

Pierre Soille[1] and Jacopo Grazzini[2]

[1] Institute for the Protection and the Security of the Citizen
Global Security and Crisis Management Unit
[2] Institute for Environment and Sustainability
Spatial Data Infrastructures Unit
European Commission, Joint Research Centre
T.P. 267, via Fermi 2749, I-21027 Ispra, Italy

Abstract. Constrained connectivity relations partition the image definition domain into maximal connected components complying to a series of input constraints such as local and global intensity variation thresholds. However, they lead to a stream of small transition regions in situations where the edge between two large homogeneous regions spans over several pixels (ramp discontinuity). In this paper, we analyse this behaviour and propose new definitions for the notions of transition pixels and regions. We then show that they provide a suitable basis for suppressing connected components originating from non ideal step edges.

1 Introduction

Owing to the natural one-to-one correspondence between the partitions of a set and the equivalence relations on it [1, p. 130] and given that connectivity relations are equivalence relations, image segmentation [2] based on logical predicates defined in terms of connectivity relations naturally leads to uniquely defined image partitions[1]. For example, the trivial connectivity relation stating that two pixels are connected if and only if they can be joined by an iso-intensity path breaks digital images into segments of uniform grey scale [4]. They are called plateaus in fuzzy digital topology [5] and flat zones in mathematical morphology [6]. In most cases, the equality of grey scale is a too strong homogeneity criterion so that it produces too many segments. Consequently, the resulting partition is *too fine*. A weaker connectivity relation consists in stating that two pixels of a grey tone image are connected if there exists a path of pixels linking them and such that the grey level difference along adjacent pixels of the path does not exceed a given threshold value. In this paper, we call this threshold value the local range parameter and denote it by α. Accordingly, we call the resulting connected components the α-connected components. This idea was introduced in image processing by Nagao, Matsuyama, and Ikeda in the late seventies [7] but was already known before in classification as the single linkage clustering method [8].

---

[1] Partial partitions relying on partial equivalence relations (i.e., symmetric and transitive relations that are not necessarily reflexive) are studied in [3].

M.H.F. Wilkinson and J.B.T.M. Roerdink (Eds.): ISMM 2009, LNCS 5720, pp. 59–69, 2009.
© Springer-Verlag Berlin Heidelberg 2009

α-connected components are called quasi- or λ-flat zones [9,10] in mathematical morphology. Although α-connected components often produce adequate image partitions, they fail to do so when distinct image objects (with variations of intensity between adjacent pixels not exceeding α) are separated by one or more transitions going in steps having a magnitude less than or equal to α. Indeed, in this case, these objects appear in the same α-connected component so that the resulting partition is *too coarse*. This problem is sometimes referred to as the 'chaining effect' of the single linkage clustering method [11]. A natural solution to this problem is to limit the difference between the maximum and minimum values of each connected component by introducing a second threshold value called hereafter global range parameter and denoted by ω. This led to the notion of constrained connectivity introduced in [12]. A more general framework where constraints are defined in terms of logical predicates is put forward in [13]. Constrained connectivity solves the chaining effect of α-connectivity because appropriate global constraints prohibit α-connected components to grow too much. However, it may create a series of small undesirable regions in situations where the edge between two homogeneous regions spans over several pixels in the form of a ramp discontinuity [14]. This motivated us to analyse this behaviour and propose appropriate definitions for the notions of transition pixels and regions. They provide a basis to suppress all connected components originating from non ideal step edges[2], see also preliminary results in [15].

Section 2 briefly recalls the notion of constrained connectivity expressed in terms of logical predicates. Transitions pixels and regions are studied in Sec. 3. Before concluding, experimental results are presented in Sec. 4.

2   Logical Predicate Connectivity

We define α-connectivity for multichannel images following [7] (see also [16] where $\alpha = (\alpha_1, \ldots, \alpha_m)$ is referred to as the differential threshold vector): if, for every channel j, the difference of the values between two adjacent pixels is less than or equal to α_j, then the two pixels belong to the same component. Equivalently, denoting by f_j a scalar image, two pixels p and q of a multichannel image $\mathbf{f} = (f_1, \ldots, f_m)$ are α-connected if there exists a path $\mathcal{P} = (p = p_1, \ldots, p_n = q)$, $n > 1$, such that $|f_j(p_i) - f_j(p_{i+1})| \leq \alpha_j$ for all $1 \leq i < n$ and for all $j \in \{1, \ldots, m\}$. By definition, a pixel is α-connected to itself. If necessary, a more general definition whereby the component-wise grey level difference is replaced by an arbitrary distance on vectors [10,17,18] can be considered. For instance, the Mahalanobis distance provides us with a scale-invariant distance taking into account the correlations between the different channels. Alternatively, one could define the α-connected component of a pixel by computing the intersection of the α_j-connected components obtained for each channel separately and then extract the connected component of this intersection that contains this pixel. In practice, because the connections between pixels may follow different paths in each channel, it may be necessary to consider the intersection of graphs rather

---

[2] An ideal step edge consists of a discontinuous jump from one value to another.

than just an intersection on the nodes of the connected components obtained for each channel. This will be detailed in a future paper.

To determine whether two vectors are ordered, we use the marginal ordering [19] and denote by $\preceq$ the resulting relation 'less than or equal to': $\mathbf{f}(p) \preceq \mathbf{f}(q) \Leftrightarrow \mathbf{f}_j(p) \leq \mathbf{f}_j(q)$ for all $j \in \{1, \ldots, m\}$. Nevertheless, the family of partitions into $\boldsymbol{\alpha}$-connected components, for varying $\boldsymbol{\alpha}$, is not totally ordered because of the absence of total ordering between the vectors $\boldsymbol{\alpha}_i$ such that $\boldsymbol{\alpha}_i \preceq \boldsymbol{\alpha}$. We solve this problem by authorising only local vector range parameters having the same range α for all channels (so that it comes down to a unique scalar value α). In practice, this is not a limitation because the individual channels of the input multichannel image can be transformed beforehand.

Recall that a logical predicate P returns true when its argument satisfies the predicate, false otherwise. In general, provided that a given predicate returns true on iso-intensity connected components, we may look for the largest α_i-connected component satisfying this predicate such that every α_j-connected component where $\alpha_j \leq \alpha_i$ also satisfies it. This leads to the following general definition when considering a series of n predicates returning true when applied to iso-intensity connected components:

$$(P_1, \ldots, P_n)\text{-CC}(p) = \tag{1}$$
$$\bigvee \left\{ \alpha_i\text{-CC}(p) \;\middle|\; P_k\left(\alpha_i\text{-CC}(p)\right) = \text{true for all } k \in \{1, \ldots, n\} \text{ and} \right.$$
$$\left. P_k\left(\alpha_j\text{-CC}(q)\right) = \text{true for all } j \leq i \text{ and all } q \in \alpha_i\text{-CC}(p) \right\}.$$

Note that in [13] the condition 'and all $q \in \alpha_i$-CC(p)' was missing. The first author proposed the additional condition following a counter-example provided to him by Christian Ronse. If all predicates P_k are decreasing[3], this procedure is equivalent to checking the predicates on the α_i-CC of p only, see Eq. 4 in [13]. Indeed, once they are satisfied by the α_i-CC of p, they will be satisfied by the α_j-CC of q, because the latter are smaller or equal to the former.

Examples of decreasing predicates are 'is the difference between the largest and smallest value of the α_i-connected component below a threshold value ω?' and 'is the α_i-connected component strongly[4] connected?'. The predicate 'is the variance of the intensity values of the α_i-connected component below a given threshold level σ?' is an example of non-decreasing predicate. Additional examples of decreasing and non-decreasing predicates (similar to attribute thinnings [20]) can be found in [12,13].

From a theoretical point of view, constrained connectivity is related to the framework of the lattice approach to image segmentation proposed by Serra [21] and further investigated by Ronse [3]. More precisely, in [21] and [3], one considers

---

[3] A (logical) predicate is said to be decreasing if and only if every subset of any set satisfying this predicate also satisfies it [13].

[4] A α-connected component is strongly connected if and only if all edges of the connected components have a weight less than or equal to α (the weight of an edge between two adjacent pixels of the CC is defined as the range of the intensity values of the nodes it links).

connective segmentation, that is, segmentation methods where a connective criterion is associated to the image so that the final segmentation consists of a partition of connected components. However, connective segmentation as per [21] and [3] does not constrain the connected components in the sense of [12]. Therefore, while constrained connectivity uses a connective criterion (the α-connectivity), not all connected components resulting from this criterion are allowed but only the largest ones satisfying a series of constraints. From a theoretical point of view, the extension of the connective segmentation paradigm in order to integrate the constrained connectivity paradigm is studied mathematically in [22]. In particular, it is shown that constrained connectivity fits the framework of open-overcondensations [23].

3   Transition Pixels and Regions

The constrained connectivity paradigm partitions the image definition domain into maximal connected components satisfying a series of constraints expressed in terms of logical predicates. In practice, large connected components are often surrounded by a stream of small, usually one pixel thick, connected components. Indeed, given the general low pass filtering nature of imaging systems, ideal step edges are actually transformed (blurred) into ramp discontinuities. In the worst case, the ramp is made of steps of equal intensity so that it is pulverised into as many connected components as the number of steps (unless the creation of a large connected component encompassing the whole ramp would not violate the input connectivity constraints). Ramp pixels are sometimes called stairs [24], transition regions [25], or simply transitions [26] because they establish a path between nearby bright and dark regions. They are analogous to the notion of samples of intermediate (transitional) characteristics lying between actual clusters, see for example [27]. They are also related to the concept of mixed pixels since the intensity of a ramp pixel can be viewed as a mixture of the intensity of the regions separated by the considered ramp.

A definition of transition regions using topological and size criteria is proposed in [28]. Methods based solely on an area criterion are investigated in [10,29]. However, small regions do not necessarily correspond to transition regions. In [25], transition regions are defined as one-pixel regions having neighbouring brighter and darker regions after a preprocessing stage based on alternating sequential filters by reconstruction. In [15], transition regions are defined as connected components disappearing with an elementary erosion and not containing any regional extremum. Here, we first introduce the notion of *transition pixel*. It is then used for determining whether a given connected component correspond to a *transition region*.

A pixel of a grey level image f is a *local* extremum if and only if all its neighbours have a value either greater or lower than that of the considered pixel. That is, in morphological terms, a pixel is a local extremum if and only if the minimum between the gradients by erosion ρ^ε and dilation ρ^δ of f at position p is equal to 0:

$$p, \text{ local extremum of } f \Leftrightarrow \left[\rho^\varepsilon(f) \wedge \rho^\delta(f)\right](p) = 0. \tag{2}$$

The local extremum map LEXTR of a grey level image f is simply obtained by thresholding the pointwise minimum of its gradients by erosion and dilation for all values equal to 0:

$$\text{LEXTR}(f) = T_{t=0}[\rho^{\varepsilon}(f) \wedge \rho^{\delta}(f)]. \tag{3}$$

The LEXTR map corresponds to the indicator function returning 1 for local extrema pixels and 0 otherwise. We define transition pixels of a grey level image f as those image pixels that are *not* local extrema:

$$p, \text{ transition pixel of } f \Leftrightarrow \left[\rho^{\varepsilon}(f) \wedge \rho^{\delta}(f)\right](p) \neq 0. \tag{4}$$

The value of the morphological gradient of a transition pixel indicates the largest intensity jump that occurs when crossing this pixel. It corresponds to the intensity difference between its highest and lowest neighbours. We call *transition map* the grey tone image obtained by setting each transition pixel to the value of this intensity difference:

$$[\text{TMAP}(\mathbf{f})](p) = \begin{cases} 0, & \text{if } p \in \text{LEXTR}^1(f), \\ [\rho(f)](p), & \text{otherwise,} \end{cases} \tag{5}$$

where ρ denotes the morphological gradient operator (i.e., sum of the gradients by erosion and dilation). Interestingly, for an ideal image with regions of constant intensity levels separated by ideal step edges (and assuming that one pixel thick regions must correspond to local extrema) the transition map is equal to zero everywhere. Note that this would not be the case if regional instead of local extrema would have been considered. Finally, a transition region is defined as a constrained connected component containing only transition pixels. We present hereafter formal definitions suitable for the processing of multichannel images.

The search for local extrema in multichannel images is not straightforward because there exists no total ordering between vectors of more than one dimension. To circumvent this problem, we adopt a marginal ordering procedure whereby each channel is processed separately. Hence, when considering a multichannel image $\mathbf{f} = (f_1, \ldots, f_m)$, we define the operator LEXTR$^{\Sigma}$ summing the outputs of the indicator function LEXTR applied to each channel f_j of the input image:

$$\text{LEXTR}^{\Sigma}(\mathbf{f}) = \sum_{j=1}^{j=m} \text{LEXTR}(f_j). \tag{6}$$

We then define the local extrema of order n as those pixels of the image that are local extrema in at least n channels of the input image. They are denoted by LEXTR^n:

$$\text{LEXTR}^n(\mathbf{f}) = \{p \mid [\text{LEXTR}^{\Sigma}(\mathbf{f})](p) \geq n\}. \tag{7}$$

In the following, we are interested in the local extrema of lowest order LEXTR^1. A pixel p is a transition pixel if and only if, in all channels of the input image, it has at least one lower and one higher neighbours:

$$p, \text{ transition pixel of } \mathbf{f} \Leftrightarrow p \notin \text{LEXTR}^1(\mathbf{f}) \Leftrightarrow \vee_{j=1}^{j=m} \text{LEXTR}(f_j) = 0. \tag{8}$$

That is, a pixel of a multichannel is a transition pixel if and only if it is a transition pixel in *each* individual channel.

The calculation of the maximal amplitude of the grey level difference between neighbours of each transition pixel and over all channels leads to the notion of transition map for multichannel image. Formally, it is denoted by TMAP and obtained by setting non transition pixels to 0 and transition pixels to the pointwise maximum of the morphological gradient computed for each channel:

$$[\text{TMAP}(\mathbf{f})](p) = \begin{cases} 0, & \text{if } p \in \text{LEXTR}^1(\mathbf{f}), \\ \vee_{j=1}^{j=m} [\rho(f_j)](p), & \text{otherwise.} \end{cases} \quad (9)$$

Finally, a connected component is deemed to be a transition region if it contains only transition pixels or, equivalently, if it does *not* contain any local extremum pixel appearing in any channel of the input image:

$$(P_1, \ldots, P_n)\text{-CC}(p) \text{ is a transition region}$$
$$\Leftrightarrow (P_1, \ldots, P_n)\text{-CC}(p) \cap \text{LEXTR}^1 = \emptyset. \quad (10)$$

This latter test can be achieved efficiently by performing the reconstruction of the labelled connected components using the complement of LEXTR^1 as seed pixels.

Once transition regions are detected, they can be removed from the partition so that the latter becomes a partial partition. The remaining non-transition regions are then expanded so as to cover again the whole image definition domain in order to obtain again a partition. This is illustrated on actual image data in the next section.

4   Experimental Results

A sample of a Landsat satellite image is displayed in Fig. 1a (only the true colour channels among the 6 available channels are considered in this experiment). The output of LEXTR^Σ is displayed in Fig. 1b. The grey level value of this image indicates how may times each pixel is a local extremum when considering each channel separately. It follows that the grey level values of Fig. 1b range from 0 (never a local extremum) to 3 (local extremum in all 3 channels). Figure 1c shows the output of LEXTR^1, i.e., the local extrema of order 1 of the colour image of Fig. 1a. It corresponds to the pixels of Fig. 1b having a value greater than 0.Transition pixels correspond to the white pixels of Fig. 1c. The transition map TMAP with transition pixels set to the pointwise maximum of the morphological gradient of each channel is displayed in Fig. 1d.

Figure 2a shows the constrained connected components of Fig. 1a using identical contrast values for the local $\boldsymbol{\alpha}$ and global $\boldsymbol{\omega}$ range vectors [12]: $\boldsymbol{\alpha} = \boldsymbol{\omega} = (32, 32, 32)$. The transition regions as per Eq. 10 are displayed in Fig. 2b. Once the transition regions have been detected, they are removed from the partition. The gaps of the resulting partial partition are filled thanks to a seeded region procedure [30] initiated by the non-transition regions. This procedure ensures that all transition regions are suppressed but at the cost of some arbitrary decisions in the

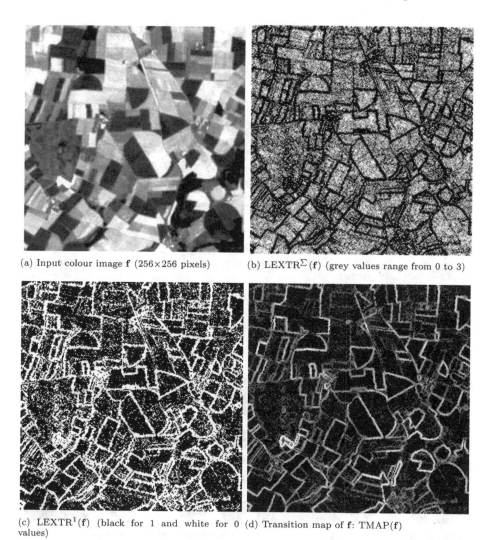

(a) Input colour image **f** (256×256 pixels)   (b) LEXTR$^\Sigma$(**f**) (grey values range from 0 to 3)

(c) LEXTR^1(**f**) (black for 1 and white for 0 values)   (d) Transition map of **f**: TMAP(**f**)

Fig. 1. Local extrema of a multichannel image (true colour channels of a Landsat image) and corresponding transition map

presence of transition regions whose smallest distance (in the spectral domain) to their neighbouring regions is obtained for more than one region. The resulting filtered partition is shown in Fig. 2c. The idea of filling the gaps of a partial partition using seeded region growing originates from [31]. In this latter case, the partial partition was obtained by removing all image iso-intensity connected components whose area is less than a given threshold value (this idea is further expanded in [29] to multichannel images and α-connected components). Figure 2d shows the simplification of Fig. 1a by setting each segment of Fig. 2c to the mean RGB value

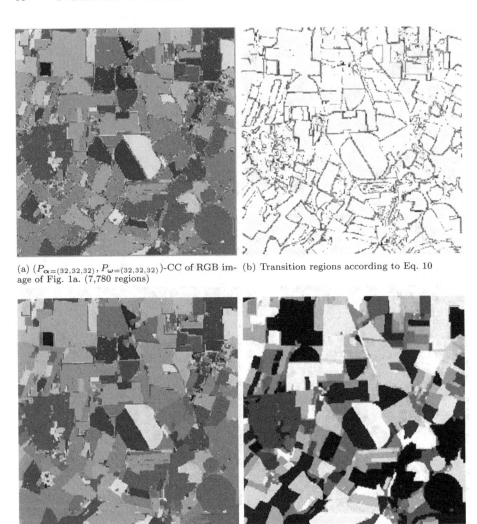

(a) $(P_{\alpha=(32,32,32)}, P_{\omega=(32,32,32)})$-CC of RGB image of Fig. 1a. (7,780 regions)

(b) Transition regions according to Eq. 10

(c) Filtered partition (1,669 regions)

(d) Resulting simplified RGB image

Fig. 2. From constrained connected components to transition regions and resulting edge preserving simplification (see input image in Fig. 1a)

of the input image pixels falling within this segment. The resulting image can be viewed as a sharpened image through ramp width zeroing based on region growing. Similarly, our approach can be linked to the discrete sharpening filters proposed in [32] and further developed in [33]. Indeed, the underlying ideas of the so-called Kramer-Bruckner filters consists in setting each pixel to its dilated or eroded value depending on which one is closer to its value. Consequently, for grey tone images, the sharpening transformation modifies only non-local extrema, i.e., transition pixels. More generally, a relationship can be established with the class of morphological image enhancement methods [34].

Note that transition regions also occur when generating a partition by automatically combining fine to coarse partition hierarchy using lifetime measurements [12]. Indeed, transition regions usually persist for a wide range of scale so that their lifetime is high. The concepts presented in this paper can also be used for removing these transitions regions.

5   Concluding Remarks

We have shown that the detection of constrained connected components corresponding to transition regions is of interest for segmentation and edge preserving simplification purposes. It is also useful for unsupervised classification techniques. Indeed, only those pixels not corresponding to transition regions should be considered when performing cluster analysis in a feature space to avoid overlap between clusters [35]. Similarly, the detection of transition pixels could be exploited by methods aiming at separating pure from mixed pixels. We have advocated the use of *local* extrema for marking non-transition regions. Local extrema instead of regional minima or maxima could form a basis for an interesting new type of jump connection [21,3]. Finally, we plan to analyse whether edge-preserving smoothing using a similarity measure in adaptive geodesic neighbourhoods [36,37] provides us with a valid pre-processing transformation to reduce the occurence of transition regions.

Acknowledgments

We wish to thank Christian Ronse and Jean Serra for their comments on a preliminary version of this paper as well as for email correspondence about connectivity.

References

1. Davey, B., Priestley, H.: Introduction to Lattices and Order, 2nd edn. Cambridge University Press, Cambridge (2002)
2. Zucker, S.: Region growing: childhood and adolescence. Computer Graphics and Image Processing 5, 382–399 (1976)
3. Ronse, C.: Partial partitions, partial connections and connective segmentation. Journal of Mathematical Imaging and Vision 32, 97–125 (2008)
4. Brice, C., Fennema, C.: Scene analysis using regions. Artificial Intelligence 1, 205–226 (1970)
5. Rosenfeld, A.: Fuzzy digital topology. Information and Control 40, 76–87 (1979)
6. Serra, J., Salembier, P.: Connected operators and pyramids. In: Dougherty, E., Gader, P., Serra, J. (eds.) Image Algebra and Morphological Image Processing IV. SPIE, vol. 2030, pp. 65–76 (1993)
7. Nagao, M., Matsuyama, T., Ikeda, Y.: Region extraction and shape analysis in aerial photographs. Computer Graphics and Image Processing 10, 195–223 (1979)
8. Gower, J., Ross, G.: Minimum spanning trees and single linkage cluster analysis. Applied Statistics 18, 54–64 (1969)

9. Meyer, F., Maragos, P.: Morphological scale-space representation with levelings. In: Nielsen, M., Johansen, P., Olsen, O., Weickert, J. (eds.) Scale-Space 1999. LNCS, vol. 1682, pp. 187–198. Springer, Heidelberg (1999)

10. Zanoguera, F., Meyer, F.: On the implementation of non-separable vector levelings. In: Talbot, H., Beare, R. (eds.) Proceedings of VIth International Symposium on Mathematical Morphology, Sydney, Australia, Commonwealth Scientific and Industrial Research Organisation, pp. 369–377 (2002)

11. Everitt, B., Landau, S., Leese, M.: Cluster Analysis, 4th edn. Oxford University Press, Oxford (2001)

12. Soille, P.: Constrained connectivity for hierarchical image partitioning and simplification. IEEE Transactions on Pattern Analysis and Machine Intelligence 30, 1132–1145 (2008)

13. Soille, P.: On genuine connectivity relations based on logical predicates. In: Proc. of 14th Int. Conf. on Image Analysis and Processing, Modena, Italy, pp. 487–492. IEEE Computer Society Press, Los Alamitos (2007)

14. Arora, H., Ahuja, N.: Analysis of ramp discontinuity model for multiscale image segmentation. In: Proc. of 18th International Conference on Pattern Recognition, pp. 99–103. IEEE, Los Alamitos (2006)

15. Soille, P., Grazzini, J.: Advances in constrained connectivity. In: Coeurjolly, D., Sivignon, I., Tougne, L., Dupond, F. (eds.) DGCI 2008. LNCS, vol. 4992, pp. 423–433. Springer, Heidelberg (2008)

16. Baraldi, A., Parmiggiani, F.: Single linkage region growing algorithms based on the vector degree of match. IEEE Transactions on Geoscience and Remote Sensing 34, 137–148 (1996)

17. Szczepanski, M., Smolka, B., Plataniotis, K., Venetsanopoulos, A.: On the distance function approach to color image enhancement. Discrete Applied Mathematics 139, 283–305 (2004)

18. Noyel, G., Angulo, J., Jeulin, D.: On distances, paths and connections for hyperspectral image segmentation. In: Banon, G., Barrera, J., Braga-Neto, U., Hirata, N. (eds.) Proceedings of International Symposium on Mathematical Morphology. São José dos Campos, Instituto Nacional de Pesquisas Espaciais (INPE), vol. 1, pp. 399–410 (2007)

19. Barnett, V.: The ordering of multivariate data (with discussion). Journal of the Royal Statistical Society (A) 139, 318–355 (1976)

20. Breen, E., Jones, R.: Attribute openings, thinnings, and granulometries. Computer Vision and Image Understanding 64, 377–389 (1996)

21. Serra, J.: A lattice approach to image segmentation. Journal of Mathematical Imaging and Vision 24, 83–130 (2006)

22. Ronse, C.: Block splitting operators on (partial) partitions for morphological image segmentation. Technical report, Université Louis Pasteur, Strasbourg, France (2009)

23. Ronse, C.: A lattice-theoretical morphological view on template extraction in images. Journal of Visual Communication and Image Representation 7, 273–295 (1996)

24. Crespo, J., Serra, J.: Morphological pyramids for image coding. In: Haskell, B., H.M. (eds.) Proc. Visual Communications and Image Processing. SPIE, vol. 2094, pp. 159–170 (1993)

25. Crespo, J., Schafer, R., Serra, J., Gratin, C., Meyer, F.: The flat zone approach: a general low-level region merging segmentation method. Signal Processing 62, 37–60 (1997)

26. Guimarães, S., Leite, N., Couprie, M., Araújo, A.: A directional and parametrized transition detection algorithm based on morphological residues. In: XV Brazilian Symposium on Computer Graphics and Image Processing, pp. 261–268 (2002)

27. Hodson, F., Sneath, P., Doran, J.: Some experiments in the numerical analysis of archaeological data. Biometrika 53, 311–324 (1966)

28. Strong, J., Rosenfeld, A.: A region coloring technique for scene analysis. Communication of the ACM 16, 237–246 (1973)

29. Brunner, D., Soille, P.: Iterative area filtering of multichannel images. Image and Vision Computing 25, 1352–1364 (2007)

30. Adams, R., Bischof, L.: Seeded region growing. IEEE Transactions on Pattern Analysis and Machine Intelligence 16, 641–647 (1994)

31. Soille, P.: Beyond self-duality in morphological image analysis. Image and Vision Computing 23, 249–257 (2005)

32. Kramer, H., Bruckner, J.: Iterations of a non-linear transformation for enhancement of digital images. Pattern Recognition 7, 53–58 (1975)

33. Meyer, F., Serra, J.: Contrasts and activity lattice. Signal Processing 16, 303–317 (1989)

34. Schavemaker, J., Reinders, M., Gerbrands, J., Backer, E.: Image sharpening by morphological filtering. Pattern Recognition 33, 997–1012 (2000)

35. Poggio, L., Soille, P.: Land cover classification with unsupervised clustering and hierarchical partitioning. In: 11th Int. Conf. of the International Federation of Classification Societies (IFCS): Classification as a Tool for Research, Dresden (2009), http://www.ifcs2009.de

36. Grazzini, J., Soille, P.: Adaptive morphological filtering using similarities based on geodesic time. In: Coeurjolly, D., Sivignon, I., Tougne, L., Dupond, F. (eds.) DGCI 2008. LNCS, vol. 4992, pp. 519–528. Springer, Heidelberg (2008)

37. Grazzini, J., Soille, P.: Edge-preserving smoothing using a similarity measure in adaptive geodesic neighbourhoods. Pattern Recognition 42, 2306–2316 (2009), http://dx.doi.org/10.1016/j.patcog.2008.11.004

Surface-Area-Based Attribute Filtering in 3D

Fred N. Kiwanuka[1,2], Georgios K. Ouzounis[3], and Michael H.F. Wilkinson[1]

[1] Institute for Mathematics and Computing Science, University of Groningen,
P.O. Box 407, 9700 AK Groningen, The Netherlands
[2] Faculty of Computing and Information Technology, Makerere University,
P.O. Box 7062 Kampala, Uganda
[3] School of Medicine, Democritus University of Thrace,
University General Hospital of Alexandroupoli, 68100 Alexandroupoli, Greece
F.N.Kiwanuka@rug.nl, gouzoun@med.duth.gr, m.h.f.wilkinson@rug.nl

Abstract. In this paper we describe a rotation-invariant attribute filter based on estimating the sphericity or roundness of objects by efficiently computing surface area and volume of connected components. The method is based on an efficient algorithm to compute all iso-surfaces of all nodes in a Max-Tree. With similar properties to moment-based attributes like sparseness, non-compactness, and elongation, our sphericity attribute can supplement these in finding blood-vessels in time-of-flight MR angiograms. We compare the method to a discrete surface area method based on adjacency, which has been used for urinary stone detection. Though the latter is faster, it is less accurate, and lacks rotation invariance.

1   Introduction

Connected filters [1,2] are members of the larger family of morphological operators, that act on the flat zone level of gray-scale images. Connected filters have the capacity to precisely identify and extract connected components in their entirety and without distorting their boundaries. This property, critical in many applications such as medical imaging, increases their popularity and makes them a suitable tool for problems in which accurate shape analysis is of importance. Connected components can either be removed or remain intact but new ones cannot emerge. Emergence and distortion of components is an existing problem in many other filtering methods.

Attribute filters are a subset of connected operators [3,4], that access connected components of threshold sets based on their attributes. Examples are attribute openings, closings, thinnings and thickenings [3,4,5]. Attribute filters often rely on either size or shape criteria such as volume, simplicity, complexity, moment of inertia, non-compactness, etc. [3,4,6,7]. Attributes are called *shape descriptors* provided they satisfy three key properties: translation, scale and rotation invariance [8]. Including scale invariance comes at the expense of increasingness [8], and rotation invariance can come at the cost of a high computational overhead, and reducing this is the topic of ongoing research. Exceptions are moment invariants [7,8], which are rotation invariant by definition, and can

M.H.F. Wilkinson and J.B.T.M. Roerdink (Eds.): ISMM 2009, LNCS 5720, pp. 70–81, 2009.

be computed efficiently. Description by moment invariants is limited, hence the need for new shape descriptors. In this paper we present a non-increasing 3D shape descriptor to measure how spherical (round) objects are, emphasizing rotation invariance. The *sphericity* attribute relies on accurate computation of surface area for which a new method is developed. It is compared to a simpler method based on the 3D equivalent of the city-block perimeter length. The performance of both methods is evaluated in terms of speed and accuracy.

Outline. The remainder of this article is organized as follows. A short introduction on attribute filters is given in Section 2 followed by a brief description of the Max-Tree algorithm in Section 3, used for computing the sphericity attribute. The latter is described in further detail in Section 4. Finally, Section 5 presents the results of the sphericity attribute used for 3D medical image enhancement, comparing it to other methods.

2   Theoretical Background

In the following, binary images X are subsets of some non-empty, universal set E. The set of all subsets of E is denoted $\mathcal{P}(E)$.

2.1   Attribute Filters

Attribute filters are based on connectivity openings. A connectivity opening $\Gamma_x(X)$ yields the connected component containing the point $x \in X$, and $\emptyset$ otherwise. A connectivity opening is anti-extensive i.e. $\Gamma_x(X) \subseteq X$, increasing i.e. $X \subseteq Y \Rightarrow \Gamma_x(X) \subseteq \Gamma_x(Y)$ and idempotent i.e. $\Gamma_x(\Gamma_x(X)) = \Gamma_x(X)$. Furthermore, for all, $X \subseteq E$, $x, y \in E$, $\Gamma_x(X)$ and $\Gamma_x(Y)$ are equal or disjoint. Attribute filters are defined using a family of connectivity openings, by imposing constraints on the connected components they return. Such constraints are expressed in the form of binary criteria which decide to accept or to reject components based on some attribute measure. Let $\Delta : \mathcal{P}(E) \rightarrow \{\text{false}, \text{true}\}$ be an attribute criterion; then Γ_Δ is a trivial opening returning the connected component C if $\Delta(C)$ is true and $\emptyset$ otherwise. Moreover, $\Gamma_\Delta(\emptyset) = \emptyset$. Attribute criteria can be represented as: $\Delta(C) = Attr(C) \geq \lambda$ where $Attr(C)$ is some real-value attribute of C and λ is an attribute threshold. A binary attribute opening Γ^Δ of a set X with an increasing criterion can be defined as

$$\Gamma^\Delta(X) = \bigcup_{x \in X} \Gamma_\Delta(\Gamma_x(X)) \tag{1}$$

For non-increasing criteria Δ we obtain an attribute thinning. The sphericity attribute $S(C)$ is an example and is represented in terms of surface area($A(C)$) and volume($V(C)$) of each component as

$$S(C) = \frac{\pi^{\frac{1}{3}}(6V(C))^{\frac{2}{3}}}{A(C)} \tag{2}$$

3   The Max-Tree

The Max-Tree [4] data structure was designed for morphological attribute filtering in image processing. The nodes (C_h^k, k is the node index, h the gray level) of the Max-Tree represent connected components for all threshold levels in a data set (see Fig. 1). These components are referred to as *peak components* and are denoted as P_h^k. Each node C_h^k contains only those pixels of peak component P_h^k which have gray level h. The root node represents the set of pixels belonging to the background, that is the set of pixels with the lowest intensity in the image and each node has a pointer to its parent. The nodes corresponding to the components with the highest intensity are the leaves. The filtering process is separated into three stages: construction, filtering and restitution. During the construction phase, the Max-Tree is built from the flat zones of the image, collecting auxiliary data used for computing the node attributes at a later stage. The auxiliary data can be used to compute one or more attributes, that describe certain properties of the peak components represented by those nodes. The filtering process is based on certain rules like the *Direct, Min, Max,* and *Viterbi* rules [3,4], and more recently the *Subtractive* rule [8], and *Branches* rule [9,10]. These filtering rules are all designed to deal with *non-increasing* attributes, in which accept and reject decisions might alternate along any path from leaf to root. In the following we will only use the Subtractive rule, which works best for blood-vessel enhancement and image decomposition based on shape [6,7,8].

Whatever filtering rule used, a key problem when computing Max-Trees lies in efficient computation of the attributes. Following the approach in Breen and Jones [3], most methods use a recursive procedure in which data are collected during the construction phase, and passed down to the parents. This approach is fine for any attribute which can be computed incrementally, such as those based on moments, histograms, or on the minimum and maximum coordinate values. It has however tended to limit the attributes used. In [7] several augmentations to the Max-Tree were made for interactive visualization purposes, including fast iso-surface rendering. We will use these extensions to compute surface area efficiently.

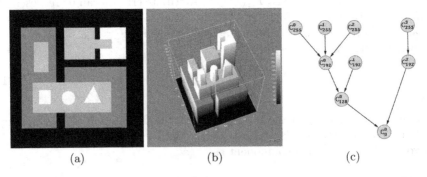

(a)   (b)   (c)

Fig. 1. Example of a Max-Tree: (a) original image; (b) viewed as 3D surface; (c) corresponding Max-Tree

4 Sphericity Attribute Computation

A simple algorithm (called adjacency method in the following) to compute an approximate surface area is based on 6 connectivity in 3D [11]. Surface area computation begins by detecting all surface voxels, i.e. object voxels that are 6-connected to background voxels. For a three-dimensional connected component that is represented as a set of voxels, one can easily identify the voxels that are 6-connected to the background and view them as the border of the object. The boundary of the object is the set of voxel faces that separates the object from the background. Now the concept of 6-connectedness stems from the fact that, for any two voxels (x_1, x_2, x_3) and (y_1, y_2, y_3) are called 6 face adjacent if $\Sigma_{i=1}^3 (x_i - y_i)^2 = 1$ [12]. To solve the problem of boundary extraction, one approach is to visit each voxel in each object and determine whether it is 6-connected to a background voxel or not.

Once the set of voxel faces that separate the object from the background are identified their areas are estimated separately and added up to constitute the surface area. In short, for each voxel in a connected component, simply compute the number of 6-connected neighbors *outside* the component. This is equal to the number of faces of each voxel on the boundary. The sum of these values over the component is the surface area of the discrete representation of the object.

In gray scale, things are more complicated, as the voxels cannot readily be classified as object and background. Using the depth-first construction algorithm for Max-Tree construction from [4] (or for that matter the union-find approach of [13]), the surface area of the current node C_h^k is initialized at zero, and the surface areas of any child node at higher grey level is added to it. For each voxel at level h within C_h^k we compute the number of adjacent voxels of *lower* gray level, and add this to the surface area of the component. We also compute the number of adjacent voxels with gray level *higher* than h. This number is subtracted from the surface area of the current node C_h^k, because it represents part of the boundary between the current node and one of its children. This part has previously been added to the surface area of the node, but is *not* part of the boundary of the peak component P_h^k represented by node C_h^k, and must therefore be subtracted.

The adjacency method can be implemented in the usual recursive way, using auxiliary data as proposed by [3,4]. The computation of the fast sphericity from the image data is performed as follows:

- Compute Max-Tree according to the algorithm in [4].
- As the Max-Tree is built compute the Volume($V(C_h^k)$) of each node using the existing voxel-based algorithm
- For each node (C_h^k)
    - initialize surface area $A(C_h^k)$ to zero,
    - add surface areas of children (components at higher levels) to $A(C_h^k)$,
    - add number of 6-adjacent voxels to C_h^k with gray level $< h$ to $A(C_h^k)$

- subtract number of 6-adjacent voxels to C_h^k with gray level $> h$ from $A(C_h^k)$
- compute sphericity $\pi^{\frac{1}{3}}(6V(C_h^k))^{\frac{2}{3}}/A(C_h^k)$.

The above algorithm assumes isotropic voxels with unit surface area faces. To deal with anisotropic voxels, we compute separate sums of neighbours in x, y and z directions, multiply each sum with the appropriate surface area, and add the individual results together to obtain the final surface area of each node. The time complexity of this algorithm (disregarding Max-Tree construction) is $O(N)$, with N the number of voxels. This is because for each voxel, we need to inspect 6 neighbors, perform at most one addition and one subtraction. The surface area of each node is added only once to its parent, and because the number of nodes is bounded by N this part is also $O(N)$.

The problem, as many have noted, is that this does not yield an accurate measure of surface area [14]. Worse still, as resolution increases, the estimate does not improve. We therefore adapt a method for fuzzy sets [14], based on iso-surface estimation to grey-scale volumes and attribute filtering.

Rendering of an object often requires iso-surface detecting algorithms such as marching cubes [15] which produce a list of triangles approximating the surface. Iso-surfaces are formed from each level set of the function f for which points (x, y) have a constant intensity. To find these triangle meshes we need to process all cells of the volume. A cell is a cube with the voxel centers at its corners. These triangle meshes are quickly obtained from *active cells* that intersect the surface during visualization process, using either range based search or seed set generation methods [16], but also using an augmented Max-Tree [7]. There can be up to 5 triangles per each active cell.

In each active cell we have n triangles. Let v_i be vertices of the triangles, such that v_i, v_{i+1}, v_{i+2} form a triangle. Then the sum of the areas of each triangle approximates the surface area $A(S)$ of the iso-surface patch S intersecting the cell. This is computed as

$$A(S) = \frac{1}{2}\Sigma_{i=0}^n |(v_{3i+1} - v_{3i}) \times (v_{3i+2} - v_{3i})| \tag{3}$$

where $\times$ denotes the vector cross product. We could recursively go through the Max-Tree, and for each node C_h^k obtain triangle meshes from the augmented Max Tree representation and compute their area $A(C_h^k)$. This would lead to repeated visits of cells in the volume leading to cash thrashing. Here we purpose a different approach, as explained below.

In our *iso-surface* method we use the augmented Max-Tree that adds visualization data to the Max-Tree [7]. For the purposes of this paper the most important additions are the `Dilated` and `Eroded` arrays. These contain the maximum and minimum node along the root path passing through each cell. Note that we use 26 connectivity so that all eight voxels at the corners of each cell are in the same root path. The aim is to compute the surface areas of all iso-surfaces at all levels in the Max-Tree corresponding to the nodes.

The surface area attribute computation takes the following procedure:

- Compute the augmented Max Tree.
- Compute the volume $V(C_h^k)$ of each node using the existing voxel-based algorithm.
- For each cell
  - find maximum and minimum nodes from Dilated and Eroded arrays
  - set current node to maximum
  - while current node is not minimum
    * compute grey level intermediate between current node and its parent
    * compute triangle mesh of iso-surface at that level for the current cell
    * calculate area of the triangles in the mesh and add to area of corresponding node.
    * set current node to its parent
- compute the surface area of the root node from the volume dimensions.
- For each node C_h^k compute sphericity $\pi^{\frac{1}{3}}(6V(C_h^k))^{\frac{2}{3}}/A(C_h^k)$.

Sphericity for a sphere is 1.0 and for a cube it is approximately 0.806. For our iso-surface method computation of surface area and volume for data objects should be as close as possible for specific objects' true values, and retain rotation invariance. Thus we compare our sphericity computation with the adjacency method in [11], which is the only other 3-D surface-area-based attribute filter implementation to date.

The computational complexity of this algorithm is $O(N\Delta G)$ with N being the number of cells or voxels, and ΔG is the mean grey level range within each cell. Thus the algorithm should slow down as grey level resolution is increased. Memory costs of both algorithms are the same, requiring storage of two doubles per node: one for volume and one for surface area.

5   Results and Discussion

5.1   Computational Cost

After computing the Max-Tree representation of a data set, the attribute of each node is computed from its auxiliary data. In the filtering stage and based on each node's attribute value, a decision is made and its level is modified according to the rule chosen. To measure the algorithm's computational performance, we ran some timing experiments on a Core 2 Duo E8400 at 3.0 GHz. Attribute computation times in seconds are shown in the Table. 1. We ran tests on the mrt16_angio2 and mrt8_angio2 time-of-flight magnetic resonance angiograms (MRA) from http://www.volvis.org and the fullhead CT data set included with VTK at 8 bit and 12 bit grey-level resolution for several attributes. As expected, the iso-surface based attribute is slower than the other attributes.

As expected the iso-surface algorithm performs more slowly on 12-bit resolution data sets than the equivalent 8-bit data sets. The effect is most pronounced on the fullHead data set, in which the grey level range is 4095 for the 12-bit data. The 12-bit mrt16_angio2 has only a 576 grey level range, which explains why it only doubles in computing time with respect to the 8-bit version.

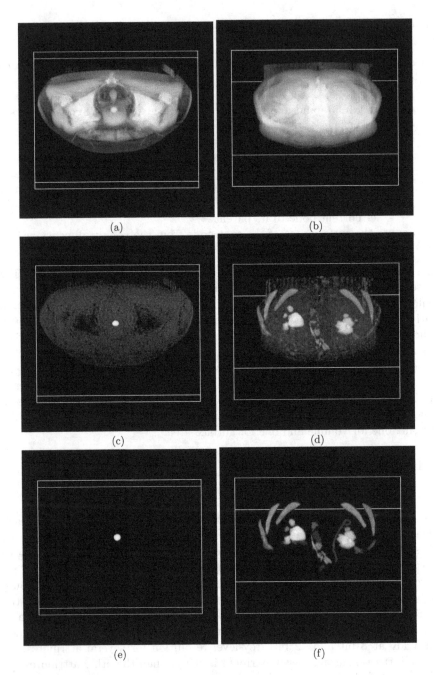

Fig. 2. Sphericity filtering of CT scans along the urinary tract in X-ray rendering mode: (a) the unfiltered view of a bladder calculus and (b) of kidney calculi; the results of the sphericity filter for each set with $\lambda = 0.94$ (c), and $\lambda = 0.39$ (d) respectively; the results of the volume filter following the sphericity filter with $\lambda = 78$ (e) and $\lambda = 450$ (f) respectively

Table 1. Computing time (in seconds) of various attributes

Attribute	fullHead 12 bits	8 bits	mrt*angio2 12 bits	8 bits	vessels 8 bits
Sphericity(Iso.)	123.13	8.64	26.77	14.10	2.94
Sphericity(Adj.)	1.44	0.98	1.44	1.30	0.85
Non Compactness	0.70	0.57	0.66	0.62	0.83
Sparseness	0.98	0.83	0.92	0.85	1.18
Volume	0.15	0.08	0.14	0.12	0.12

5.2 Performance Evaluation

CT Scans of the Urinary Tract. To evaluate the performance of the sphericity filter in isolating compact structures, we used two 3D CT data sets of patients suffering from urolithiasis. Both are courtesy of the Department of Radiology and Medical Imaging, University General Hospital of Alexandroupolis, Greece. The first one shown at the top of left column of Fig. 2 is of a patient diagnosed with a bladder calculus. The stone is rather compact allowing for high values of λ. Fig. 2 (c) shows the sphericity filter output for $\lambda = 0.94$, for which all other anatomical structures are suppressed. Followed by a volume opening with $\lambda = 78$, we can completely isolate the stone from the remaining noise - Fig. 2 (e).

The second data set shows a severe case of nephrolithiasis, in which a stent is inserted to one of the kidneys to bypass the obstruction of urine flow. In each kidney there exists one major calculus accompanied by other smaller ones. Due to their arbitrary shape, extracting them all (Fig. 2 (d)) requires a low value for the sphericity threshold – $\lambda = 0.39$. To enhance the scene further, a volume opening is applied with $\lambda = 450$, shown in Fig. 2 (f).

Time-of-Flight Angiography. Time-of-Flight angiograms are very difficult to filter. On the mrt16_angio2 data set (from http://www.volvis.org/, the performance of the two methods is hard to distinguish as shown in Fig. 3(c) and (d). Both methods perform equally in terms of retaining or discarding specific features. However, both struggle in suppressing the background without removing the vessels. By contrast, the moment-based non-compactness attribute from [6,7] removes background much more effectively (see Fig. 3(b)). However, it too retains unwanted features. In Fig. 3(e) and (f) we see the effect of applying the sphericity filters to the volume in Fig. 3(b). Both show a distinct improvement on background suppression. Evidently, the blood vessels have a larger sphericity than most unwanted features retained by the non-compactness filter.

Accuracy and Rotational Invariance. To measure the accuracy of the methods we measure the sphericity for several synthetic objects of known sphericity. The first volume is an inverted Euclidean distance map from the Volume Library (http://www9.informatik.uni-erlangen.de/External/vollib/). This yields a series of Euclidean spheres with increasing radii. A separate volume of cubes and blocks of various sizes was made. Finally, a volume consisting of a rod of $4 \times 4 \times 32$

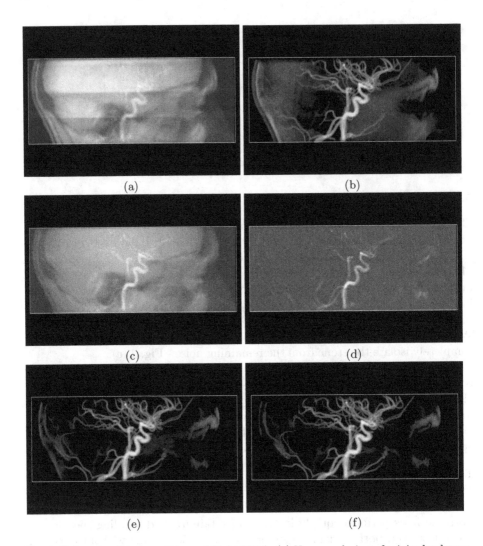

Fig. 3. Sphericity filtering of time-of-flight MRA: (a) Xray rendering of original volume; (b) filtered with non-compactness attribute at $\lambda = 3.7$; (c) original filtered with iso-surface sphericity attribute at $\lambda = 0.1320$; (d) original filtered with adjacency sphericity attribute at $\lambda = 0.1124$; (e) volume (b) filtered with iso-surface sphericity attribute at $\lambda = 0.1320$; (f) volume (b) filtered with adjacency sphericity attribute at $\lambda = 0.1124$

rotated from $0°$ to $45°$ in $5°$ steps. For both algorithms we computed the sphericity attributes. For an axis aligned cube of 32^3 the sphericity of the iso-surface method yielded 0.821 versus 0.806 for the adjacency method. The latter is the theoretical value. The difference is due to the rounding of the corners by the iso-surface method.

The results for the Euclidean distance map are in Fig. 4. This shows that the iso-surface method approaches the theoretical value of 1.0 as radius r increases.

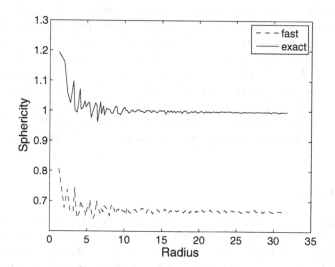

Fig. 4. Attribute measure as a function of sphere diameter for Euclidean spheres, for both the iso-surface (solid) and adjacency method (dashed)

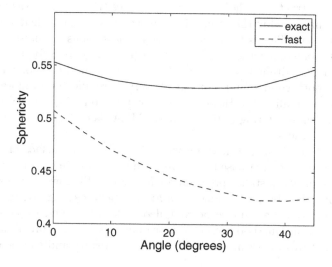

Fig. 5. Sphericity of a $4 \times 4 \times 32$ rod rotated from $0°$ to $45°$ in steps of $5°$, for both the iso-surface (solid) and adjacency method (dashed)

The values larger than 1 at small r are caused by the fact that the volume is computed voxel-wise, and the surface from the iso-surface. This means that the iso-surface can enclose a smaller volume than that measured by the voxel-based volume computation. By contrast the adjacency method consistently approaches $2/3$ asymptotically, which is what is expected from theory.

Fig. 5 shows the results of the rotation invariance test. Clearly, as we rotate the rod the iso-surface method shows more rotation invariance than the adjacency method. Between 45° and 90° the results mirror those in the graph. In a second test we compared the sphericity measure of a $32 \times 32 \times 32$ cube (axis aligned) with the same cube rotated by 45° over the z-axis. For the axis aligned cube we have 0.806 for the adjacency method and 0.821 for the iso-surface method as noted above. When we rotate over 45 degrees, the values become 0.637 for the adjacency method (theoretically approaching 0.632 or 22% off), versus 0.818 for ours (or 0.4% lower than unrotated) again proving our improved rotation invariance.

6   Conclusions

Rotation invariance is an important property for object detection, and it is a challenge to achieve it in the case of surface-area-based attributes. In this paper we presented a rotation invariant, iso-surface method whose performance was compared to the adjacency method from [11]. The performance of both methods is shown for images rotated at different angles. While the iso-surface method performs well in terms of rotation invariance, the adjacency method [11] is faster. The accuracy of the iso-surface method was better than that of the adjacency method for large structures, though on small spheres the accuracy was reduced. This should be improved in the future. Both methods struggled when applied to very noisy data sets. This is because sphericity is less robust to noise than the moment-based non-compactness measure from [6]. The reason for this is that surface area changes dramatically as small holes appear in a component due to noise, whereas both volume and moment of inertia are affected in more or less the same way by small holes. However, the combination of sphericity filtering and non-compactness filtering in time-of-flight MRAs is clearly superior to only using non-compactness.

Though we focused on sphericity measures in this paper, surface area itself, and many other surface-area-based attributes can be derived in the same way. All code, data sets and an installer program for Microsoft Windows demonstrating these possibilities can be downloaded from `http://www.cs.rug.nl/~michael/MTdemo`. This source code can also be compiled under Linux and Apple's OS-X. Finally, the iso-surface method presented here only uses a single core of the CPU. Parallelization of this algorithm should be straightforward, and will be done in the future.

References

1. Heijmans, H.J.A.M.: Connected morphological operators for binary images. Comp. Vis. Image Understand. 73, 99–120 (1999)
2. Salembier, P., Serra, J.: Flat zones filtering, connected operators, and filters by reconstruction. IEEE Trans. Image Proc. 4, 1153–1160 (1995)
3. Breen, E.J., Jones, R.: Attribute openings, thinnings and granulometries. Comp. Vis. Image Understand. 64(3), 377–389 (1996)

4. Salembier, P., Oliveras, A., Garrido, L.: Anti-extensive connected operators for image and sequence processing. IEEE Trans. Image Proc. 7, 555–570 (1998)
5. Meijster, A., Wilkinson, M.H.F.: A comparison of algorithms for connected set openings and closings. IEEE Trans. Pattern Anal. Mach. Intell. 24(4), 484–494 (2002)
6. Wilkinson, M.H.F., Westenberg, M.A.: Shape preserving filament enhancement filtering. In: Niessen, W.J., Viergever, M.A. (eds.) MICCAI 2001. LNCS, vol. 2208, pp. 770–777. Springer, Heidelberg (2001)
7. Westenberg, M.A., Roerdink, J.B.T.M., Wilkinson, M.H.F.: Volumetric attribute filtering and interactive visualization using the max-tree representation. IEEE Trans. Image Proc. 16, 2943–2952 (2007)
8. Urbach, E.R., Roerdink, J.B.T.M., Wilkinson, M.H.F.: Connected shape-size pattern spectra for rotation and scale-invariant classification of gray-scale images. IEEE Trans. Pattern Anal. Mach. Intell. 29, 272–285 (2007)
9. Naegel, B., Passat, N., Boch, N., Kocher, M.: Segmentation using vector-attribute filters: Methodology and application to dermatological imaging. In: Proc. Int. Symp. Math. Morphology (ISMM 2007), pp. 239–250 (2007)
10. Purnama, K.E., Wilkinson, M.H.F., Veldhuizen, A.G., van Ooijen, P.M.A., Lubbers, J., Sardjono, T.A., Verkerke, G.J.: Branches filtering approach for max-tree. In: Proc. 2nd Int. Conf. Comput. Vision Theory Applic (VISAPP 2007), Barcelona, March 8-11, pp. 328–332 (2007)
11. Ouzounis, G.K., Giannakopoulos, S., Simopoulos, C.E., Wilkinson, M.H.F.: Robust extraction of urinary stones from CT data using attribute filters. In: Proc. Int. Conf. Image Proc. (in press, 2009)
12. Udupa, J.K.: Multidimensional digital boundaries. CVGIP: Graphical Models and Image Processing 56(4), 311–323 (1994)
13. Najman, L., Couprie, M.: Building the component tree in quasi-linear time. IEEE Trans. Image Proc. 15, 3531–3539 (2006)
14. Sladoje, N., Nyström, I., Saha, P.K.: Measurements of digitized objects with fuzzy borders in 2d and 3d. Image Vis. Comput. 23, 123–132 (2005)
15. Kaufman, A., Mueller, K.: Overview of volume rendering. In: The Visualization Handbook. Academic Press, London (2005)
16. Cignoni, P., Marino, P., Montani, C., Puppo, E., Scopigno, R.: Speeding up isosurface extraction using interval trees. IEEE Trans. Visualization Comp. Graph. 3, 150–170 (1997)

Levelings and Geodesic Reconstructions

Jose Crespo

GIB - LIA
DLSIIS
Facultad de Informática
Universidad Politécnica de Madrid
28660 Boadilla del Monte (Madrid), Spain
jcrespo@fi.upm.es

Abstract. This paper investigates some geodesic implementations that have appeared in the literature and that lead to connected operators. The focus is on two so-called self-dual geodesic transformations. Some fundamental aspects of these transformations are analyzed, such as whether they are actually levelings, and whether they can treat each grain or pore independently from the rest (connected-component locality). As will be shown, one of the geodesic self-dual reconstructions studied appears to be not a leveling. Nevertheless, it possesses a distinctive characteristic: it can process grains and pores in a connected-component local manner. The analysis is performed in the set or binary framework, although results and conclusions extend to (flat) gray-level operators.

Keywords: Levelings, geodesic operators, geodesic reconstructions.

1 Introduction

Levelings are a class of operators that are connected and that satisfy certain constraints. In the set or binary framework, levelings are called set or binary levelings. The analysis of this paper is performed in the set or binary framework, although results and conclusions extend to (flat) gray-level operators that commute with thresholding.

Geodesic transformations are a usual way to implement levelings. This paper investigates two so-called self-dual geodesic operations presented in the literature. In the analysis, the sequence of performing an under-reconstruction followed by an over-reconstruction (and vice-versa) is also considered.

In particular, this work investigates the following fundamental properties of them: (a) their leveling nature (i.e., if they are levelings), and (b) whether they treat each grain or pore independently from the rest (whether they are *connected-component local*). A significant finding, as will be shown, is that one self-dual geodesic reconstruction would not be a leveling. This fact does not necessarily implies that it is not useful; it has a distinctive characteristic that can make it interesting in certain applications. Researchers and users of geodesic reconstructions should know the properties of them.

M.H.F. Wilkinson and J.B.T.M. Roerdink (Eds.): ISMM 2009, LNCS 5720, pp. 82–91, 2009.

The paper is organized as follows. Section 2 provides some background, including some definitions of concepts related to connected operators and levelings. The two self-dual reconstructions (as well as the elementary geodesic transformations they are based on) are described in Section 3, and are analyzed in Section 4. Then, Section 5 concludes the paper.

2 Background

2.1 Basic Definitions

Some general references in the field of Mathematical Morphology are the following [1][2][3][4][5][6][7].

In the theoretical expressions in this paper, we will be working on the lattice $\mathcal{P}(E)$, where E is a given set of points (the space) and $\mathcal{P}(E)$ denotes the set of all subsets of E (i.e., $\mathcal{P}(E) = \{A : A \subseteq E\}$). Nevertheless, results can be extended to gray-level functions by means of the so called flat operators [2][6].

In this work we will deal later with the duality concept, which has a precise definition in Mathematical Morphology. Two morphological operators ψ_1 and ψ_2 are *dual* of each other if $\psi_1 = \mathsf{C}\psi_2\mathsf{C}$, where C symbolizes the complement operator. As a particular case, a morphological operator ψ is said to be *self-dual* if $\psi = \mathsf{C}\psi\mathsf{C}$.

Connected operators do not introduce discontinuities and extend partitions in the sense that the partition of the output is *coarser* that that of the input [8] [9]. For binary images (or sets), they treat the connected components of the input and its complement in an all or nothing way. The operator that extracts the connected-component a point x belongs to is the opening γ_x [3]. In this work, the space connectivity is assumed to be a strong connectivity [10][11], which avoids the existence of isolated grains and pores. More particularly, connected subsets of $\mathbb{Z}^2$ with four- or eight-connectivity are used as the space E of points in this paper. Connected operations can be considered as graph operations. Image representations based on inclusion trees can be useful [12].

The following two sections define two constraints that are particularly useful for studying the connected operator class: *connected-component (c.c.) locality* and *adjacency stability*.

2.2 Connected-Component Locality

Definition 1. *[13][14][15] Let E be a space endowed with γ_x, $x \in E$. An operator $\psi : \mathcal{P}(E) \longrightarrow \mathcal{P}(E)$ is said to be **connected-component** local (or **c.c.-local**) if and only if, $\forall A \in \mathcal{P}(E)$:*

- *ψ preserves (or, respectively, removes) a non-empty grain G of A in operation $\psi(A)$ if and only if ψ preserves (respectively, removes) grain G in operation $\psi(G)$.*
- *ψ preserves (or, respectively, fills) a non-empty pore P of A in operation $\psi(A)$ if and only if ψ preserves (respectively, fills) pore P in operation $\psi(E \setminus P)$.*

Where "\" denotes the set subtraction operation. (Note that connected operators just preserve or remove grains, and preserve or fill pores.)

Thus, a c.c.-local connected operator is one that (1) fills grains and/or remove pores, and (2) treats each grain or pore independently from the rest of grains and pores. The connected-component local operator concept was also later discussed in [11], where the term "grain-operator" is used.

2.3   Adjacency Stability

The adjacency stability constraint restrains in some way the behavior of adjacent flat zones, in particular the switch from grain to pore and vice-versa. The adjacency stability concept was introduced in [13][14], and was further studied in [15]. A related concept and formulation are discussed in [11].

Definition 2. *Let E be a space endowed with γ_x, $x \in E$. An operator ψ : $\mathcal{P}(E) \longrightarrow \mathcal{P}(E)$ is **adjacency stable** if, for all $x \in E$:*

$$\gamma_x(\text{id} \bigvee \psi) = \gamma_x \bigvee \gamma_x\psi. \tag{1}$$

Property 1. Extensive and anti-extensive mappings are adjacency stable.

The next property states the composition laws of adjacency stable connected operators, and, conjointly with Property 1, provides a way to build operators of this class.

Property 2. The class of adjacency stable connected operators is closed under the sup, the inf and the sequential composition operations.

2.4   Levelings

Definition 3. *An image g is a leveling of an input image f if and only if:*

$$\forall\ (p,q) \text{ neighboring pixels} : g_p > g_q \ \Rightarrow \ f_p \geq g_p \text{ and } g_q \geq f_q \tag{2}$$

The previous definition of leveling is that in [16, Definition 4, p. 193] [17, Definition 2.2, p. 4]. A more general definition is introduced in [18, Definition 10, p. 62].

Set levelings are those defined in the set or binary framework. As discussed in [19], the leveling concept is equivalent to the adjacency stability connected operator concept that was presented in [13][14][15], which therefore constitute the origin of the leveling concept in the set or binary framework. Composition laws of set levelings (which are extended to and satisfied by flat gray-level levelings as well) can be found in [13][14][15] (see Property 1 and Property 2). Regarding some clarifications about whether levelings satisfy the strong property, see [19][20]. Levelings are useful operators for image filtering that simplify an image while imposing input-output restrictions, and that can be computed using compositions of morphological connected operators (from Properties 1 and 2).

Other works about levelings are [21][22][23][24].

3 Geodesic Reconstructions: Definitions and Formulae

The geodesic reconstructions that will be investigated in this work are defined in the following. This work will focus specially on the R_ν and $R_{\nu'}$ self-dual geodesic reconstructions.

– **Reconstruction R_ν**

  The R_ν self-dual reconstruction is based on the elementary self-dual geodesic operator ν. (Note: the description follows the presentation in [6, Section 6.1.3], although the notation varies in minor details.)

$$[\nu^g_{(1)}(f)](x) = \begin{cases} (\delta^g_{(1)}(f))(x), & \text{if } f(x) \leq g(x) \\ (\varepsilon^g_{(1)}(f))(x), & \text{otherwise} \end{cases} \tag{3}$$

  where $\delta^g_{(1)}(f) = \delta_1(f) \wedge g$, and $\varepsilon^g_{(1)}(f) = \varepsilon_1(f) \vee g$. The mask image is denoted by g, and the marker image is symbolized by f.

  The $\nu^g_{(1)}(f)$ operator can be equivalently expressed [6] as $\varepsilon_1(f) \vee \delta^g_{(1)}(f)$ or $\delta_1(f) \wedge \varepsilon^g_{(1)}(f)$ (which would follow from leveling expressions presented in [16][25]).

  The corresponding transformation of size n is $\nu^g_{(n)}(f) = \nu^g_{(1)}(\nu^g_{(n-1)}(f))$. Reconstruction R_ν denotes the iteration of ν until idempotence:

$$R_\nu(g; f) = \nu^g_{(i)}(f) \tag{4}$$

  where i is such that $\nu^g_{(i+1)}(f) = \nu^g_{(i)}(f)$.

– **Reconstruction $R_{\nu'}$**

  In [6, Section 6.1.3], the next self-dual transformation variant is also presented:

$$[\nu'^g_{(1)}(f)](x) = \begin{cases} (\delta^g_{(1)}(f \wedge g))(x), & \text{if } f(x) \leq g(x) \\ (\varepsilon^g_{(1)}(f \vee g))(x), & \text{otherwise} \end{cases} \tag{5}$$

  $\nu'^g_{(n)}(f) = \nu'^g_{(1)}(\nu'^g_{(n-1)}(f))$, and the reconstruction based on $\nu'^g_{(1)}$ until idempotence is:

$$R_{\nu'}(g; f) = \nu'^g_{(i)}(f) \tag{6}$$

  where i is such that $\nu'^g_{(i+1)}(f) = \nu'^g_{(i)}(f)$.

– **Under-reconstruction $\underline{R}$ and over-reconstruction $\overline{R}$**

  Let $\underline{R}(g; f)$ denote the normal under-reconstruction or reconstruction by dilation (i.e., the iteration of $\delta^g_{(1)}(f)$ until idempotence), where $f \leq g$. Let $\overline{R}(g; f)$ symbolize the normal over-reconstruction or reconstruction by erosion (i.e., the iteration of $\varepsilon^g_{(1)}(f)$ until idempotence), where $f \geq g$.

  In the work of the paper, we will also consider the sequential compositions of $\underline{R}$ and $\overline{R}$ (i.e., $\overline{R} \circ \underline{R}$ and $\overline{R} \circ \underline{R}$) to complete the analysis and comparisons.

(a) Case study 1   (a) Case study 2

Fig. 1. Case studies. Input functions (continuous line) and marker functions (dotted line) used in the analysis. Note: the space of points is discrete.

4   Analysis of Geodesic Reconstructions R_ν and $R_{\nu'}$

In this section, we will analyze the geodesic reconstructions defined in the previous section on two different cases: one that can be referred to as "normal or non-problematic", and another one that has some adjacency issues. The two cases are displayed in Fig. 1, where the input functions are displayed as continuous lines, and the marker function as dotted lines:

4.1   On the Leveling Nature

We will first use a simple example of the application of the geodesic reconstructions defined in Section 3 to a 1-D gray-level function and an associated marker, as displayed in Fig. 2. In fact, we will perform a thresholding operation to operate on a section to better illustrate the behaviors of the geodesic reconstructions regarding adjacency matters.

We can observe that all geodesic reconstructions considered compute the same result in the example of Fig. 2 at each level. In the case at the left of Fig. 2, a grain is marked and reconstructed (see Fig. 2(d.1)-(g.1)); in the right part, that same grain is removed (see Fig. 2(d.2)-(g.2)). No differences are observed.

We apply next the geodesic reconstructions considered in this paper to a case that shows some adjacency issues in Fig. 3. Two levels are considered: the first one, at the left part of Fig. 3, does not present any problem and all geodesic reconstructions compute the same result. However, the level at the right part of the figure poses some adjacency issues, and, as can be observed in Fig. 3(d.2)-(g.2), *not all geodesic reconstructions considered behave the same.*

In fact, the result shown in Fig. 3(c.2) computed by $R_{\nu'}$ shows the behavior that is in fact excluded by the leveling nature: a grain has been removed and an adjacent pore has been filled. The adjacency stability equation (1) (or expression (2)) is not satisfied. Thus, based on expression (5) by itself[1], **the $R_{\nu'}$ reconstruction is**

---

[1] We mean without imposing restrictions on the marker. As a matter of fact, in relation to this issue, it can be mentioned that the marker-based formulation (which does not get into the details of concrete geodesic implementations) of a set leveling in [22] poses some constraints on the marker (more exactly, on the markers, since it considers two) to take into account the adjacency constraints derived from the adjacency stability equation (1) (or expression (2)).

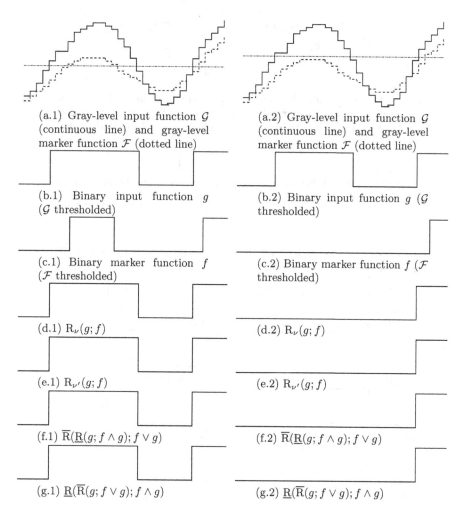

(a.1) Gray-level input function $\mathcal{G}$ (continuous line) and gray-level marker function $\mathcal{F}$ (dotted line)

(a.2) Gray-level input function $\mathcal{G}$ (continuous line) and gray-level marker function $\mathcal{F}$ (dotted line)

(b.1) Binary input function g ($\mathcal{G}$ thresholded)

(b.2) Binary input function g ($\mathcal{G}$ thresholded)

(c.1) Binary marker function f ($\mathcal{F}$ thresholded)

(c.2) Binary marker function f ($\mathcal{F}$ thresholded)

(d.1) $R_\nu(g; f)$

(d.2) $R_\nu(g; f)$

(e.1) $R_{\nu'}(g; f)$

(e.2) $R_{\nu'}(g; f)$

(f.1) $\overline{R}(\underline{R}(g; f \wedge g); f \vee g)$

(f.2) $\overline{R}(\underline{R}(g; f \wedge g); f \vee g)$

(g.1) $\underline{R}(\overline{R}(g; f \vee g); f \wedge g)$

(g.2) $\underline{R}(\overline{R}(g; f \vee g); f \wedge g)$

Fig. 2. Geodesic reconstructions for case study 1 at two levels. Part (a) shows the input gray-level function $\mathcal{G}$ (continuous line) and marker $\mathcal{F}$ (dotted line), along with the thresholding level (horizontal discontinuous line) used for parts (b) and (c), which show, respectively, the binary input function g and marker f employed for the reconstructions displayed in parts (d) and (e). Note: the left and right parts of the figure refer to different thresholding levels; the space of points is discrete.

not generally a leveling. The previous statement can depend on the particular space: it could be the case that a certain self-dual grain removing and pore filling operation is not a leveling in a certain space but it is in a subset or a superset of it (see further discussion about some of these aspects in [15]).

The other reconstructions considered, $R_\nu(g; f)$, $\overline{R}(\underline{R}(g; f \wedge g); f \vee g)$ and $\underline{R}(\overline{R}(g; f \vee g); f \wedge g)$ do not show those adjacency issues. Nevertheless, in the situation at the right of Fig. 3 $\overline{R}(\underline{R}(g; f \wedge g); f \vee g)$ is not equal to $\underline{R}(\overline{R}(g; f \vee g); f \wedge g)$

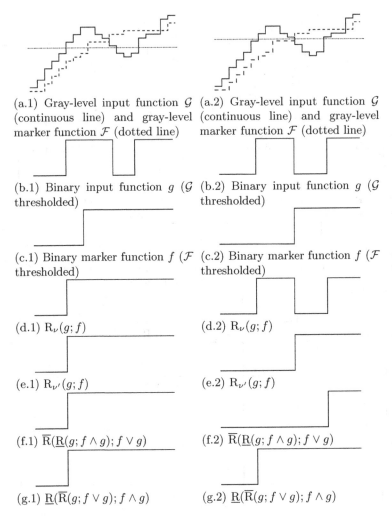

(a.1) Gray-level input function $\mathcal{G}$ (continuous line) and gray-level marker function $\mathcal{F}$ (dotted line)

(a.2) Gray-level input function $\mathcal{G}$ (continuous line) and gray-level marker function $\mathcal{F}$ (dotted line)

(b.1) Binary input function g ($\mathcal{G}$ thresholded)

(b.2) Binary input function g ($\mathcal{G}$ thresholded)

(c.1) Binary marker function f ($\mathcal{F}$ thresholded)

(c.2) Binary marker function f ($\mathcal{F}$ thresholded)

(d.1) $R_\nu(g; f)$

(d.2) $R_\nu(g; f)$

(e.1) $R_{\nu'}(g; f)$

(e.2) $R_{\nu'}(g; f)$

(f.1) $\overline{R}(\underline{R}(g; f \wedge g); f \vee g)$

(f.2) $\overline{R}(\underline{R}(g; f \wedge g); f \vee g)$

(g.1) $\underline{R}(\overline{R}(g; f \vee g); f \wedge g)$

(g.2) $\underline{R}(\overline{R}(g; f \vee g); f \wedge g)$

Fig. 3. Geodesic reconstructions for case study 2 at two levels. Part (a) shows the input gray-level function $\mathcal{G}$ (continuous line) and marker $\mathcal{F}$ (dotted line), along with the thresholding level (horizontal discontinuous line) used for parts (b) and (c), which show, respectively, the binary input function g and marker f employed for the reconstructions displayed in parts (d) and (e). Note: the left and right parts of the figure refer to different thresholding levels; the space of points is discrete.

(see Fig. 3(f.2) and Fig. 3(g.2)), and they should generally be avoided if self-dual processing is desired.

4.2   On Connected-Component Locality

A complete study of the c.c.-locality of a marker-based connected operator implemented using reconstruction transformations (self-dual or not) would obviously

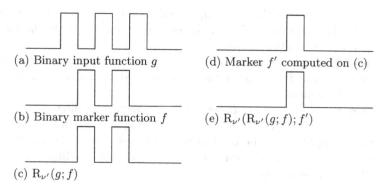

(a) Binary input function g

(b) Binary marker function f

(c) $R_{\nu'}(g; f)$

(d) Marker f' computed on (c)

(e) $R_{\nu'}(R_{\nu'}(g; f); f')$

Fig. 4. Iteration of $R_{\nu'}$. Note that part (c) is different from part (e) (i.e., $R_{\nu'}(g; f) \neq R_{\nu'}(R_{\nu'}(g; f); f')$).

need to take into account the marker computation. In the following, to simplify the treatment, we will focus on the reconstruction transformations themselves (expressions (3) and (5)), except when a second sequential application is commented where the marker computation stage is also considered.

By examining the behaviors of the R_ν and $R_{\nu'}$ self-dual reconstructions in the example of the right part of Fig. 3, we can see that the treatment of a grain (or respectively, a pore) in Fig. 3(d.2) has been influenced by an adjacent pore (respectively, a grain). Thus, reconstruction R_ν is not c.c.-local. Note that the geodesic operations themselves of R_ν make it non c.c.-local (even when the marker is c.c.-local).

Regarding the alternative reconstruction $R_{\nu'}$, those issues do not arise: a grain that has not been marked would be removed by $R_{\nu'}$ disregarding what happens at the adjacent pores. Analogously for pores. Thus, **$R_{\nu'}$ can be used as a basis for c.c.-local connected processing**.

We will briefly comment that the switching of adjacent grains and pores that happens in non levelings (such as $R_{\nu'}$) is normally linked to non-idempotent behavior when c.c.-local processing is desired. This is illustrated in Fig. 4, where two iterations of $R_{\nu'}$ are applied to the input image Fig. 4(a) using Fig. 4(b) as marker, where three grains of Fig. 4(a) are signaled to be removed and two pores are signaled to be filled. Let us assume that the marker criterion is c.c.-local. Fig. 4(c) displays $R_{\nu'}(g; f)$. If we apply the previous operation again, the marker computed on Fig. 4(c) is Fig. 4(d). We can see that $R_{\nu'}(R_{\nu'}(g; f); f')$ (Fig. 4(e)) is different from $R_{\nu'}(g; f)$ (Fig. 4(c)), in other words, $R_{\nu'}$ does not show idempotence. As was the case when considering the leveling nature, this can depend on the particular space.

5   Conclusion

This paper has investigated some fundamental issues that exist with geodesic reconstructions, and, particularly, the emphasis of the analysis has been on a pair of self-dual reconstructions that have appeared in the literature. It is important that researchers and users of geodesic reconstructions know the distinctive properties and characteristics of them.

The focus of the analysis has been on: (a) whether the geodesic transformations are levelings; and (b) whether they can be used for building connected-component local operators.

As has been found out, one of them is not generally a leveling. This operator possesses a characteristic that makes it interesting for certain situations, as discussed in the paper: it can be used as basis for processing grains and pores independently from the rest of grains and pores, i.e., for building connected-component local connected operators.

Acknowledgements

This work has been supported in part by "Ministerio de Ciencia e Innovación" of Spain (Ref.: TIN2007-61768).

References

1. Matheron, G.: Random Sets and Integral Geometry. Wiley, New York (1975)
2. Serra, J.: Mathematical Morphology, vol. I. Academic Press, London (1982)
3. Serra, J. (ed.): Mathematical Morphology. Theoretical Advances, vol. II. Academic Press, London (1988)
4. Heijmans, H.: Morphological Image Operators. In: Hawkes, P. (ed.) Advances in Electronics and Electron Physics. Academic Press, Boston (1994)
5. Banon, G.: Formal introduction to digital image processing. INPE, São José dos Campos (2000)
6. Soille, P.: Morphological Image Analysis, 2nd edn. Springer, Heidelberg (2003)
7. Dougherty, E., Lotufo, R.: Hands-on Morphological Image Processing. SPIE Press, Bellingham (2003)
8. Serra, J., Salembier, P.: Connected operators and pyramids. In: Proceedings of SPIE, Non-Linear Algebra and Morphological Image Processing, San Diego, July 1993, vol. 2030, pp. 65–76 (1993)
9. Salembier, P., Serra, J.: Flat zones filtering, connected operators, and filters by reconstruction. IEEE Transactions on Image Processing 4(8), 1153–1160 (1995)
10. Ronse, C.: Set-theoretical algebraic approaches to connectivity in continuous or digital spaces. Journal of Mathematical Imaging and Vision 8(1), 41–58 (1998)
11. Heijmans, H.: Connected morphological operators for binary images. Computer Vision and Image Understanding 73, 99–120 (1999)
12. Monasse, P., Guichard, F.: Fast computation of a contrast-invariant image representation. IEEE Trans. on Image Proc. 9(5), 860–872 (2000)
13. Crespo, J., Serra, J., Schafer, R.W.: Image segmentation using connected filters. In: Serra, J., Salembier, P. (eds.) Workshop on Mathematical Morphology, Barcelona, May 1993, pp. 52–57 (1993)
14. Crespo, J.: Morphological Connected Filters and Intra-Region Smoothing for Image Segmentation. PhD thesis, School of Electrical and Computer Engineering, Georgia Institute of Technology (December 1993)

15. Crespo, J., Schafer, R.W.: Locality and adjacency stability constraints for morphological connected operators. Journal of Mathematical Imaging and Vision 7(1), 85–102 (1997)
16. Meyer, F.: From connected operators to levelings. In: Heijmans, H.J.A.M., Roerdink, J.B.T.M. (eds.) Mathematical morphology and its applications to image and signal processing, pp. 191–198. Kluwer Academic Publishers, Dordrecht (1998)
17. Meyer, F., Maragos, P.: Nonlinear scale-space representation with morphological levelings. Journal of Visual Communication and Image Representation 11(3), 245–265 (2000)
18. Meyer, F.: Levelings, image simplification filters for segmentation. Journal of Mathematical Imaging and Vision 20(1-2), 59–72 (2004)
19. Crespo, J.: Adjacency stable connected operators and set levelings. In: Banon, G.J.F., Barrera, J., Braga-Neto, U.d.M., Hirata, N.S.T. (eds.) Proceedings of the 8th International Symposium on Mathematical Morphology 2007 - ISMM 2007, October 2007. São José dos Campos, Universidade de São Paulo (USP), Instituto Nacional de Pesquisas Espaciais (INPE), vol. 1, pp. 215–226 (2007)
20. Crespo, J., Maojo, V.: The strong property of morphological connected alternated filters. Journal of Mathematical Imaging and Vision 32(3), 251–263 (2008)
21. Matheron, G.: Les nivellements. Technical Report N-54/99/MM, Report Centre de Morphologie Mathmatique, E.N.S. des Mines de Paris (February 1997)
22. Serra, J.: Connections for sets and functions. Fundamenta Informaticae 41(1-2), 147–186 (2000)
23. Maragos, P.: Algebraic and PDE approaches for lattice scale-spaces with global constraints. International Journal of Computer Vision 52(2/3), 121–137 (2003)
24. Serra, J., Vachier-Mammar, C., Meyer, F.: Nivellements. In: Najman, L., Talbot, H. (eds.) Morphologie mathématique 1: approches déterministes, pp. 173–200. Lavoisier, Paris (2008)
25. Meyer, F.: The levelings. In: Heijmans, H.J.A.M., Roerdink, J.B.T.M. (eds.) Mathematical morphology and its applications to image and signal processing, pp. 199–206. Kluwer Academic Publishers, Dordrecht (1998)

A Comparison of Spatial Pattern Spectra

Sander Land and Michael H.F. Wilkinson

Institute for Mathematics and Computing Science, University of Groningen,
P.O. Box 407, 9700 AK Groningen, The Netherlands

Abstract. Pattern spectra have frequently been used in image analysis.
A drawback is that they are not sensitive to changes in spatial distribution of features. Various methods have been proposed to address this
problem. In this paper we compare several of these on both texture classification and image retrieval. Results show that Size Density Spectra
are most versatile, and least sensitive to parameter settings.

1 Introduction

In image analysis, pattern spectra are a way of extracting information about the
amount of image content at various sizes from an image. The original area pattern
spectra, first introduced by Maragos [1], are insensitive to position information.
This can cause very different images to have nearly identical pattern spectra.
In response to this drawback, many variants have been developed to include
sensitivity to spatial arrangement of the image content at each scale [2, 3, 4].
Apart from choosing a variant, there are many parameters settings to choose.

This paper presents a thorough comparison of four types of spatial size spectra, with as main goal to investigate the differences between the various spatial
pattern spectra, and which ones are most suitable for certain tasks. We test generalized pattern spectra [3], multi-scale connectivity pattern spectra [5], spatial
size distributions [2], and size-density spectra [4]. This latter method has not
been used much yet, and this paper includes the first direct comparison of this
method against other pattern spectra. In all cases we will use area openings [6],
for which fast algorithms are available [7]. Choosing a single type of granulometry
ensures a fair comparison of the underlying strategies to add spatial information to pattern spectra. Connected granulometries were chosen for their speed
advantage over structural granulometries, especially when large numbers of bins
are needed in the pattern spectrum [8].

The methods will be compared on their performance in classifying simple
objects from the COIL-20 database and classifying textures from the Brodatz
database. Finally, there will be a test comparing their performance on contentbased image retrieval, that is, the problem of finding images similar to some
query images, in a database of images. This last task is especially interesting in
the context of online image searches, and is part of continuing research to extend
image searches to include visual information instead of performing queries just
based on the text surrounding the images.

M.H.F. Wilkinson and J.B.T.M. Roerdink (Eds.): ISMM 2009, LNCS 5720, pp. 92–103, 2009.
© Springer-Verlag Berlin Heidelberg 2009

2 Theory

Before going into the theory on spatial size spectra, we give some basic definitions. A binary image X is defined as some subset of the image domain $\mathbf{M}$ (usually $\mathbf{M} \subset \mathbb{Z}^2$), while a grey-scale image is a map $f : \mathbf{M} \to \mathbb{Z}$. Dilation, erosion and opening using a structuring element B are indicated by $\delta_B, \epsilon_B, \gamma_B$ respectively. The area A of an image X is defined as $A(X) = \#X$ for binary images, with $\#$ denoting the cardinality. In the case of grey scale images f area A must be substituted by the sum of grey levels S in pattern spectra, i.e. $S(f) = \sum_{x \in M} f(x)$.

We will also use the area opening [6,7], based on the connected opening $\Gamma_x(X)$, which yields the connected component of X that contains x, and the empty set if $x \notin X$. The area opening is defined as

$$\Gamma_\lambda^a(X) = \{x \in X | A(\Gamma_x(X)) \geq \lambda\}, \tag{1}$$

i.e., the area opening returns all connected components which have an area $\geq \lambda$. Parameter λ is the area threshold. For grey-scale images threshold decomposition can be used.

$$\gamma_\lambda^a(f)(x) = \max \{h | x \in \Gamma_\lambda^a(T_h(f))\}, \tag{2}$$

where $T_h(f) = \{x \in \mathbf{M} | f(x) \geq h\}$ is the threshold set at grey level h, and γ_λ^a is the area opening for grey-scale images. Fast algorithms can be found in [7].

2.1 Granulometries and Pattern Spectra

A granulometry [1,9] is a collection of openings $\{\gamma_\lambda\}_{\lambda \geq 0}$ with a size parameter λ such that $\gamma_\lambda \gamma_\mu = \gamma_{\max(\lambda,\mu)}$. Informally, it can be considered a collection of sieves, with the previous criterion indicating that using two sieves of different sizes has the same effect as only using the coarser sieve. The pattern spectra described in this paper all use area openings for the granulometry. A pattern spectrum [1] is a way to describe the result of applying a granulometry. It stores the amount of extra details $\gamma_{\lambda+1}$ removes compared to γ_λ. More formally the pattern spectrum PS_X is defined as

$$PS_X(\lambda) = A(\gamma_\lambda(X) \setminus \gamma_{\lambda+1}(X)), \tag{3}$$

in the binary case. For a grey-scale image f, the pattern spectrum becomes

$$PS_f(\lambda) = S(\gamma_\lambda(f)) - S(\gamma_{\lambda+1}(f)) = S(\gamma_\lambda(X) - \gamma_{\lambda+1}(f)), \tag{4}$$

where S denotes the summation of all grey values.

Generalized pattern spectra. A generalized pattern spectrum [3] is similar to a normal pattern spectrum, but uses a general parameterization with a vector $M(f)$ instead of the area $A(f)$. This ensures differences in spatial arrangement are noticed. One possibility for M is to use central moments μ_{ij}, of order $i + j$, with $i, j \in \mathbb{Z} \setminus \mathbb{Z}^-$, which are defined as

$$\mu_{ij} = \sum_{(x,y)} (x - \bar{x})^i (y - \bar{y})^j f(x, y), \tag{5}$$

where $(\bar{x}, \bar{y})$ is the center of gravity of the image. These can be normalized using the the sum of grey levels $S = \mu_{00}$, yielding the normalized central moments η_{ij} which are invariant under scaling:

$$\eta_{ij} = \frac{\mu_{ij}}{\mu_{00}^{1+(i+j)/2}}, \text{ for } i + j \geq 2. \tag{6}$$

Using these, it is also possible to define some quantities which are invariant under scaling, translation and rotation, such as Hu's moment invariants [10].

Spatial size distributions. Another way to capture the spatial distribution of applying a granulometry is to use spatial size distributions denoted SSD_f. Originally proposed by Ayala and Domingo [2], it is defined as:

$$SSD_f(\lambda, \mu) = \frac{1}{A(f)^2} \sum_{h \in \mu U} \sum_{x \in \mathbf{M}} f(x) f(x + h) - \gamma_\lambda(f)(x) \gamma_\lambda(f)(x + h) \tag{7}$$

where μU is some convex set U containing the origin scaled by a size parameter μ and $\mathbf{M} \subset \mathbb{Z}^2$ is an image domain. A typical choice for U would be a disc or square, and any granulometry can be used for γ_λ. We will use the area opening γ_λ^a for γ_λ. This function produces the difference between the overlap of $f(x)$ and its translates $f(x + h)$, and the overlap of the opened image $\gamma_\lambda(f)$ and its translates.

Multi-scale connectivity. Multi-scale connectivity is yet another extension to normal pattern spectra [5]. Instead of considering the spatial layout of applying a granulometry, it looks at how the connectivity of clusters in an image changes under a change of scale. One way to simulate such a change of scale is to use dilations. When zooming out on a scene, smaller clusters seem to merge. Applying a dilation also merges these clusters into one larger cluster, and is therefore somewhat similar to zooming out for the changes in connectivity. The clustering based connected opening using some extensive operator ψ is defined for a binary image as

$$\Gamma_\lambda^\psi(X) = \{x \in X \mid A(\Gamma_x(\psi(X)) \cap X) \geq \lambda\}. \tag{8}$$

This is an area opening which effectively looks at connected components in the extended image $\psi(X)$ while still restricting the area of these components to the original image X. The method extends to grey-scale as before as in (2). By using a series of dilation of the form δ_{iB}, which denotes i−times dilation with a structuring element B we can obtain a connectivity pyramid [5], which is sensitive to the mutual proximity of structures in the image. Using these openings, a multi-scale connectivity pattern spectrum MSC_f can be defined as:

$$MSC_f(\lambda, i) = S(\gamma_{\lambda+1}^{\delta_i B}(f) - \gamma_\lambda^{\delta_i B}(f)). \tag{9}$$

Because $\delta_{0B}(X) = X$, this spectrum includes a normal area pattern spectrum.

Size-density spectra. Size-density spectra combine sensitivity to spatial information with low sensitivity to noise. Originally proposed by Zingman et al. [4], they are significantly different from normal pattern spectra, and not based on the usual granulometries, but on the density opening $\theta_{B,d}$, defined as:

$$\theta_{B,d}(f) = f \wedge \bigvee_{x \in M} \{B_x \mid D(B_x, f) \geq d\} \text{ with } D(B_x, f) = S(B_x \wedge f)/S(B) \tag{10}$$

where B_x is a *grey-scale* structuring element B shifted by a vector x. Effectively, it is an opening which uses whether the overlap $D(B_x, f)$ exceeds some density d, allowing imperfect overlap, thereby making it less sensitive to noise. However, even for $d = 1$ it is still different from the usual structural opening. In the binary case it is identical to the rank-max opening [11].

This is an opening (anti-extensive, increasing and idempotent), but it lacks the absorption property with respect to the size of B. However, the opening does have the absorption property with respect to the density, i.e., $\theta_{B,d_1}\theta_{B,d_2} = \theta_{B,\max(d_1,d_2)}$. This can be used to define a normal pattern spectrum for a constant structuring element $PS(d) = A(\theta_{B,d(n)}(f) - \theta_{B,d(n+1)}(f))$. Although top-hat spectra, which do not need the absorption property, can be used instead, using normal pattern spectra is more efficient. The resulting size-density spectra used here are a concatenation of several pattern spectra (with density as parameter) for a range of sizes, also known as "type 2 size-density signatures". The structuring elements B are symmetric and completely at the maximum intensity, allowing fast calculation of density by convolution $D(B_x, f) = \frac{S(f*B)}{S(B)}$.

3   Comparison

In this section, the different pattern spectra will be compared on how well they perform on a variety of tasks.

3.1   COIL-20

The first test is on classification performance using the COIL-20 database [12]. This database consists of 20 objects photographed from 72 different angles, for a total of 1440 images of size 128×128. The nearest neighbour classifier is used with the L_1 distance. In most tests the features will be normalized to a standard deviation of 1, eliminating the large differences in average size between features. Table 1 contains the results, showing the mean precision $P(1)$, $P(10)$, $P(20)$, $P(50)$ for a variety of methods and settings. The mean precision $P(n)$ is defined

Table 1. Results for the COIL-20 database

Method	#features	P1	P10	P20	P50
Generalized Pattern Spectra					
gps$_{\lambda<2000,\text{lin}}$: $A,\overline{x},\overline{y},\eta_{20},\eta_{02},\eta_{11}$	120	0.950	0.776	0.679	0.516
gps$_{\lambda<2000,\text{lin}}$: Hu's invariants	140	0.673	0.471	0.394	0.280
gps$_{\lambda<4000,\text{lin}}$: $A,\overline{x},\overline{y},\eta_{20},\eta_{02}$	400	0.938	0.788	0.703	0.553
gps$_{\lambda<4000,\text{quad}}$: A	80	0.930	0.786	0.704	0.539
gps$_{\lambda<4000,\text{quad}}$: $A,\overline{x},\overline{y},\eta_{20},\eta_{02}$	400	0.980	0.854	0.770	**0.631**
Spatial Size Distributions					
ssd$_{\lambda<2000,\text{lin}}$	520	0.891	0.703	0.592	0.431
ssd$_{\lambda<4000,\text{quad}}$	520	0.948	0.816	0.713	**0.527**
Multiscale Pattern Spectra					
msc$_{\lambda<2000,\text{lin}}$	500	0.863	0.744	0.665	0.518
msc$_{\lambda<4000,\text{quad}}$	500	0.978	0.875	0.810	**0.661**
Size Density Spectra					
sds$_{r\%\leq10}$	70	0.983	0.844	0.750	0.606
sds$_{r\%\leq20}$	380	0.995	0.915	0.835	**0.664**
Combinations of pattern spectra					
gps + msc	900	0.996	0.927	0.860	0.723
sds + msc	880	0.997	0.931	0.867	0.731
sds + gps	780	0.999	0.941	0.876	0.740
gps + msc + sds	1280	1.000	0.952	0.891	**0.759**
Results with optimized parameters for generalized pattern spectra.					
gps$_{r\leq58}$	400	0.978	0.890	0.818	0.662
msc$_{r\leq58,i=0,3...57}$	500	0.990	0.916	0.837	0.682
ssd$_{r\leq58}$	400	0.969	0.870	0.777	0.586
sds$_{r\%\leq0.45..}$	495	0.983	0.883	0.802	0.630
sds$_{r\%\leq20}$ + msc$_{r\leq58}$ + gps$_{r\leq58}$	1280	1.000	0.954	0.897	0.750
Results with non-normalized features					
gps$_{\lambda<4000,\text{quad}}$: $A,\overline{x},\overline{y},\eta_{20},\eta_{02}$ (nn)	400	0.880	0.714	0.627	0.472
sds$_{r\%\leq20}$ (nn)	380	0.972	0.862	0.772	0.602

as the fraction of the first n results retrieved that belong to the correct class, averaged over all images. Normalization almost always yields an improvement of about 10%, so only a few results using non-normalized features are shown here to show typical differences.

The 'gps' results are from generalized pattern spectra [3]. A smaller test using 20 evenly spaced $\lambda < 2000$ showed that the best properties to use were area, center of gravity, and η_{20},η_{02}, giving a 15-20% improvement over area pattern spectra. Hu's invariants were shown to be not very suitable, yielding the lowest overall performance. Best results were obtained using quadratically spaced $\lambda = \pi r^2$ for evenly spaced r.

The 'ssd' results are from spatial size distributions [2], using $U = B(0,1)$, $\mu = 0,1,\ldots25$ and 20 values of λ spaced linearly or quadratically, as indicated.

The 'msc' results are from multi-scale connectivity spectra [5]. The number of dilations used is $i = 0,1,\ldots19$ and 25 values of λ spaced are linearly or quadratically, resulting in 500 features in total. This method performs well, and is better than the spatial size distributions and generalized pattern spectra, although the latter difference is not significant.

The 'sds' results were generated using size-density spectra, with settings $r_\% = 1, 2, \ldots 10, d = 0.2, 0.3, \ldots 0.8$ for $\mathrm{sds}_{r_\% \leq 10}$ and $r_\% = 1, 2, \ldots 20, d = 0.05, 0.15, \ldots 0.95$ for $\mathrm{sds}_{r_\% \leq 20}$, where $r_\%$ indicates the radius of the structuring element B as a percentage of the minimum of the image width and height. These size-density spectra perform best, although again the differences compared to generalized pattern spectra and multiscale connectivity spectra are very small. The various settings used show that even a very small size-density spectrum of only 70 features performs very well, and the method seems to be less sensitive to the exact parameters used than the other spectra are.

Combining several spectra also significantly improves the classifier in most cases. Much better results were obtained using a combination of a generalized pattern spectrum and a size-density spectrum, which manages to return a correct first result for 1438 of the 1440 images, and this combination also has the best $P(50)$ of all combinations of two pattern spectra. Combining all three yields a slightly higher $P(50)$ of 0.759 and returns a correct first result for all images.

A detailed analysis showed that the distribution of $P(50)$ over different classes is quite skewed, consisting mostly of almost perfect and very poor results. Although there are some differences between the various pattern spectra, overall they are fairly similar, failing mostly on the same objects. Also, among the objects which are hard to classify are several cars and several blocks. This could be caused by the classifier being unable to distinguish them from each other.

When compared to the results of shape-size spectra [8], the combinedspatially sensitive spectra outperform the shape-spectra at the $P(1)$ level (0.989 vs 1.000), despite the fact that Urbach et al. [8] use a more advanced classifier.

Optimizing the size parameters. As the generalized pattern spectra were shown to be quite sensitive to parameter settings it is interesting to see how much the result depends on the choice of the λ values.

We measured the classification performance as a function of the lower and upper r (with $\lambda = \pi r^2$ as usual), where 80 different r values were used, equally spaced between the lower and upper values in each run. The results showed that the smaller size scales are very important. We found that the choice of the upper r is important too, although overestimating the optimal value is better than underestimating it. The best results were obtained for $r < 58$, which corresponds to $\lambda \leq 10500$, or about 64% of the total image size.

Also interesting is to consider whether this choice of size parameter is also relevant for the other methods, these results are shown in table 1 under 'optimized results'. There is also significant improvement in the classification using spatial size distributions here. Surprisingly, performance of the similar multi-scale connectivity spectra only improves after also increasing the number of dilations used. The size-density spectra once again show completely different behaviour as the other spectra, with an equivalent an increase to $r_\% \leq 0.45$ significantly

decreasing the performance. This is probably because the blurring involved in
the density estimation becomes too strong for large r.

The few small improvements obtained here for the generalized and multi-scale
pattern spectra all disappear when once again combining pattern spectra, where
performance is similar to the results in the previous section, which shows that
spending much time optimizing these choices is probably not worth it.

In conclusion, the most important thing in choosing parameters for a generalized pattern spectrum is not to underestimate the maximum size and not to
overestimate the minimum size. When in doubt, using $0 - 66\%$, or even $0 - 90\%$
of the total image size is probably best. Also, normalizing features and using a
quadratically spaced size parameter is especially important in using these pattern spectra, with the best results being obtained this way.

3.2   Brodatz

For the next test, the performance of each method for texture classification
is compared. The Brodatz data set consists of 113 textures of size 640×640.
We selected 100 textures shown in Fig. 1 to eliminate flat textures. Next, 25
randomly selected squares of size 128×128 were taken from these and used as
the images, for a total of 2500 images. The results for the Brodatz test can be
seen in table 2. These results were obtained in the same way as with the COIL-20
database. Note that because there are only 25 images in each class, the $P(50)$ is
not shown here.

The generalized pattern spectra perform very poorly, with a normal pattern
spectrum outperforming any use of extra moments. This can be explained by the
fact this spectrum, unlike the rest is sensitive to *absolute* positional information,
not relative. Focussing on the smaller size scales ($\lambda \leq 400$) for a normal pattern
spectrum does not improve results compared to the $\lambda \leq 4000$ settings.

The multiscale connectivity pattern spectra are not very good at classifying these textures, although they perform better than the generalized pattern
spectra. This is possibly because dilating a texture will quickly make everything connected, after which the pattern spectra contain little useful information,
making their performance similar to normal pattern spectra. Overall, these methods simply does not seem suitable for use on textures.

Fig. 1. A random 128×128 image from each of Brodatz database classes, sorted from
worst to best classification using the best classifier found. The value of the mean $P(20)$
are indicated for each class in the outer rows.

Table 2. Results for the images generated using the Brodatz database

Method	#features	P1	P10	P20
histogram	25	0.921	0.724	0.569
Generalized Pattern Spectra				
$\text{gps}_{\lambda<4000,\text{quad}}: A$	80	0.320	0.209	**0.173**
$\text{gps}_{\lambda<4000,\text{quad}}: A, \eta_{20}, \eta_{02}$	240	0.288	0.165	0.134
$\text{gps}_{\lambda<400,\text{quad}}\ : A$	95	0.386	0.212	0.171
Spatial Size Distributions				
$\text{ssd}_{\lambda\leq1000,\text{lin}}$	520	0.555	0.374	0.290
$\text{ssd}_{\lambda\leq1250,\text{quad}}$	520	0.665	0.479	0.380
$\text{ssd}_{\lambda\leq400,\text{quad}}$	520	0.684	0.485	**0.385**
Multiscale Pattern Spectra				
$\text{msc}_{\lambda<2000,\text{quad}}$	500	0.495	0.270	**0.214**
Size Density Spectra				
$\text{sds}_{r_\%\leq10}$	70	0.808	0.569	0.442
$\text{sds}_{r_\%\leq20}$	380	0.803	0.542	0.413
$\text{sds}_{r_\%\leq10}$ (2)	380	0.869	0.627	**0.495**
Combinations of pattern spectra				
$\text{ssd}_{\lambda\leq400,\text{quad}} + \text{sds}_{r_\%\leq10}$ (2)	900	0.92	0.725	**0.587**

Spatial size distributions are reasonably good at distinguishing different textures, and performance improves considerably when using quadratically spaced λ. Again, focussing on the small details only barely improves the classifier. The size-density spectra are once again the best performing method. Looking at the difference between the $\text{sds}_{r_\%\leq10}$ and $\text{sds}_{r_\%\leq20}$ shows that looking at smaller size scales improves the result. Indeed, calculating some more features in this size range ($\text{sds}_{r_\%\leq10}$ (2), $r_\% = 0.5, 1, \ldots 10$) yields the best results. Again, the method is not very sensitive to the choice of parameters.

Combining the best spectra, spatial size distributions and size-density spectra, for a final round of classification yields the best results. However, even this elaborate combination of several pattern spectra has trouble matching the performance of a simple grey-value histogram. Also, previous work shows that shape-size spectra [8] can obtain significantly better results ($P(1) = 0.965$) even on more dificult tests based on this database.

Overall, spatial pattern spectra are not very suitable for use with textures, as textures usually do not contain the type of global spatial information that spatial pattern spectra describe. This is in stark conttrast to size-shape spectra [8], where a classification performance (comparable to our $P(1)$ result) of 0.965 was achieved. This suggests that shape is more important in classifying texture than positional information.

3.3   ImageCLEF 2007 Photographic Retrieval Task

In this section, the pattern spectra will be tested to see how well they perform on an content-based image retrieval task. The task in question is the ImageCLEF

2007 photographic retrieval task[1]. The IAPR TC-12 photographic collection is a database of 20,000 still natural images taken from locations around the world, including pictures of different sports and actions, photographs of people, animals, cities, landscapes and many others. Three query images are provided for each of the 60 queries, and the task is to find the 1000 most similar images as in [13]. Evaluation is done using the provided list of relevant results for each query. As performance measures we use the mean average precision (MAP) as in [13]. The average precision is calculated by taking the mean of the precision values of the list truncated after each relevant document. Unlike a single precision value $P(n)$, which yields the same value regardless of whether the relevant documents are the first k or the last k in the first n, the average precision rewards returning the relevant documents earlier. Because there are multiple query images, there are several ways to determine the nearest neighbour even when just using the L_1 distance using normalized features. The most important distance measures tried were minimum, maximum and average distance to a query image. Of these the minimum distance outperformed the others by far, and is the only one shown.

Results. Because of the computational cost of working with so many images the faster generalized pattern spectra and multi-scale connectivity spectra will be investigated first, and tests are limited to the best performing ones in the COIL-20 case. Three different pattern spectra of each type were computed, one for each 30% size range, with 80 features per property in each 30% interval. Table 3 shows that the smaller details are the most important. This differs from the results in [13], where larger details were most important. This can be explained by considering that the shape of large background objects like mountains may be more important than their size or spatial distribution. Even though the larger scales perform poorly on their own, adding these features may improve results.

For the multi-scale connectivity spectra, the settings for iterations used in the COIL-20 tests were first tested. To avoid the curse of dimensionality , a

Table 3. Results for generalized and multiscale pattern spectra for the ImageCLEF 2007 photographic retrieval task. Shown are the MAP (mean average precision) values using the minimum L_1 distance to the query images. The 2^n columns use $i = 0, 1, 2, 4, 8, 16, 32, 64$.

	Generalized Pattern Spectra			Multiscale Pattern Spectra			
	$A, \bar{x}, \bar{y}$	A (rgb)	$A, \bar{x}, \bar{y}$ (rgb)	≤ 19	$0, 6 \ldots 54$	2^n	2^n (rgb)
0-30%	0.0125	0.0165	**0.0169**	0.0144	0.0148	0.0190	0.0254
30-60%	0.0052	0.0063	0.0058	0.0074	0.0076		
60-90%	0.0050	0.0054	0.0062	0.0065	0.0062		
0-60%	0.0125	0.0161	0.0152	0.0165	0.0190		
30-90%	0.0071	0.0082	0.0098	0.0103	0.0096	0.0144	0.0195
0-90%	0.0160	0.0153	0.0169	0.0193	0.0211	0.0265	**0.0318**

---

[1] See http://eureka.vu.edu.au/~grubinger/ImageCLEFphoto2007/adhoc.htm

shorter feature vector was computed, in which the radius of the dilations was distributed exponentially as $i = 0, 1, 2, 4, 8, 16, 32, 64$. This was done because earlier tests showed the benefits of having high resolution at low scales. Also, because the larger features perform poorly, they were put together with 25 bins for size in one 60% interval. These settings perform significantly better that all others. Because fewer features are used, extension to color images is easier, yielding another 20% improvement.

Finally, a limited number of tests were done on the spatial size distributions and size-density spectra. These results are shown in table 4, together with more detailed performance measures for the best results obtained in previous tests. This shows the size-density spectra performing better than the generalized pattern spectra, although they are also significantly slower. They are also significantly more sensitive to choice of parameters than in previous tests, with only the smaller size scales performing well. Extending the best performing spectra to use color data yields a 60% improvement, much higher than the improvements shown in the other pattern spectra and competitive with the multi-scale spectra.

Table 4. Detailed results for the ImageCLEF 2007 photographic retrieval task. The bold column heading indicates the performance measure used in table 3, and "hp" indicates classification by distance to a hand-picked "best" query image, as in [13].

Method	#features	time	$P(20)$ min	$P(100)$ min	**MAP** **min**	MAP hp
gps (best,0-90%)	720	0.55s	0.055	0.021	0.0160	0.0132
msc (best, 0-90%)	400	1.0s	0.080	0.029	0.0265	0.0235
ssd 0-30%	520	10s	0.034	0.016	0.0091	0.0091
$sds_{r_\% \leq 10}$	70	5.7s	0.058	0.021	0.0176	0.0155
gps (best rgb,0-30%)	720	0.58s	0.060	0.024	**0.0169**	0.0165
msc (best rgb, 0-90%)	1200	3.1s	0.097	0.034	**0.0318**	0.0293
$sds_{r_\% \leq 10}$ (rgb)	210	17s	0.090	0.032	**0.0285**	0.0305
Tushabe 0-100%	600					0.0215
Tushabe 30-100%	600					0.0273
Tushabe 30-100% (rgb)	1800					0.0337

Spatial size distributions perform worse, and are much slower, than generalized pattern spectra. Given this extremely poor performance on the $0-30\%$ size scale, no more tests were done to extend this to larger scales.

The multi-scale pattern spectra give the best results at a reasonable computational cost. However, their sensitivity to choice of parameters means that large performance gains might be lost by making poor choices for these. Results for the best size-density spectrum are similar. Combining both pattern spectra yields a significant improvement, giving a MAP of 0.0396. Compensating for the difference in number of features by including the size-density spectra two or three times yields the best result of 0.0416.

Grubinger et al. [14] report an average MAP of 0.0681 over all purely image-based methods in ImageCLEF 2007. These methods mostly involve relevance feedback by users during a training stage, or query expansion by providing more query images, or combinations of the two, whereas neither were used here. The comparatively low performance does not mean that these pattern spectra do not work well. In combination with other methods, and using query expansion and relevance feedback the performance is expected to increase. Indeed, in a later extension, using multiple shape-size pattern spectra and more advanced retrieval method, Tushabe and Wilkinson [15] already improved their results to a MAP value of 0.0571 (70% improvement).

4   Conclusions

The three different problems considered all have quite different dynamics, and few conclusions can be made that apply to all of them. Generaly, normalizing features and using quadratically spaced binning for the size parameter improved performance. Also, combining several pattern spectra, especially when size-density spectra are included, tends to work very well.

Classification using the COIL-20 and Brodatz tests both show that considerable effort is needed to determine the optimal parameters. It is safer to overestimate than underestimate the maximum scale. Size-density spectra are more robust in this sense. For the COIL-20 test, only the spatial size distributions performed worse than the other pattern spectra, whereas in the Brodatz test, the generalized pattern spectra and multi-scale spectra performed poorly. This suggests that *local* spatial pattern spectra are probably not the best choice for texture segmentation. Size-shape spectra are probably better [8].

In the test for the ImageCLEF 2007 photographic retrieval task, the multi-scale connectivity spectra are best. However, the size-density spectra almost matched their performance and improved performance when combined with them. Unlike the COIL-20 test and also unlike earlier work using shape-size spectra [13], the larger size scales were shown being less important for spatial pattern spectra, although they often still contributed to better performance.

The spatial size distributions [2] were only useful in the texture classification test, where they finished second. They are, however, very slow. Only the size-density spectra [4] performed well in all three tests, but are also very slow compared to some of the other pattern spectra. The current implementation uses about 10-20 seconds to process a single 0.2 megapixel image, and this time is unlikely to be improved much, with about 50% of this time needed for the highly optimized density calculation.

Overall, there are inherent limits in only using area and spatial information at various size scales. In the future we will combine spatial pattern spectra with techniques capturing different information, such as shape [13, 8]. Furthermore, all pattern spectra are based on the notion of image extrema. This paradigm is not always the best when dealing with colour images. In our case we simply used marginal processing, which worked best on size-shape spectra. It would be interesting to see whether methods based on, e.g., binary partition trees [16] could be adapted to compute pseudo-pattern spectra efficiently.

References

1. Maragos, P.: Pattern spectrum and multiscale shape representation. IEEE Trans. Pattern Anal. Mach. Intell. 11(7), 701–716 (1989)
2. Ayala, G., Domingo, J.: Spatial size distributions: Applications to shape and texture analysis. IEEE Trans. Pattern Anal. Mach. Intell. 23(12), 1430–1442 (2001)
3. Wilkinson, M.H.F.: Generalized pattern spectra sensitive to spatial information. In: ICPR 2002: Proceedings of the 16th International Conference on Pattern Recognition, vol. 1, pp. 21–24. IEEE Computer Society, Los Alamitos (2002)
4. Zingman, I., Meir, R., El-Yaniv, R.: Size-density spectra and their application to image classification. Pattern Recogn. 40(12), 3336–3348 (2007)
5. Braga-Neto, U., Goutsias, J.: Object-based image analysis using multiscale connectivity. IEEE Trans. Pattern Anal. Mach. Intell. 27(6), 892–907 (2005)
6. Cheng, F., Venetsanopoulos, A.N.: An adaptive morphological filter for image processing. IEEE Trans. Image Proc. 1, 533–539 (1992)
7. Meijster, A., Wilkinson, M.H.F.: A comparison of algorithms for connected set openings and closings. IEEE Trans. Pattern Anal. Mach. Intell. 24(4), 484–494 (2002)
8. Urbach, E.R., Roerdink, J.B.T.M., Wilkinson, M.H.F.: Connected shape-size pattern spectra for rotation and scale-invariant classification of gray-scale images. IEEE Trans. Pattern Anal. Mach. Intell. 29(2), 272–285 (2007)
9. Breen, E.J., Jones, R.: Attribute openings, thinnings, and granulometries. Comput. Vis. Image Underst. 64(3), 377–389 (1996)
10. Hu, M.K.: Visual pattern recognition by moment invariants. IRE Transactions in Information Theory IT-8, 179–187 (1962)
11. Ronse, C.: Erosion of narrow image features by combination of local low rank and max filters. In: Proc. Int. Conf. Image Proc. 1986, London, pp. 77–81 (1986)
12. Nene, S.A., Nayar, S.K., Murase, H.: Columbia object image library (COIL-20). Technical Report CUCS-005-96, Columbia University (1996)
13. Tushabe, F., Wilkinson, M.H.F.: Content-based image retrieval using combined 2D attribute pattern spectra. In: In Working Notes of the 2007 CLEF Workshop (2007),
    http://www.clef-campaign.org/2007/working_notes/TushabeCLEF2007.pdf
14. Grubinger, M., Clough, P., Hanbury, A., Müller, H.: Overview of the ImageCLEF-photo 2007 photographic retrieval task. In: Peters, C., Jijkoun, V., Mandl, T., Müller, H., Oard, D.W., Peñas, A., Petras, V., Santos, D. (eds.) CLEF 2007. LNCS, vol. 5152, pp. 433–444. Springer, Heidelberg (2007)
15. Tushabe, F., Wilkinson, M.H.F.: Content-based image retrieval using combined 2D attribute pattern spectra. In: Peters, C., Jijkoun, V., Mandl, T., Müller, H., Oard, D.W., Peñas, A., Petras, V., Santos, D. (eds.) CLEF 2007. LNCS, vol. 5152, pp. 554–561. Springer, Heidelberg (2008)
16. Salembier, P., Garrido, L.: Binary partition tree as an efficient representation for image processing, segmentation and information retrieval. IEEE Trans. Image Proc. 9(4), 561–576 (2000)

Differential Equations for Morphological Amoebas

Martin Welk, Michael Breuß, and Oliver Vogel

Mathematical Image Analysis Group
Faculty of Mathematics and Computer Science, Campus E1.1
Saarland University, 66041 Saarbrücken, Germany
{welk,breuss,vogel}@mia.uni-saarland.de
http://www.mia.uni-saarland.de

Abstract. This paper is concerned with amoeba median filtering, a structure-adaptive morphological image filter. It has been introduced by Lerallut et al. in a discrete formulation. Experimental evidence shows that iterated amoeba median filtering leads to segmentation-like results that are similar to those obtained by self-snakes, an image filter based on a partial differential equation. We investigate this correspondence by analysing a space-continuous formulation of iterated median filtering. We prove that in the limit of vanishing radius of the structuring elements, iterated amoeba median filtering indeed approximates a partial differential equation related to self-snakes and the well-known (mean) curvature motion equation. We present experiments with discrete iterated amoeba median filtering that confirm qualitative and quantitative predictions of our analysis.

Keywords: morphological amoebas, median filtering, partial differential equations.

1 Introduction

Morphological amoebas are a class of morphological image filters in which structuring elements adapt to image structures with a maximum of flexibility. They have been introduced by Lerallut et al. [11,12]. In the amoeba construction, the structuring elements adapt locally to the variation of grey (or colour) values, also taking into account the distance to the origin pixel. Thereby, large deviations in the image values are penalised, so that the amoebas may grow around corners or along anisotropic image structures. Using the resulting shape as a structuring element, many filtering procedures can be applied on it. In this paper, we are particularly interested in the use of the median filter.

Iterated application of amoeba median filtering (AMF) can be carried out in different ways. In [11], a *pilot image* is used to steer the iterated processes via an alternating procedure. This works as follows. A smoothed version of the original image f is used for constructing amoebas for all pixels. Then, the median filter is applied using the corresponding structuring elements. The filtered image is

M.H.F. Wilkinson and J.B.T.M. Roerdink (Eds.): ISMM 2009, LNCS 5720, pp. 104–114, 2009.

in turn used for constructing new amoebas, and these amoebas are then used as structuring elements to filter the original image f. We concentrate for this paper on more straightforward iterative procedures for AMF, using pixelwise the following steps subsequently: (i) amoeba construction, and (ii) median filtering using the amoeba as structuring element.

For iterated median filtering with a fixed structuring element, work by Guichard and Morel [7] has brought out that, in the continuous-scale limit, it approximates the *partial differential equation (PDE)* $u_t = |\nabla u| \operatorname{div}(\nabla u / |\nabla u|)$, known as (mean) curvature motion [1]. In this sense, iterated discrete median filtering with a fixed structuring element can be understood as a specific discretisation of that PDE.

Iterated AMF simplifies images towards a cartoon-like appearance with homogeneous regions separated by sharp contours. Even corners are preserved fairly well, in contrast to median filtering with a fixed structuring element. Using PDE approaches, similar segmentations can be achieved e.g. by so-called self-snakes [14,18]. These are filters that stand in close relationship to curvature motion, with the difference that the evolution is modulated by an edge-stopping function depending on the local image gradient. Thereby the displacement of edges is avoided, and edges are sharpened. In the light of Guichard and Morel's above-mentioned result it is therefore natural to ask whether there exists a similar correspondence between a continuous-scale limit case of amoeba filters and a self-snakes-like PDE.

In the present paper, we address this question. We prove that iterated amoeba filtering can indeed be understood as a discrete approximation of a PDE which is related to curvature motion. We discuss how different choices for the distance measures involved in the amoeba definition influence the limit case.

Our results extend the framework of known correspondences between discrete and PDE formulations of morphological filters. The study of these relationships helps to gain a unified view on image filtering methods and to combine advantages of both approaches.

Related work. Median filtering in its non-adaptive form goes back to Tukey [16] and became common as a structure-preserving image filter in the 90s [6,9].

On the PDE side, (mean) curvature motion for image smoothing has been proposed by Alvarez et al. [1], already together with the generalisation of the basic PDE by multiplying the right-hand side with a decreasing function of the image gradient. Sapiro [14] proposed a variant of this idea, named self-snakes, in which the edge-stopping factor is placed *within* the divergence expression. While curvature motion smoothes in level-line direction only, Caselles et al. [3] defined for image interpolation purposes a process that smoothes exclusively in gradient direction, called *adaptive monotone Lipschitz extension* (AMLE). The representation of an image as a manifold embedded in the product space of image domain and greyvalue range has been introduced in PDE-based image filtering with the so-called Beltrami framework by Kimmel et al. [8] and Yezzi [19].

Since the seminal paper by Guichard and Morel [7] further cross-relationships between discrete and PDE-based image filters have been studied. For example, van den Boomgaard [17] proved a PDE approximation result for the Kuwahara-Nagao operator [10,13]. Didas and Weickert [5] studied correspondences between adaptive averaging and a class of generalised curvature motion filters. Barash [2] and Chui and Wang [4] considered PDE limits of bilateral filters [15].

Structure of the paper. The paper is organised as follows. In Section 2 we describe the discrete algorithm. Our main contribution, namely the derivation of a PDE corresponding to AMF, follows in Section 3. In Section 4, we show some test results. The paper is finished with a conclusion in Section 5.

2   The Discrete Amoeba Construction

The basic procedure is described in Lerallut et al.'s papers [11,12]. Here, we give a brief account of the algorithm in the form we have implemented, which is slightly modified in a few points that will be pointed out in the sequel.

In the following, we work with images f whose pixels are numbered by integers, such that f_i denotes the grey value of the pixel with index i. The coordinates of this pixel are denoted by (x_i, y_i). We distinguish the initial image f from the iterated images $u^{(n)}$, where n denotes the iteration number. For starting the iterative process, we set $u^{(0)} := f$. On the amoebas whose construction is described below the standard median filter is applied.

Description of the algorithm. For each pixel i_0 with (x, y)-coordinates (x_{i_0}, y_{i_0}), an adaptive structuring element is determined as follows. We consider pixels i^* within a prescribed maximal Euclidean distance ϱ of pixel i_0. The number ϱ represents the maximal size of the shape of the amoeba, since it will also be used for limiting the allowed *amoeba distance*. For the so pre-selected pixels we consider paths $(i_0, i_1, \ldots, i_k \equiv i^*)$ that connect i_0 with i^* via a sequence of pixels in which each two subsequent pixels i_j, i_{j+1} are neighbours. Among all these, we determine the shortest path P with respect to the amoeba distance $L(P)$. If the amoeba distance is below ϱ for P, the pixel i^* is accepted as a member of the amoeba structuring element.

It remains to specify the amoeba distance as well as the neighbourhood relation between subsequent pixels. In [11,12], the amoeba distance is given by

$$L_L^{(n)}(P) = \sum_{m=0}^{k-1} 1 + \sigma \sum_{m=0}^{k-1} \left| u_{i_{m+1}}^{(n)} - u_{i_m}^{(n)} \right|, \tag{1}$$

where $\sigma > 0$ is a parameter that penalises large deviations in grey value data, and each pixel is required to be in the 4-neighbourhood of its predecessor, i.e. a horizontal or vertical neighbour. Note that this definition involves the measurement of spatial distances by the city-block metric, since the first sum in (1) counts the pixels in the path P (without the starting pixel i_0). Moreover, spatial and tonal distances (i.e. greyvalue differences) are combined via an l_1 sum.

In our implementation, we use a metric that better approximates the Euclidean distance in space. To this end, we use 8-neighbourhoods that include horizontal, vertical, and diagonal neighbours, and use the Euclidean distance on these pixel pairs. This results in shorter paths compared to the procedure of Lerallut et al., as well, conceptually, in an improvement in terms of rotational invariance. For the way how spatial and tonal distances are combined we consider either a Euclidean sum, or an l_1 sum like in (1), which leads finally to two alternative amoeba distance measures L_2 and L_1 given by

$$L_2^{(n)}(P) = \sum_{m=0}^{k-1} \sqrt{\left(x_{i_{m+1}} - x_{i_m}\right)^2 + \left(y_{i_{m+1}} - y_{i_m}\right)^2 + \sigma^2 \left(u_{i_{m+1}}^{(n)} - u_{i_m}^{(n)}\right)^2}, \quad (2)$$

$$L_1^{(n)}(P) = \sum_{m=0}^{k-1} \left(\sqrt{\left(x_{i_{m+1}} - x_{i_m}\right)^2 + \left(y_{i_{m+1}} - y_{i_m}\right)^2} + \sigma \left| u_{i_{m+1}}^{(n)} - u_{i_m}^{(n)} \right| \right). \quad (3)$$

3 Space-Continuous Analysis

For our further investigation, we need a space-continuous formulation of AMF. We base this on the representation of a (smooth) image u by its graph $\Gamma = \Gamma_{u,\sigma} := \{p(x,y) = (x, y, \sigma u(x)) \mid (x,y) \in \Omega\}$ where $\Omega \subset \mathbb{R}^2$ is the image domain, and σ a scaling parameter for grey-values as in (1)–(3). Note that this embedding is analogous to the Beltrami framework, compare [19]. The surface Γ is equipped with a metric d which can be obtained by restricting the Euclidean metric of the embedding space $\mathbb{R}^3$, i.e.

$$d(p_1, p_2) \equiv d_2(p_1, p_2) = \min \int_0^1 \sqrt{x'(s)^2 + y'(s)^2 + \sigma^2 u'(s)^2} \, ds \quad (4)$$

where the minimum is taken over all curves $[0,1] \to \Gamma$ that start in $p_1 := p(x_1, y_1)$ and end in $p_2 := p(x_2, y_2)$. Alternatively, and closer to the setting of [11], one can use an l_1 sum of the Euclidean distance in space and the greyvalue distance,

$$d(p_1, p_2) \equiv d_1(p_1, p_2) = \min \int_0^1 \left(\sqrt{x'(s)^2 + y'(s)^2} + \sigma |u'(s)| \right) ds. \quad (5)$$

One step of amoeba filtering then reads as follows. For a given location (x_0, y_0) in the image domain, an amoeba structuring element $\mathcal{A}(x_0, y_0)$ is constituted by all locations (x,y) for which $d(p(x_0, y_0), p(x, y))$ does not exceed a given radius ϱ. Typical shapes of amoeba structuring elements with both metrics are shown in Figure 1. It is worth noticing that with the metric (4) the boundary of $\mathcal{A}(x_0, y_0)$ crosses the level line through (x_0, y_0) orthogonally and smoothly, while with (5) it has kinks at the intersection points, giving the structuring element a digonal overall shape in contrast to the elliptical contour with (4).

Once the structuring element has been constructed, the median of all grey-values within the structuring element is taken, i.e. the value μ whose level line

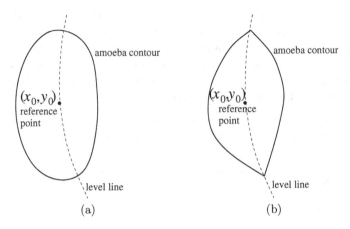

Fig. 1. Amoeba structuring elements. **(a)** Typical amoeba with metric $d \equiv d_2$ from (4). **(b)** Typical amoeba with metric $d \equiv d_1$ from (5).

(the curve along which $u(x, y) = \mu$ holds) cuts $\mathcal{A}(x_0, y_0)$ into two parts of equal area. In the filtered image, μ becomes the new grey-value at location (x_0, y_0).

We analyse this filter now in a manner similar to Guichard and Morel's approach [7]. We focus first on the case of the metric $d \equiv d_2$, see (4). Without loss of generality, we assume that we are dealing with the location $(x_0, y_0) = (0, 0)$. We assume further that $u(x_0, y_0) = 0$, and that the image gradient at (x_0, y_0) is given by $\nabla u(x_0, y_0) = (\alpha/\sigma, 0)^{\mathrm{T}}$ with some positive α. Then σu possesses the Taylor expansion

$$\sigma u(x, y) = \alpha x + \beta x^2 + \gamma xy + \delta y^2 + \mathcal{O}(\varrho^3) \tag{6}$$

within $\mathcal{A} = \mathcal{A}(x_0, y_0)$, where we have used that $x, y = \mathcal{O}(\varrho)$.

Consider now a value $z = \mathcal{O}(\varrho)$. We are interested in the level line of u corresponding to the grey-value z/σ, restricted to $\mathcal{A}$. On this line, $\sigma u(x, y) = z$ holds. Due to the prescribed gradient direction of u, level lines of u within $\mathcal{A}$ are roughly oriented in y direction. We can therefore express the level line by writing x as a function of y. Resolving the equation $\sigma u(x) = z$ for x yields

$$x = x(y) = \left(\frac{z}{\alpha} - \frac{z^2\beta}{\alpha^3} \right) - \frac{z\gamma}{\alpha^2} y - \frac{\delta}{\alpha} y^2 + \mathcal{O}(\varrho^3) . \tag{7}$$

(As a quadratic equation needs to be solved, there is a second solution which is, however, outside $\mathcal{A}$ if ϱ is small enough.) The length of the level line segment within $\mathcal{A}$ acts as a weight with which the value $u = z/\sigma$ enters the computation of the median μ. The end points of this segment are obtained by equating $d_2(p(x_0, y_0), p(x(y), y))$ to ϱ. Approximating d_2 by the Euclidean distance within $\mathbb{R}^3$, this equation becomes $x(y)^2 + y^2 + z^2 = \varrho^2$, a quadratic equation for y with two solutions y_1, y_2. The length $L(z)$ of the level line segment within $\mathcal{A}$ equals up to $\mathcal{O}(\varrho^3)$ the difference $|y_1 - y_2|$. We compute therefore

$$L(z) = 2\varrho \sqrt{1 - \frac{z^2(\alpha^2+1)}{\varrho^2\alpha^2}} \left(1 + \frac{z\delta}{\alpha^2} + \frac{z^3\beta}{\alpha^2(\alpha^2\varrho^2 - z^2(\alpha^2+1))} \right) + \mathcal{O}(\varrho^3) . \quad (8)$$

The median μ is now determined by the equality

$$\int_{Z_-}^{\sigma\mu} L(z)\,\mathrm{d}z = \int_{\sigma\mu}^{Z_+} L(z)\,\mathrm{d}z , \quad (9)$$

where Z_+ and Z_- are the smallest positive and largest negative values for which $L(Z_+) = L(Z_-) = 0$. One has $Z_+, Z_- = Z^* + \mathcal{O}(\varrho^3)$ with $Z^* = \varrho\alpha/\sqrt{\alpha^2+1}$. Provided that $\mu = \mathcal{O}(\varrho^2)$, the equality (9) can be transformed into

$$\int_0^{Z^*} (L(z) - \dot{L}(-z))\,\mathrm{d}z = 2\sigma\mu L(0) + \mathcal{O}(\varrho^4) . \quad (10)$$

Resolving the integral on the left-hand side analytically yields $\frac{4\varrho^3\delta}{3(\alpha^2+1)} + \frac{8\varrho^3\beta}{3(\alpha^2+1)^2}$. Together with $L(0) = 2\varrho + \mathcal{O}(\varrho^3)$, this implies

$$\mu = \frac{\varrho^2}{3\sigma} \left(\frac{\delta}{\alpha^2+1} + \frac{2\beta}{(\alpha^2+1)^2} \right) + \mathcal{O}(\varrho^3) \quad (11)$$

which can be restated in terms of spatial derivatives of u as

$$\mu = \frac{\varrho^2}{6} \left(\frac{u_{yy}}{1+\sigma^2 u_x^2} + \frac{2u_{xx}}{(1+\sigma^2 u_x^2)^2} + \mathcal{O}(\varrho) \right) . \quad (12)$$

One amoeba median filter step acts therefore approximately like one time step of an explicit scheme for the PDE

$$u_t = \frac{u_{\xi\xi}}{1+\sigma^2|\nabla u|^2} + \frac{2u_{\eta\eta}}{\left(1+\sigma^2|\nabla u|^2\right)^2} \quad (13)$$

with time step size $\tau = \varrho^2/6$. On the right-hand side, second derivatives are taken in the directions of the normalised gradient vector $\eta := \nabla u/|\nabla u|$ and the perpendicular vector $\xi := \eta^\perp$, the tangential vector of the local level line of u.

When ϱ tends to zero, the iterated amoeba median filter therefore converges to the PDE (13). The first summand of the right-hand side of (13) can obviously be interpreted as curvature motion $u_t = u_{\eta\eta}$ modulated in the way proposed in [1] by an edge-stopping factor $g_1(|\nabla u|) := \left(1+\sigma^2|\nabla u|^2\right)^{-1}$. It can also be compared to the self-snakes PDE [14,18]

$$u_t = |\nabla u|\,\mathrm{div} \left(g(|\nabla u|) \frac{\nabla u}{|\nabla u|} \right) = g(|\nabla u|)u_{\xi\xi} + \langle \nabla g(|\nabla u|), \nabla u \rangle , \quad (14)$$

except that the term $\langle \nabla g, \nabla u \rangle$ is not present. As this "shock term" contributes to the edge-enhancing properties of the self-snakes evolution, the edge-enhancing effect may be less pronounced with the amoeba filter than with self-snakes.

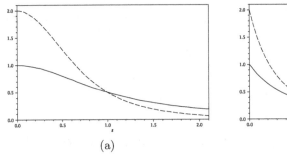 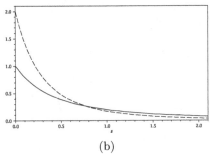

(a)   (b)

Fig. 2. Edge-stopping functions in PDEs approximated by iterated amoeba median filtering. For visualisation, σ is fixed to 1. **(a)** Weight functions $g_1 = \left(1 + |\nabla u|^2\right)^{-1}$ for the curvature motion term (solid line), $g_2 = 2\left(1 + |\nabla u|^2\right)^{-2}$ for the AMLE term (dashed line) from the PDE (13) based on the Euclidean amoeba metric (4). **(b)** Corresponding weight functions for the amoeba metric (5).

The second summand of (13) resembles the AMLE [3] evolution $u_t = u_{\eta\eta}$, but with an edge-stopping factor $g_2(|\nabla u|) := 2\left(1 + \sigma^2 |\nabla u|^2\right)^{-2}$. Note that g_2 decreases faster than g_1, with $g_1 = g_2$ for $|\nabla u| = \sigma^{-1}$, see Figure 2 (a). At all locations where the gradient is sufficiently large, the PDE (13) is therefore dominated by the self-snakes-like modulated curvature motion part. The AMLE contribution dominates in almost flat image regions.

A similar analysis applies if instead of d_2 the metric d_1 from (5) is used. The resulting equation is again of the form $u_t = g_1(|\nabla u|)u_{\xi\xi} + g_2(|\nabla u|)u_{\eta\eta}$ with decreasing functions g_1, g_2 of the gradient. Here, g_1 and g_2 are given by complicated integral expressions that are best evaluated numerically, see Figure 2 (b). The derivation for this case will be published in a forthcoming paper.

4   Experiments

We present two experiments that confirm the behaviour suggested by the analytical results from the previous section.

The *House* experiment. In this experiment we use a relatively "simple" image in order to investigate the influence of parameters, see Figure 3.

Subfigure (a) shows the original image. Figure 3(b) depicts the steady state achieved by standard median filtering employing a fixed (3×3) structuring element. As usual with median filtering, the shape of edges is rounded, and the facade of the depicted house is quite non-uniform in its grey value distribution. The use of a larger non-adaptive structuring element will distort the shape of important image features.

In Figure 3(c–f) we compare the results of iterated AMF using the L_2 amoeba distance together with varying parameters.

We start with a relatively strong penalisation of grey value differences given by $\sigma = 0.25$, see (c, d). As predicted, we observe the influence of the self-snakes

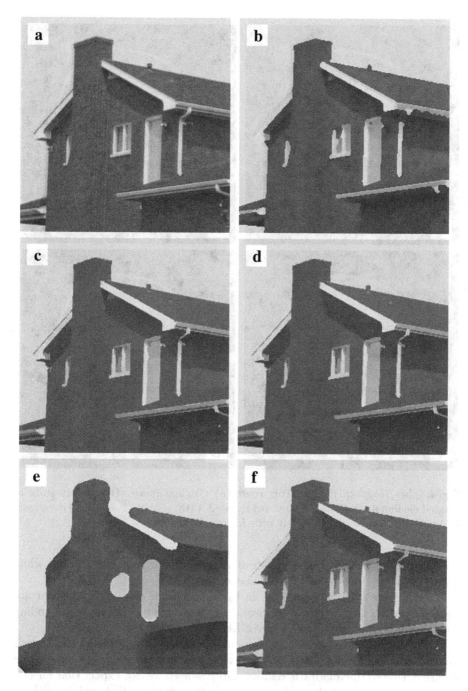

Fig. 3. The *House* experiment. **Top row:** (a) Original image. (b) Filtered with iterated median filter, 3×3 stencil, 40 iterations. **Middle row:** (c) Iterated AMF, $\varrho = 10$, $\sigma = 0.25$, 4 iterations. (d) Same as in (c) but 20 iterations. **Bottom row:** (e) Iterated AMF, $\varrho = 10$, $\sigma = 0.02$, 10 iterations. (f) Iterated AMF, $\varrho = 20$, $\sigma = 0.25$, 1 iteration.

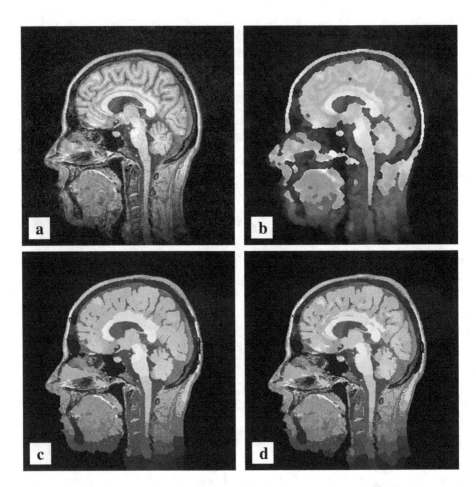

Fig. 4. The *Head* experiment. **Top row:** (a) Original image. (b) Steady state of iterated median filter. **Bottom row:** (c) Iterated AMF, $\varrho = 10$, $\sigma = 0.25$, 10 iterations, L_2 amoeba distance. (d) Same but with L_1 amoeba distance.

very clearly by the sharp transition of regions of different grey values, while nearly flat image regions are flattened even more.

When a very small σ is used, as in Figure 3(e), the size of regions that are treated as nearly flat increases significantly. Indeed, we observe the corresponding dominant blurring-like influence of AMLE.

In Figure 3(f) we increase the amoeba parameter ϱ relative to the setting from (c, d). From the analytic point of view, this corresponds to a larger time step size: Due to the quadratic relationship $\tau = \varrho^2/6$ we can expect that for two structuring elements with radii ϱ_1 and $\varrho_2 = 2\varrho_1$, four AMF iterations with ϱ_1 should roughly make up one iteration with ϱ_2. The comparison of Figure 3(c) and (f) confirms this approximate relationship: One iteration with $\varrho = 20$ has a similar outcome as four iterations with $\varrho = 10$. We observe especially that the transition zones at the shadows are located very similarly. The self-snake-like sharpening, however, appears somewhat more prominent in the image processed with four iterations.

The _Head_ experiment. In this experiment (Figure 4) we use an MR image of a human head which is rich in details of different contrast and scale. The original image is shown in Subfigure (a). In (b–d) iterated AMF results both with L_2 and L_1 amoeba distance are displayed. It can be seen that both distance measures lead to similar results. Moreover, we observe even clearer than in the _House_ experiment the good quality of segmentation that is achieved in spite of the relative simplicity of the filtering approach.

5   Conclusion

Our analysis of iterated amoeba median filtering shows that even highly adaptive discrete image filters can be interpreted in terms of PDE-based evolutions. This viewpoint leads to clear explanations of qualitative properties of iterated AMF, and predictions that can be confirmed by experiments. At the same time, the cross-relation sheds new light on well-known PDE filters and may inspire the development of new discretisations of PDE filters. Continuing this direction of research, we believe that it will not only expedite the development of both classes of image filters, but also bring forward a fusion between formerly distinct branches of image processing.

References

1. Alvarez, L., Lions, P.-L., Morel, J.-M.: Image selective smoothing and edge detection by nonlinear diffusion. II. SIAM Journal on Numerical Analysis 29, 845–866 (1992)
2. Barash, D.: Bilateral filtering and anisotropic diffusion: towards a unified viewpoint. In: Kerckhove, M. (ed.) Scale-Space 2001. LNCS, vol. 2106, pp. 273–280. Springer, Heidelberg (2001)
3. Caselles, V., Morel, J.-M., Sbert, C.: An axiomatic approach to image interpolation. IEEE Trans. Image Proc. 7(3), 376–386 (1998)
4. Chui, C.K., Wang, J.: PDE models associated with the bilateral filter. Advances in Computational Mathematics (2008)
5. Didas, S., Weickert, J.: Combining curvature motion and edge-preserving denoising. In: Sgallari, F., Murli, F., Paragios, N. (eds.) SSVM 2007. LNCS, vol. 4485, pp. 568–579. Springer, Heidelberg (2007)
6. Dougherty, E.R., Astola, J. (eds.): Nonlinear Filters for Image Processing. SPIE Press, Bellingham (1999)
7. Guichard, F., Morel, J.-M.: Partial differential equations and image iterative filtering. In: Duff, I.S., Watson, G.A. (eds.) The State of the Art in Numerical Analysis, pp. 525–562. Clarendon Press, Oxford (1997)
8. Kimmel, R., Sochen, N., Malladi, R.: Images as embedding maps and minimal surfaces: movies, color, and volumetric medical images. In: Proc. 1997 IEEE Computer Society Conference on Computer Vision and Pattern Recognition, pp. 350–355 (1997)

9. Klette, R., Zamperoni, P.: Handbook of Image Processing Operators. Wiley, New York (1996)
10. Kuwahara, M., Hachimura, K., Eiho, S., Kinoshita, M.: Processing of RI-angiocardiographic images. In: Preston, J.K., Onoe, M. (eds.) Digital Processing of Biomedical Images, pp. 187–202. Plenum, New York (1976)
11. Lerallut, R., Decencière, E., Meyer, F.: Image processing using morphological amoebas. In: Ronse, C., Najman, L., Decencière, E. (eds.) Mathematical Morphology: 40 Years On, pp. 13–22. Springer, Dordrecht (2005)
12. Lerallut, R., Decencière, E., Meyer, F.: Image filtering using morphological amoebas. Image and Vision Computing 25(4), 395–404 (2007)
13. Nagao, M., Matsuyama, T.: Edge preserving smoothing. Computer Graphics and Image Processing 9(4), 394–407 (1979)
14. Sapiro, G.: Vector (self) snakes: a geometric framework for color, texture and multiscale image segmentation. In: Proc. IEEE International Conference on Image Processing 1996, vol. 1, pp. 817–820 (1996)
15. Tomasi, C., Manduchi, R.: Bilateral filtering for gray and color images. In: Proc. Sixth International Conference on Computer Vision, pp. 839–846. Narosa Publishing House (1998)
16. Tukey, J.W.: Exploratory Data Analysis. Addison–Wesley, Menlo Park (1971)
17. Van den Boomgaard, R.: Decomposition of the Kuwahara–Nagao operator in terms of linear smoothing and morphological sharpening. In: Mathematical Morphology: Proc. Sixth International Symposium, pp. 283–292. CSIRO Publishing (2002)
18. Whitaker, R.T., Xue, X.: Variable-conductance, level-set curvature for image denoising. In: Proc. IEEE International Conference on Image Processing 2001, pp. 142–145 (2001)
19. Yezzi Jr., A.: Modified curvature motion for image smoothing and enhancement. IEEE Trans. Image Proc. 7(3), 345–352 (1998)

Spatially-Variant Anisotropic Morphological Filters Driven by Gradient Fields

Rafael Verdú-Monedero[1], Jesús Angulo[2], and Jean Serra[3]

[1] Department of Information Technologies and Communications, Technical University of Cartagena, 30202, Cartagena, Spain
rafael.verdu@upct.es

[2] CMM-Centre de Morphologie Mathématique, Mathématiques et Systèmes, MINES Paristech, 35, rue Saint Honoré, 77305 Fontainebleau Cedex, France
jesus.angulo@ensmp.fr

[3] Laboratoire A2SI - ESIEE, B.P. 99, 93162 Noisy-le-Grand, France
serraj@esiee.fr

Abstract. This paper deals with the theory and applications of spatially-variant mathematical morphology. We formalize the definition of spatially variant dilation/erosion and opening/closing for gray-level images using exclusively the structuring function, without resorting to complement. This sound theoretical framework allows to build morphological operators whose structuring elements can locally adapt their orientation across the dominant direction of image structures. The orientation at each pixel is extracted by means of a diffusion process of the average square gradient field, which regularizes and extends the orientation information from the edges of the objects to the homogeneous areas of the image. The proposed filters are used for enhancement of anisotropic images features such as coherent, flow-like structures.

1 Introduction

The expression "spatially variant" encompasses both ideas of *i*) the two level structure of a space E, and of all subsets $\mathcal{P}(E)$ and functions on E, and of *ii*) some variable processing over space E. Concerning mathematical morphology, the two founding texts about point *i*) are [16] (ch.2,3, and 9) and [7]. In [16] ch.2, devoted to the set case, Serra introduces the structuring function, with the derived four basic operations of dilation, erosion, and their two products, and the three dualities (adjunction, reciprocal and complement); it is shown that a compact structuring function may have a reciprocal version infinite everywhere. In ch.3, Matheron gives topological conditions for limiting such an expansion. Ch.9 is a first introduction to the function case, which is actually treated for the first time in [7]. In particular, Heijmans and Ronse develop the key approach by pulses sup-generators. More recent advances, due to Bouaynaya and Schonfeld can be found in [3], due to Soille in [19], and concerning Roerdink group morphology in [13]. Other papers focus specifically on efficient implementations of spatially variant morphological operators, such as those of Cuisenaire [5], Lerallut *et al.* [11] and Dokladal and Dokladalova [6].

M.H.F. Wilkinson and J.B.T.M. Roerdink (Eds.): ISMM 2009, LNCS 5720, pp. 115–125, 2009.

Point *ii*) involves two branches. All examples in the founding papers refer to some geometrical deformation of the Euclidean space, by perspective [16] Ch.4, or by rotation invariance [7]. The perspective case corresponds to an actual application to traffic control, by Beucher *et al.* [2]. But one can imagine another mode of variability, not given by a geometrical law, but by the images under study themselves. In [17] for example, in the description of a forest fire, the structuring elements are discs whose variable radii are drawn from a so-called spread map, and they act on another image, that of the fuel map.

Rather often, spatially variant morphology is associated with the search of directions. For example, fast implementation of morphological filters along discrete lines at arbitrary angles have been reported by Soille and Talbot in [18]. Other more sophisticated algorithms for morphological operators on thin structures are the path openings of Heijmans [9]. Closer to our study, Breuß *et al.* [4] considered a PDE formulation for adaptive morphological operators and Tankyevych *et al.* [20] proposed also locally orientated operators.

In this paper we focus on linear orientated structuring elements which vary over the space according to a vector field. The originality of our approach lies in that we draw the information on the structuring elements from the image under study itself. The morphological processing is thus locally adapted to some features that already exist in the image, but that this processing aims to emphasize. Evolved filters (based on successive openings and closings) can be then used for enhancement of anisotropic images features such as coherent, flow-like structures.

In the work by Tankyevych *et al.* [20], the orientation information is computed from Hessian matrix (i.e., second-order derivatives) and the curvilinear operators, such as the morphological closing, are computed by means of reciprocal structuring functions. In this paper we show how spatially-variant anisotropic numerical openings/closings can also be computed from their direct geometric definitions. In addition, we prefer to use first-order derivatives and a diffusion-like regularization step in order to calculate the directional vector field.

The paper is organized in four parts. Section 2 sets up the morphological background. We build and describe the directional vector field from which a structuring function is generated in Section 3. With this tool in hand, we perform experiments on a series of numerical images in Section 4, measurements that are followed by the conclusion.

2   Spatially-Variant Morphology

In mathematical morphology, many usual notions are dual from each other under complement. When the variation of a structuring function follows a geometrical law, then the complement of the dilation and of the adjoint opening can be theoretically calculated. But that is no longer true for data based variation, and this drawback obliges us to express the four basic operations by means of the structuring function exclusively, without resorting to complement, or equivalently, to reciprocal dilation.

Notation. Letter E denotes an arbitrary set, which can be a digital or continuous space, or any graph. The points of E are given in bold small letters (e.g. $\mathbf{x} \in E$), and their coordinates are represented by small letters (e.g. $\mathbf{x} = (x, y)$). The subsets of E are given in capital letters (e.g. $X \subseteq E$), and the set of all these subsets (including the empty set $\varnothing$) is denoted by $\mathcal{P}(E)$. The points $\mathbf{x}$ of E, considered as elements $\{\mathbf{x}\}$ of $\mathcal{P}(E)$, are called singletons and form the sub-class $\mathcal{S}(E)$ of $\mathcal{P}(E)$. A structuring function $\delta : E \to \mathcal{P}(E)$, or equivalently from $\mathcal{S}(E)$ into $\mathcal{P}(E)$, is an arbitrary family $\{\delta(\mathbf{x})\}$ of sets indexed by the points of E. One also writes $\delta(\mathbf{x}) = B(\mathbf{x})$, for emphasizing that the transform of a point is a set.

The numerical axis $\mathcal{T}$ is an arbitrary family closed sequence of non negative numbers between two extreme bounds, 0 and M say. They can be $[0, +\infty]$, or the integers $[0, 255]$, etc. The family of all numerical functions $f : E \to \mathcal{T}$ is denoted by $\mathcal{F}(E, \mathcal{T})$. Both sets $\mathcal{P}(E)$ and $\mathcal{F}(E, \mathcal{T})$ are complete lattices, i.e., posets whose any family of elements admits a supremum (a smaller upperbound) and an infimum (a larger lowerbound) [16], [7]. For $\mathcal{P}(E)$, they are union and intersection, and for $\mathcal{F}(E, \mathcal{T})$ the pointwise sup and inf.

Dilation, erosion. Since supremum and infimum do characterize a lattice, the two basic operations that map lattice $\mathcal{P}(E)$ into itself are those which preserve either union or intersection. In mathematical morphology, they are called dilation and erosion, and denoted by δ and ε respectively:

$$\delta(\cup X_i) = \cup \delta(X_i) \quad ; \quad \varepsilon(\cap X_i) = \cap \varepsilon(X_i) \qquad X_i \in \mathcal{P}(E). \tag{1}$$

Both operations are increasing. The two families of dilations and erosions on $\mathcal{P}(E)$ correspond to each other by the Galois's relation of an adjunction, namely

$$\delta(X) \subseteq Y \quad \Leftrightarrow \quad X \subseteq \varepsilon(Y), \tag{2}$$

and given a dilation δ, there always exists one and only one erosion ε that satisfies Equivalence (2) [7].

As a set X is the union of its singletons, i.e.

$$X = \cup \{\{\mathbf{x}\} \,|\, \{\mathbf{x}\} \subseteq X\}$$

and as dilation commutes under union, this operation is generated by its restriction $\delta : \mathcal{S}(E) \to \mathcal{P}(E)$ which associates the structuring function $\delta(\mathbf{x})$ with each singleton $\{\mathbf{x}\}$ (or equivalently with each point $\mathbf{x}$ of E). The dilation of Rel.(1) becomes

$$\delta(X) = \cup \{\delta\{\mathbf{x}\} | \{\mathbf{x}\} \subseteq X\} = \cup \{\delta(\mathbf{x}) | \ \mathbf{x} \in X\} = \cup \{B(\mathbf{x}) | \ \mathbf{x} \in X\}. \tag{3}$$

We then draw from adjunction (2) the expression of the erosion, namely

$$\varepsilon(X) = \{\mathbf{z} \ | \ B(\mathbf{z}) \subseteq X\}. \tag{4}$$

In spite of the name, a dilation may not be extensive (i.e. $\delta(X) \supseteq X$). Extensivity is obtained iff for all $\mathbf{x} \in E$ we have $\mathbf{x} \in B(\mathbf{x})$. Then the adjoint erosion ε is anti-extensive.

The operation dual of dilation δ under complement is the erosion

$$\varepsilon^*(X) = [\delta(X^c)]^c,$$

whose associated structuring function is the reciprocal version of δ, i.e.

$$\mathbf{y} \in \zeta(\mathbf{x}) \qquad \text{if and only if} \qquad \mathbf{x} \in \delta(\mathbf{y}) \qquad\qquad \mathbf{x}, \mathbf{y} \in E. \qquad (5)$$

Opening and closing. Though erosion ε usually admits many inverses, the composition product

$$\gamma(X) = \delta\varepsilon(X) \qquad\qquad (6)$$

results in the smallest inverse of $\varepsilon(X)$. This product γ is increasing ($X \subseteq Y \Rightarrow \gamma(X) \subseteq \gamma(Y)$), anti-extensive ($\gamma(X) \subseteq X$) and idempotent ($\gamma\gamma(X) = \gamma(X)$). In algebra, these three features define an *opening*. Similarly, by inverting δ and ε, we obtain the *closing* $\varphi = \varepsilon\delta$, which is increasing, extensive, and idempotent.

The analytical representation of dilation δ by means of the structuring function $\mathbf{x} \to B(\mathbf{x})$ extends to the associated opening and closing. We directly draw from Relations (3) and (4) that

$$\delta\varepsilon(X) = \cup\{B(\mathbf{x})|B(\mathbf{x}) \subseteq X\} \qquad\qquad (7)$$
$$\varepsilon\delta(X) = \cup\{\mathbf{x} \,|B(\mathbf{x}) \subseteq \cup [B(\mathbf{x})| \,\mathbf{x} \in X]\} \qquad\qquad (8)$$

The geometrical meaning of the first relation is clear: $\delta\varepsilon(X)$ is the region of the space swept by all structuring sets $B(\mathbf{x})$ that are included in X.

Finally, Relations (3), (4), (7), and (8) that give the four basic operations are completely determined by the datum of the structuring function $\mathbf{x} \to B(\mathbf{x}) = \delta(\mathbf{x})$ and do not involve any reciprocal function. In particular, opening $\gamma = \delta\varepsilon$ and closing $\varphi = \varepsilon\delta$ are *not* dual of each other for the complement.

Dilation for numerical functions. Note that the duality under complement works for sets only, whereas adjunction duality applies to any complete lattice, such as that $\mathcal{F}(E, \mathcal{T})$ of the numerical functions that we now consider. Associated with the numerical function $f : E \to \mathcal{T}$ under study and the set structuring function $\mathbf{x} \to B(\mathbf{x})$, we introduce the following pulse function $i_{\mathbf{x},t}$ of level t at point $\mathbf{x}$

$$i_{\mathbf{x},t}(\mathbf{x}) = t \quad ; \quad i_{\mathbf{x},t}(\mathbf{y}) = 0 \quad \text{when } \mathbf{y} \neq \mathbf{x}.$$

Dilating $i_{\mathbf{x},t}$ by the structuring function B results in the cylinder $C_{B(\mathbf{x}),t}$ of base $B(\mathbf{x})$ and height t. Now, function f can be decomposed into the supremum of its pulses, i.e.

$$f = \vee\{i_{\mathbf{x},f(\mathbf{x})}, \mathbf{x} \in E\}.$$

Since dilation commutes under supremum, the dilate of f by δ, of structuring function B is given by the supremum of the dilates of its pulses, namely

$$\delta(f) = \vee\{C_{B(\mathbf{x}),f(\mathbf{x})}, \mathbf{x} \in E\}. \qquad\qquad (9)$$

Similarly, the eroded $\varepsilon(f)$ is the supremum of those pulses whose dilated cylinders are smaller than f, i.e.

$$\varepsilon(f) = \vee\{i_{\mathbf{x},t} \mid C_{B(\mathbf{x}),t} \leq f, \mathbf{x} \in E\}. \tag{10}$$

These two operations satisfy the equalities $\delta(\vee f_i) = \vee\delta(f_i)$ and $\varepsilon(\wedge f_i) = \wedge\varepsilon(f_i)$, $f_i \in \mathcal{F}$, and Galois equivalence (2) in the lattice $\mathcal{F}(E, \mathcal{T})$ of the numerical functions. Moreover, the cross sections of $\delta(f)$ (resp. $\varepsilon(f)$) are the dilated (resp. the eroded) versions of the cross sections of f by the same structuring function. For this reason they are called "flat operations". Just as in the set case, the duality under adjunction does not coincide with that under the involution $f \to M - f$, which plays a role similar to a complement. The operation

$$\varepsilon^* = M - \delta(M - f)$$

turns out to still be an erosion, but ε^* is different from the ε of Rel.(10).

The two products $\gamma = \delta\varepsilon$ and $\varphi = \varepsilon\delta$ are opening and closing on $\mathcal{F}(E, \mathcal{T})$, and, as δ and ε, they commute under cross sectioning. Opening γ, for example, admits the following expression

$$\gamma(f) = \vee\{C_{B(\mathbf{x}),t} \leq f, \mathbf{x} \in E\}. \tag{11}$$

In the product space $E \times \mathcal{T}$ the subgraph of the opening $\gamma(f)$ is generated by the zone swept by all cylinders $C_{B(\mathbf{x}),t}$ smaller than f. Again, the closing $\varphi = \varepsilon\delta$ does not coincide with that, $M - \gamma(M - f)$, obtained by replacing f by $M - f$ in Rel.(11).

3   Directional Field Modelling

This section describes the method for estimating the orientation of the structures contained in a gray-level image. This vector field is obtained by using the average squared gradient and then applying a regularization process.

Average Squared gradient. The average squared gradient (ASG) method provides the directional field by squaring and averaging the gradient vectors [10,1]. Given an image $f(x, y)$, ASG uses the following definition of gradient

$$\mathbf{g} = \begin{bmatrix} g_1(x, y) \\ g_2(x, y) \end{bmatrix} = \text{sign}\left(\frac{\partial f(x,y)}{\partial x}\right) \begin{bmatrix} \frac{\partial f(x,y)}{\partial x} \\ \frac{\partial f(x,y)}{\partial y} \end{bmatrix}. \tag{12}$$

Then the gradient is squared (i.e., doubling its angle and squaring its magnitude) and averaged in some neighborhood using the window W:

$$\overline{\mathbf{g}_s} = \begin{bmatrix} \overline{g_{s,1}}(x, y) \\ \overline{g_{s,2}}(x, y) \end{bmatrix} = \begin{bmatrix} \sum_W \left(g_1^2(x, y) - g_2^2(x, y)\right) \\ \sum_W \left(2\, g_1(x, y)\, g_2(x, y)\right) \end{bmatrix}. \tag{13}$$

The directional field ASG is $\mathbf{d} = [d_1(x, y), d_2(x, y)]^\top$, where its angle is obtained as $\angle\mathbf{d} = \frac{\Phi}{2} - \text{sign}(\Phi)\frac{\pi}{2}$, which is in the range $[-\frac{\pi}{2}, \frac{\pi}{2}]$, being $\Phi = \angle\overline{\mathbf{g}_s}$; and the magnitude of $\mathbf{d}$, $\|\mathbf{d}\|$, can be left as the magnitude of $\overline{\mathbf{g}_s}$, or the squared root of $\overline{\mathbf{g}_s}$ or, in some applications (see e.g [12]) and in this work, it can be set to unity. In ongoing research, we consider also the use of the magnitude and coherence of directional field in order to build more general anisotropic structuring functions.

Regularization of the ASG. The vectors of the ASG field are generally different from zero only near the edges and, in homogeneous regions, where the gradient is nearly zero, the ASG is also zero. In order to extend the orientation information to pixels where the gradient is nearly zero a diffusion process is performed (similar to gradient vector flow, GVF, [21]), providing the ASG vector flow (ASGVF). The ASGVF is the vector field $\mathbf{v} = [v_1(x,y), v_2(x,y)]^\top$ that minimizes the energy functional:

$$\mathcal{E}(\mathbf{v}) = \mathcal{D}(\mathbf{v}) + \alpha \mathcal{S}(\mathbf{v}), \tag{14}$$

where $\mathcal{D}$ represents a distance measure given by the squared difference between the original and the regularized average squared gradient, weighted by the squared value of the last one,

$$\mathcal{D}(\mathbf{v}) = \frac{1}{2} \sum_{l=1}^{2} \int_E ||\mathbf{d}||^2 ||v_l - d_l||^2 \, dx \, dy, \tag{15}$$

where E is the image support and $l = 1, 2$ is the component index; the energy term $\mathcal{S}$ determines the smoothness of the directional field and represents the energy of the first order derivatives of the signal:

$$\mathcal{S}(\mathbf{v}) = \frac{1}{2} \sum_{l=1}^{2} \int_E ||\nabla v_l||^2 \, dx \, dy. \tag{16}$$

The parameter α is a regularization parameter which governs the trade-off between the smoothness and data-fidelity.

Using the calculus of variations, the ASGF field can be found by solving the following Euler equations

$$(\mathbf{v} - \mathbf{d})|\mathbf{d}|^2 - \alpha \nabla^2 \mathbf{v} = \mathbf{0}. \tag{17}$$

These equations can be solved by treating $\mathbf{v}$ as a function of time and considering the steady-state solution

$$\mathbf{v}_t + (\mathbf{v} - \mathbf{d})|\mathbf{d}|^2 - \alpha \nabla^2 \mathbf{v} = \mathbf{0}. \tag{18}$$

These equations are known as generalized diffusion equations. To set up the iterative solution, let the indices i, j, and n correspond to the discretization of x, y and t axes, respectively, and let the spacing between pixels be Δx and Δy, and the time step for each iteration be Δt. Replacing partial derivatives with its discrete approximations and considering discrete images ($\Delta x = \Delta y = 1$) gives our iterative solution to ASGF as follows:

$$v_{i,j}^{l,n} = v_{i,j}^{l,n-1} - \Delta t f_{i,j}^{l,n-1} + \frac{1}{\eta}(v_{i+1,j}^{l,n-1} + v_{i-1,j}^{l,n-1} + v_{i,j+1}^{l,n-1} + v_{i,j-1}^{l,n-1} - 4v_{i,j}^{l,n-1}) \tag{19}$$

where $f_{i,j}^{l,n-1} = (v_{i,j}^{l,n-1} - d_{i,j}^l)|\mathbf{d}_{i,j}|^2$, n is the iteration index and $\eta = (\alpha \Delta t)^{-1}$.

4   Applications and Discussion

This section shows the results of applying spatially-variant morphology operators for gray-level filtering, with a linear structuring element of fixed length λ but variable orientation. More precisely, the structuring function

$$B(\mathbf{x}) \equiv L_\lambda^{\theta(\mathbf{x})},$$

where $\theta(\mathbf{x})$ is the angle at point $\mathbf{x}$ from the regularized vector field $\mathbf{v}$. Besides erosions $\varepsilon_{L_\lambda^{\theta(\mathbf{x})}}$, dilations $\delta_{L_\lambda^{\theta(\mathbf{x})}}$, openings $\gamma_{L_\lambda^{\theta(\mathbf{x})}}$ and closings $\varphi_{L_\lambda^{\theta(\mathbf{x})}}$, we illustrate in the examples the application of alternate sequential filters (ASF): $\varphi_n \gamma_n \cdots \varphi_2 \gamma_2 \ \varphi_1 \gamma_1(f)$, or the dual version $\gamma_n \varphi_n \cdots \gamma_2 \varphi_2 \ \gamma_1 \varphi_1(f)$. ASF present excellent properties for image denoising and regularization of dominant structures.

The present adaptive morphological filters are compared with standard translation invariant linear openings/closings which are built according to the property that the supremum (resp. infimum) of openings (resp. closings) is an opening (resp. closing) as well. The translation-invariant linear opening of size λ is given by

$$\gamma_{L_\lambda^{lines}}(f)(\mathbf{x}) = \gamma_{L^{l,\theta_1}}(f)(\mathbf{x}) \vee \gamma_{L^{l,\theta_2}}(f)(\mathbf{x}) \vee \cdots \vee \gamma_{L^{l,\theta_d}}(f)(\mathbf{x}),$$

where the following directions $\{\theta_1, \theta_2, \cdots, \theta_d\}$ are considered.

In Fig. 1 are given two synthetic images and the application of spatially-variant morphological operators. In the first example, the ASG uses a flat squared

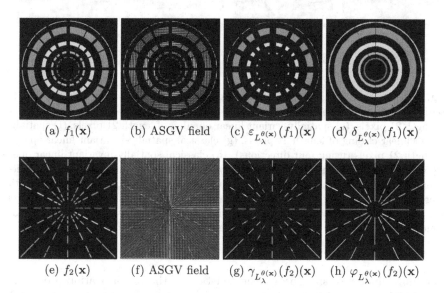

    (a) $f_1(\mathbf{x})$ (b) ASGV field   (c) $\varepsilon_{L_\lambda^{\theta(\mathbf{x})}}(f_1)(\mathbf{x})$ (d) $\delta_{L_\lambda^{\theta(\mathbf{x})}}(f_1)(\mathbf{x})$

    (e) $f_2(\mathbf{x})$ (f) ASGV field   (g) $\gamma_{L_\lambda^{\theta(\mathbf{x})}}(f_2)(\mathbf{x})$ (h) $\varphi_{L_\lambda^{\theta(\mathbf{x})}}(f_2)(\mathbf{x})$

Fig. 1. (a) Image "circles", (b) ASGV field using $\eta = 1$, (c) locally oriented linear erosion of length 9 pixels, (d) equivalent dilation. (e) Images "lines", (f) ASGV, $\eta = 1$, (g) locally oriented linear opening of length 11 pixels, (h) equivalent closing. Both images of size 256×256 pixels.

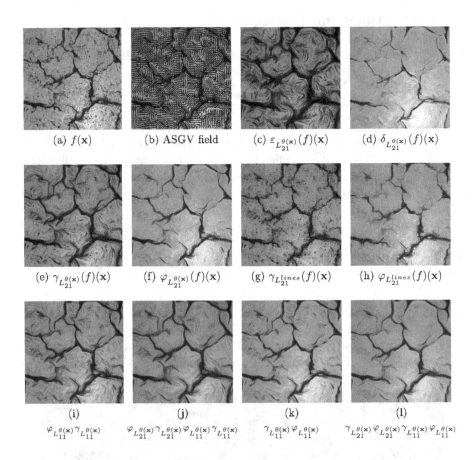

(a) $f(\mathbf{x})$ (b) ASGV field   (c) $\varepsilon_{L_{21}^{\theta(\mathbf{x})}}(f)(\mathbf{x})$ (d) $\delta_{L_{21}^{\theta(\mathbf{x})}}(f)(\mathbf{x})$

(e) $\gamma_{L_{21}^{\theta(\mathbf{x})}}(f)(\mathbf{x})$ (f) $\varphi_{L_{21}^{\theta(\mathbf{x})}}(f)(\mathbf{x})$ (g) $\gamma_{L_{21}^{lines}}(f)(\mathbf{x})$ (h) $\varphi_{L_{21}^{lines}}(f)(\mathbf{x})$

(i)   (j)   (k)   (l)

$\varphi_{L_{11}^{\theta(\mathbf{x})}}\gamma_{L_{11}^{\theta(\mathbf{x})}}$ $\varphi_{L_{21}^{\theta(\mathbf{x})}}\gamma_{L_{21}^{\theta(\mathbf{x})}}\varphi_{L_{11}^{\theta(\mathbf{x})}}\gamma_{L_{11}^{\theta(\mathbf{x})}}$ $\gamma_{L_{11}^{\theta(\mathbf{x})}}\varphi_{L_{11}^{\theta(\mathbf{x})}}$ $\gamma_{L_{21}^{\theta(\mathbf{x})}}\varphi_{L_{21}^{\theta(\mathbf{x})}}\gamma_{L_{11}^{\theta(\mathbf{x})}}\varphi_{L_{11}^{\theta(\mathbf{x})}}$

Fig. 2. (a) Original image (of size 447×447 pixels), (b) ASGV field using $\eta = 0.001$, (c) spatially-variant linear erosion of length 21 pixels, (d) spatially-variant linear dilation of length 21 pixels, (e) spatially-variant linear opening of length 21 pixels, (f) spatially-variant linear closing of length 21 pixels, (g) classical spatially-invariant linear opening of length 21 pixels (using four directions), (h) classical spatially-invariant linear opening of length 21 pixels (using four directions), (i) spatially-variant linear alternate filter (opening followed by closing) of size 11, (j) spatially-variant linear alternate sequential filter (opening followed by closing) of sizes 11 and 21, (k) spatially-variant linear alternate filter (closing followed by opening) of size 11, (l) spatially-variant linear alternate sequential filter (closing followed by opening) of sizes 11 and 21

averaging window of size 15×15, the constant of the regularization process for the ASGVF is $\eta = 1$ and the length of the structuring element is $\lambda = 9$. In the second example the window used in ASG is 11×11 pixels, η equals 1 and the structuring element is 11 pixel long. We can remark the appropriateness of very simple morphological filters for closing interrupted line structures and for the suppression of line structures of small size. As expected, the results are

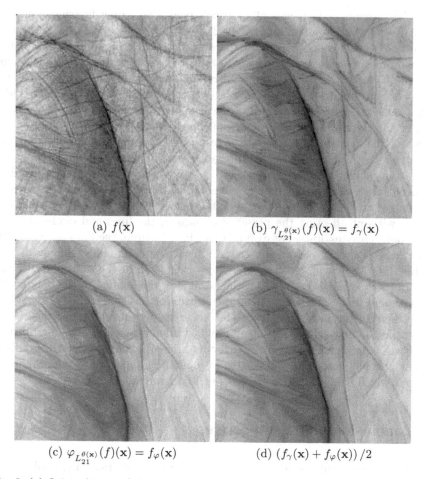

(a) $f(\mathbf{x})$

(b) $\gamma_{L_{21}^{\theta(\mathbf{x})}}(f)(\mathbf{x}) = f_\gamma(\mathbf{x})$

(c) $\varphi_{L_{21}^{\theta(\mathbf{x})}}(f)(\mathbf{x}) = f_\varphi(\mathbf{x})$

(d) $(f_\gamma(\mathbf{x}) + f_\varphi(\mathbf{x}))\,/2$

Fig. 3. (a) Original image (of size 256×256 pixels), (b) spatially-variant linear opening of length 21 pixels, (c) spatially-variant linear closing of length 21 pixels, (d) average image between the spatially-variant linear opening and closing. The ASGV field has been computed using $\eta = 0.01$.

quite spatially regular and only a negligible number of points propagate values according to wrong directions.

The example depicted in Fig. 2 is a real world image, presenting a network of coherent, flow-like structures on a very irregular background. The window used in ASG is 21×21 pixels, $\eta = 0.001$. We have studied the proposed filters for enhancement of anisotropic image features. In particular, it is shown how the ASF produce very interesting anisotropic effects. It is evident in this case that the closing (or the ASF starting by a closing) is more appropriate because the target structures are darker than their background. We can also compare the new spatially-variant linear opening/closing with their counterpart classical spatially-invariant linear opening/closing.

Fig. 3 provides a last case of morphological anisotropic filtering. For this example, the window used in ASG is 21×21 pixels, $\eta = 0.01$. The original image represents a hand palm and the aim of the filtering step is to enhance the main lines which are the most important for biometric identification. A pair of spatially-variant linear opening and closing are compared with the average images of both operators. As we can observe, the average image is a good trade-off between these two basic morphological anisotropic filters and the result is quite regular but preserves the significant linear structures.

One of the key points for the good results of these anisotropic filters is the computation of the gradient field. In the three examples, the value of the time step of the regularization process is $\Delta t = 1$. The size of the averaging window has been taken in order to preserve the orientation of the main structures in the images: circles or lines in Fig. 1, big crevices in Fig. 2) and main lines in Fig. 3), without affecting the presence of gaps in the lines and circles nor the small spots/crevices/lines. For example, in Fig. 1-bottom, the lines have a width of 3 pixels, therefore computing the ASG with an averaging window of size greater than 3×3 will preserve the main orientation of the lines ignoring the gaps. In the regularization process, the parameter η is related to the bandwidth of a low-pass filter which filters the increments of ASGVF in Eq. (19). When the gradient of the image has homogeneous areas and abrupt transitions (as it happens in the first synthetic image) a higher value is necessary to allow abrupt transitions (spatial high frequencies) exist. On the other hand, when the gradient has not abrupt transitions a low value of η gets the appropriate smoothness in the regularized vector field (as happens in Fig. 2 and 3).

5   Conclusion and Perspectives

We have clarified and solved some difficulties in the definition of spatially variant dilation/erosion and opening/closing for gray-level images. Then, we have proposed an algorithm for a reliable extraction of orientation information, which is finally used to build linear anisotropic morphological filters. The performance of derived operators has been illustrated for enhancement of anisotropic images features such as coherent, flow-like structures.

In ongoing research, we would like to address the theory of spatially-variant geodesic operators. From a practical viewpoint, we are working on a full exploitation on the directional field, including the information of magnitude and angular coherence in order to propose more general anisotropic structuring elements, not limited to orientated lines of fixed length. The extension to 3-D and more generally to n-D spaces and the application to 3-D images (e.g., denoising MRI data, enhancement of fiber networks, etc.) will be also explored in future work.

Acknowledgements. This work is partially supported by the Spanish *Ministerio de Ciencia y Tecnología*, under grant TEC2006-13338/TCM.

References

1. Bazen, A.M., Gerez, S.H.: Systematic methods for the computation of the directional fields and singular points of fingerprints. IEEE Trans. Pattern Anal. Mach. Intell. 24(7), 905–919 (2002)
2. Beucher, S., Blosseville, J.M., Lenoir, F., Motyka, V., Kraft, C.: TITAN, a traffic measurement system using image processing techniques. In: Proceedings IEE Road Traffic Congress (1989)
3. Bouaynaya, N., Schonfeld, D.: Theoretical foundations of spatially-variant mathematical morphology part II: Gray-level images. IEEE Trans. Pattern Anal. Mach. Intell. 30, 837–850 (2008)
4. Breuß, M., Burgeth, B., Weickert, J.: Anisotropic continuous-scale morphology. In: Martí, J., Benedí, J.M., Mendonça, A.M., Serrat, J. (eds.) IbPRIA 2007, Part II. LNCS, vol. 4478, pp. 515–522. Springer, Heidelberg (2007)
5. Cuisenaire, O.: Locally adaptable mathematical morphology using distance transformations. Pattern Recognition 39, 405–416 (2006)
6. Dokladal, P., Dokladalova, E.: Grey-scale Morphology with Spatially-Variant Rectangles in Linear Time. In: Blanc-Talon, J., Bourennane, S., Philips, W., Popescu, D., Scheunders, P. (eds.) ACIVS 2008. LNCS, vol. 5259, pp. 674–685. Springer, Heidelberg (2008)
7. Heijmans, H.J.A.M., Ronse, C.: The algebraic basis of mathematical morphology - part I: Dilations and erosions. Computer Vision, Graphics and Image Processing 50, 245–295 (1990)
8. Heijmans, H.J.A.M.: Morphological Image Operators. Academic Press, Boston (1994)
9. Heijmans, H., Buckley, M., Talbot, H.: Path openings and closings. Journal of Mathematical Imaging and Vision 22, 107–119 (2005)
10. Kass, M., Witkin, A.: Analyzing oriented patterns. Comput. Vision Graph. Image Process. 37(3), 362–385 (1987)
11. Lerallut, R., Decenciére, E., Meyer, F.: Image filtering using morphological amoebas. Image Vision Comput. 25, 395–404 (2007)
12. Perona, P.: Orientation diffusions. IEEE Trans. Image Processing 7(3), 457–467 (1998)
13. Roerdink, J.: Group morphology. Pattern Recognition 33, 877–895 (2000)
14. Ronse, C., Heijmans, H.J.A.M.: The algebraic basis of mathematical morphology - part II: Openings and closings. Computer Vision, Graphics and Image Processing: Image Understanding 54, 74–97 (1991)
15. Serra, J.: Image Analysis and Mathematical Morphology, vol. I. Academic Press, London (1982)
16. Serra, J.: Image Analysis and Mathematical Morphology: Theoretical Advances, vol. II. Academic Press, London (1988)
17. Serra, J., Suliman, M.D.H., Mahmud, M.: Prediction of Scars in Malaysian Forest fires by means of Random Spreads. In: Proc. of Int. ISPRS Conf. on Techniques and Applications of Optical and SAR Imagery fusion (2007)
18. Soille, P., Talbot, H.: Directional morphological filtering. IEEE Trans. Pattern Anal. Mach. Intell. 23, 1313–1329 (2001)
19. Soille, P.: Morphological image analysis. Springer, Heidelberg (1999)
20. Tankyevych, O., Talbot, H., Dokladal, P.: Curvilinear Morpho-Hessian Filter. In: Proceedings of ISBI 2008, pp. 1011–1014 (2008)
21. Xu, C., Prince, J.L.: Generalized gradient vector flow external forces for active contours. Signal Processing 71(2), 131–139 (1998)

3-D Extraction of Fibres from Microtomographic Images of Fibre-Reinforced Composite Materials

Petr Dokládal and Dominique Jeulin

Center of Mathematical Morphology, Mines-Paristech,
35, r. Saint Honoré, 77300 Fontainebleau, France

Abstract. Mechanical properties of molded components made from fibre-reinforced composite materials locally depend of the orientation of the fibres. The evaluation of the properties is done by sampling the component at known positions. The samples (of size of 1 mm^3) are scanned in a tomograph which yields 3-D images. We are interested in extracting the individual fibres, to analyze their length and local orientation.

The segmentation of the fibres is a challenging task. First, the resolution of the reconstruction being at the limits of the capabilities of the device (optics, sensor, wavelength), the images are noisy and fuzzy. Second, the fibres have a non uniform length and are heavily tangled. This paper describes the segmentation process.

1 Introduction

Starting from X-ray microtomographic images, the objective is to extract 3-D presentational maps of fibres in components manufactured by molding from fibre-composite materials. The map will serve as a support for the simulation of alignment of fibres in the flowing, liquid matrix during the molding process.

The proposed method proceeds in two steps. First step, it extracts the skeleton of the fibres by a thinning. Second, individual fibres are reconstructed from the skeleton.

The reconstruction process is formulated and implemented using the theory of graphs. It uses basic, local graph operations as the edge or vertex contraction. The graph models a real object. Its geometrical properties must not alter during the simplification. They are encoded as weights associated to the graph. During the contraction of the graph, the weights are iteratively inherited, until the ultimate state, beyond which no additional simplification is possible.

After that, we perform a statistical analysis of geometric properties of the material, such as distribution of the orientation and length of the fibres.

Previous, related work:

- Directional morphological filtering - a set of oriented morphological filters, Soille and Talbot [1], [2]. For bright objects on a dark backgroung, one would use an anti-extensive filter such as an opening.

  Letting the filter rotate, finding the maximum in the response, gives an indication on the local orientation. In addition to that, to evaluate the length of every fibre would require to let vary also the SE length.

M.H.F. Wilkinson and J.B.T.M. Roerdink (Eds.): ISMM 2009, LNCS 5720, pp. 126–136, 2009.

Even though an efficient algorithm for computing openings with multiple structuring elements has been proposed by Urbach and Wilkinson [3] this would still take considerable time.

The computational cost makes that this approach is unusable for large or n-D data, n$\geq$3.

– In the Fourier domain - the local orientation can be detected by observing the spectrum energy, Kass and Witkin 1987. Jeulin and Moreaud [4] detect local orientation of textures in 2-D and 3-D from the covariance matrix of the gradient.

The FFT allows an efficient implementation. However, the non locality of the FFT makes impossible to extract individuals to measure their length and evaluate the length histogram of the whole population of the fibres.

– Orientation space - a useful concept introduced by Chen and Hsu [5], and explored later by Van Vliet and Verbeek [6], Chen *et al* [7] or Ginkel [8]. Works well on isolated objects. Using a bank of filters infers a trade off between accuracy and locality. This drawback makes that this approach is not usable for tangled objects. Also, the computational cost, induced by applying a set of filters, is quite high.

Adding supplementatry dimensions for the orientation increases the memory requirements. The orientation in 2-D is one, and in 3-D two values. Analysing 3-D images requires working with 5-D data.

– Perona and Malik [9] or Frangi [10] use extraction of local orientation for enhancement of thin, elongated, tubular objects. The detection of orientation is local, based on second (or higher) derivatives. It is not suitable for noisy images.

– Stein *et al.* [11] use graphs to analyze the geometry of collagen gel images acquired by a confocal microscope. This work is perhaps the most similar to our approach, described below.

1.1 Preliminaries

Consider 3-D, grey-scale $\mathbf{Z}^3 \to \mathbf{Z}$ and binary images $\mathbf{Z}^3 \to \{0,1\}$. A binary object $X \subset \mathbf{Z}^3$ is a set $X = \{x_i \mid x_i{=}1\}$. All $x \in \mathbf{Z}^3$ are associated a set $N(x)$ of neighbors (or adjacent points), $N(x) \subset \mathbf{Z}^3$. Below, the same holds for all types of neighborhood most often used on the rectangular grid $\mathbf{Z}^3$, that are N_6, N_{18} or N_{26}.

A connected component of X is such a subset $\mathrm{CC} \subset X$ that for any points $x, y \in \mathrm{CC}$ there is a sequence of adjacent points all included in CC. $\mathcal{CC}(X)$ shall denote the set of all connected components of X.

The skeleton $sk(X)$ of some object X is its binary, thin representation. It is included and centered in X and it preserves its topology. In 3-D there are 2 types of skeleton, a *surface* like and a *wireframe* like. Here, given the geometric form of the objects - fibres, we use the wireframe type skeleton.

A wireframe-type skeleton sk, $sk \subset \mathbf{Z}^3$, contains three types of points, characterized according to the number of their neighbors. For any $x \in sk$ we have $\mathrm{Card}\{x_i \mid x_i \in N(x) \cap sk\} = 1, 2$ or 3 (or more), respectively denoted hereafter as *terminal, linear* and *triple* points. For convenience, n-P, with $n = 1, 2$ and 3 shall respectively denote the sets of terminal, linear and triple points of sk.

The skeleton sk is *thin* in the sense that $\forall x$, $x \in 2\text{-P} \cup 3\text{-P}$, the set $sk \setminus \{x\}$ changes the topology w.r.t. sk.

Connected components $\mathcal{CC}(2\text{-P})$ shall be called *linear segments* (or *branches*) of the skeleton.

The 3-P points constitute the *junctions*. One expects junctions be composed of triple points. Notice that the skeleton junctions may be non singleton sets, composed of more triple points. Fig. 1 shows a place where two fibres touch. The junction in the skeleton may be composed of several triple points, see Fig. 1a or Fig. 1b. Obviously, other configurations of junctions exist. Consequently, in practice, one needs to consider junctions as connected components of triple points $\mathcal{CC}(3\text{-P})$.

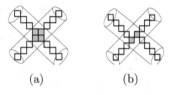

(a)   (b)

Fig. 1. Various configurations of skeleton junctions: a) four, and b) two triple points

2   Method

For several reasons, the segmentation of the fibres is a challenging task:

• The material properties - The contrast between the fibres and the matrix is low. The reconstruction is not easy. The fibres have a non uniform length and are heavily tangled.

• The image quality - The fibres are small (units μm) compared to the size of the sample (1mm^3). The resolution of the reconstruction is at the limits of the capabilities of the device elements, dependent of a number of parameters: the optics, sensor, available directional sampling, etc.

The images represent a considerable amount of data, 2000x2000x1100 pixels, coded in 16bits. The execution time of the analysis is also to take into consideration.

The used method comprises two steps: 1) From the grey-scale images we extract a 3-D binary skeleton, 2) the skeleton is then submitted to a filtering and reconstruction (performed on a graph).

2.1   Extraction of the Skeleton of the Fibres

We consider here a skeleton sk of a grayscale image I, given a binary marker M. The skeleton sk is obtained by a homotopic thinning of M, as proposed in [12], [13].

The thinning consists of an iterative deletion of *simple*, non *terminal* points from M.

Algorithm 1

Input: Priority Image F, Marker M
Output: Skeleton sk

$X^0 = M$
repeat until stability:
    select a point x, such that $x = \arg \min_{x \in X^n} \{F(x)\}$
    if x is *simple* and not 1-P then :
        $X^{n+1} = X^n \setminus \{x\}$
sk $= X^\infty$

The selection scheme is any order-generating criterion on the set of points in X, given I. It can be the intensity, distance, etc. Here the choice is inferred from a priority image F. The priorities can either be directly the input image, $F = I$, or what we have used here a pile of successive erosions

$$F = I + \varepsilon I + \varepsilon \varepsilon I + \ldots \tag{1}$$

Using a pile of erosions is advantageous when the objects itself have a poorly defined center. Accumulated erosion from both sides towards the center creates a crest line situated in the center of the object.

The marker image was obtained by thresholding. Experimentally, the most convenient threshold was found above the 80 percentile of the intensity histogram of the input image. The constant 80 comes from the a priory knowledge of the charge of the material in fibres - $\approx 20\%$. The fibres appear bright and occupy upper 20% of the intensity histogram.

A *simple* point is a point that can be deleted from some object without modifying its topology [14]. The decision whether some point is simple is local. It can be done by examination of the neighborhood of the point [15]. Optimal implementations in $\mathcal{O}(1)$ have been proposed, based on a Look-Up-Table in 2-D and a Binary Decision Diagram in 3-D. Here we use a BDD-based scheme that decides in at mosts 26 tests whether a point is simple or not.

In order to run efficiently, the selection scheme is implemented with priority queues, for details see [12].

2.2   Extraction of fibres

The extraction of individual fibres is done in several steps. All these steps are easily formulated and implemented using common operations from the graph theory.

A graph G is a pair (V, E), with V a set of vertices, and E a set of *edges* or *links*. E is a subset of $V \times V$. Furthermore, we consider undirected graphs, i.e. graphs for which the relations between pairs of vertices are symmetric.

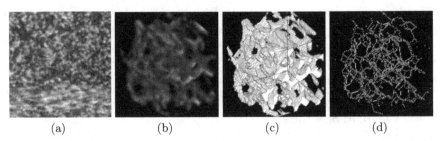

(a)   (b)   (c)   (d)

Fig. 2. Fibre-reinforced composite material: a) a 256x256 illustration crop from original 2000x2000x1100 data, b) priority image (a 3-D 60x60x60 crop) used for the thinning (Eq. 1), c) marker image, d) skeleton

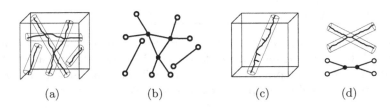

(a)   (b)   (c)   (d)

Fig. 3. Construction of the graph, a) a wire-frame skeleton of fibres, b) its graph, c) spurious skeleton branches due to noise, d) doubled skeleton junction and its graph

For a given skeleton sk, the graph $G = (V, E)$ is constructed by taking $V \leftrightarrow CC(\text{3-P}) \cup \text{1-P}$ and $E \leftrightarrow CC(\text{2-P})$, see Fig. 3. The black and the unfilled vertices in Fig. 3b represent the junctions and the terminal points of the skeleton, respectively.

On undirected graphs, the rank of a vertex v_i denotes the number of vertices linked to it by an edge, given by $d(v_i) = \text{Card}\{v_j \mid e_{ij} \in E, j \neq i\}$. It represents here the number of linear segments meeting in the corresponding triple point.

Both sets V and E are associated weights. The vertices are associated coordinates of extremal points or barycenters for junctions: $b_0 : V \to \mathbf{Z}^3$.

Every branch ls_i of the skeleton is approximated by an oriented, undirected line. Therefore, the set of edges E is also associated the barycenter $b_0 : E \to \mathbf{Z}^3$ of the line. The orientation is given as a pair of angles, azimuth and elevation $(\theta, \varphi) : E \to [0, \pi[\times [0, \pi[$ (radians). The length is $l : E \to \mathbf{R}^+$.

Two edges of a graph are called adjacent if they are incident to a common vertex.

A pair of edges e_{ij}, e_{jk}, that are incident to the same vertex v_j, bound an angle $\alpha(e_{ij})$, in $0 < \alpha(e_{ij}, e_{jk}) \leq \pi$. α defines an order on $E \times E$.

Definition 1. *Using α allows us to define on the set E the relation* closest neighbor. *For any edge $e_{ij} \in E$, the closest neighbor is the adjacent edge that delimits with e the smallest angle $\pi - \alpha$.*

This order induced by α is used below to conceive an algorithm extracting from the graph salient, co-linear structures representing individual fibres. This algorithm proceeds in two steps, i) filters spurious edges induced by noise and ii) extracts the fibres by reconstruction.

2.3   Filtering

The objective is to filter spurious branches, due principally to noise.

Wherever fibres touch, the skeleton contains intersections. The fibres being thick, the intersections may contain several triple points.

Filter 1. The filter 1 is to filter spurious skeleton branches (shorter than L) that can be deleted without modifying the topology of the object. Coded in terms of graphs, the filter sums to Algorithm 1.

Algorithm Filter 1

Input: $G(V, E)$, Length $L \in \mathbf{R}+$
Output: $G(V, E)$

---

for all $e_{ij} \in E$ do:
    if $l(e_{ij}) < L$ and $d(v_i) = 1$ and $d(v_j) = 3$:
        delete e_{ij}

The deletion *delete* is done in two steps:

1. contraction of e_{ij}: $E \leftarrow E \setminus \{e_{ij}\}$, $V \leftarrow V \setminus \{v_i\}$
2. contraction of the attaching 3-P vertex v_j:

$$V \leftarrow V \setminus \{v_j\}$$
$$E \leftarrow E \setminus \{e_{jk}, e_{jl}\}$$
$$E \leftarrow E \cup \{e_{kl}\}$$

The new edge is associated weights. The length is the distance of the barycenters of the attaching vertices $l\{e_{kl}\} = ||b(v_k) - b(v_l)||$. The orientation and the barycenter are obtained as weighted means, weighted by the length of the ancestors $l(e_{jk})$ and $l(e_{jl})$. In this way, the global geometric properties of the object are preserved during the filtering process.

Remark 1: The case $d(v_i) = 3$ and $d(v_j) = 1$ is geometrically identical and processed similarly; it differs only in the indices i, j.

Remark 2: We consider here only T-junctions occuring in such v_j that have $d(v_j) = 3$. The crossings are processed by Filter 2 below.

Filter 2. Junctions in real wire-frame type skeletons are often erroneously doubled, see Fig. 3d. These doubled junctions prevent from correct simplification of the skeleton, because the skeleton branches that belong to one fibre are actually not incident to a unique vertex of the graph. The filter type 2 is to eliminate these doubled junctions. In terms of graph operations this filter is an edge contraction of all non terminal edges shorter than L. Notice, that geometric properties of contracted edges need to be inherited.

It is well known that edge contraction may result in creating multiple edges or loops. Simple graphs, that are used here, can not contain multiple edges. Adding an edge to already connected vertices of a simple graph is identity operator. On the other hand, loops do not have a real physical meaning for coding adjacency relation of straight fibres and are therefore deleted.

This filter is coded by Algorithm 2.

Algorithm Filter 2

Input: Graph $G(V, E)$, Length $L \in \mathbf{R}+$
Output: Graph $G(V, E)$
———

for all $e_{ij} \in E$ do:
    if $l(e_{ij}) < L$ and $d(v_i) > 1$ and $d(v_j) > 1$:
        closest neighbor of e_{ij} inherits its geometric properties
        contract e_{ij}

Recall the edge contraction operation: for some given graph $G(V, E)$, the contraction of some edge e_{ij}, does the following: It removes e_{ij} from the set of edges E. It removes the vertex v_j, and attaches to v_i all edges previously incident to v_j, except e_{ij}.

The closest neighbor inherits the geometric properties of e_{ij} as in the case of filter 1.

Reconstruction of fibres. The fibres to extract are supposed to not bend (made from some rigid material). This assumption is being used during the reconstruction of the fibres from the skeleton. A straight object will comprise line segments attached in junctions and almost aligned.

Proposition 2. *Using the Definition 1 above, extracting rigid structures is equivalent to extraction of shortest mean graph walks.*

Proof: An open graph walk (defined as an alternating sequence of vertices and edges, starting and ending by a vertex) describes a continuous object such as a fibre stem.

The term shortest expressed in terms of angle $\pi-\alpha$, see Definition 1, with α being the angle between two adjacent edges, gives preference to straight continuous structures.

Mean makes the overall weight express the mean curvature of the walk. Hence shortest mean graph walk prefers straight continuous structures with the lowest mean curvature.

Notice that this approach is - to some extent - equivalent to minimal path finding for extraction of fine structures used in [16]. Here, the algorithm extracts from the graph the shortest mean walks by the from-closest-to-closest reconstruction strategy.

Algorithm Reconstruction of fibres

> Input: Graph $G(V, E)$
> Output: Graph $G(V, E)$
> ──────
> repeat until stability:
>   select a pair of edges (e_{ij}, e_{jk})
>   unlink e_{ij} and e_{jk} from v_j
>   contract e_{ij} and e_{jk}
>   if rank(v_j)=0 then
> $V \leftarrow V \setminus \{v_j\}$

The selection criterion is done on the following basis:

1. the edges (e_{ij}, e_{jk}) are adjacent, incident to a common vertex v_j.
2. The pair (e_{ij}, e_{jk}) are the currently closest neighbors in $E \times E$,

$$(e_{ij}, e_{jk}) = \arg \min_{e_{ij}, e_{jk} \in E} \alpha(e_{ij}, e_{jk})$$

The angle α is the difference between the bound edge and π. The *min* selection criterion therefore chooses pairs of edges geometrically closest to the straight line.

The stopping criterion is the maximum authorized angle bound by two edges, i.e. the cycle *repeat* may stop as soon as there are no more adjacent edges bounding and angle inferior to some given bound. We have used 30 degrees. This value has been determined experimentally as a trade-off between the over- and under-reconstruction.

During the reconstruction, the fibres lose their common vertex v_j. One can also iterate the *repeat* cycle until there are no more adjacent edges. Experiments have shown practically no influence of the value of this criterion on the result.

Note: In order to run efficiently, the select criterion needs to be implemented with a priority waiting list containing all pairs of adjacent edges of the graph $G(V, E)$, ordered with the priority $\pi - \alpha$. This list is filled once before, and progressively emptied during the reconstruction algorithm.

3   Experimental Results

The method has been proven on synthetic images. A full validation on real data is currently not feasible because of the absence of the ground truth data.

A set of fibres have been randomly drawn in a test volume according to a known distribution law. Several synthetic images have been generated, with uniform and normal distribution, and one image where two normal distributions meet.

The validity was proved by comparison of the measured parameters to those used to generate the images.

Several simulations were done to validate the algorithm. For all, 100 fibres were drawn in a 3-D image 100^3. The thickness of the fibres is 4 pixels.

Simulation 1: Normal distribution of orientations with mean of azimuth and elevation equal $\pi/6$, and standard deviation 0.1 rad, see Fig. 4.

Simulation 2: Uniform distribution of orientations in $[0, \pi]$, see Fig. 5.

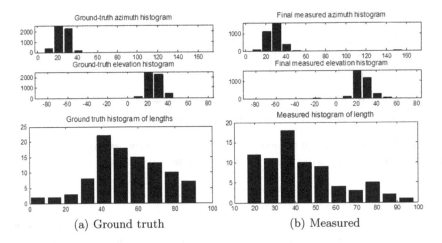

(a) Ground truth   (b) Measured

Fig. 4. Simulation 1. Anisotropic bundle of random fibres with normal distribution of orientations.

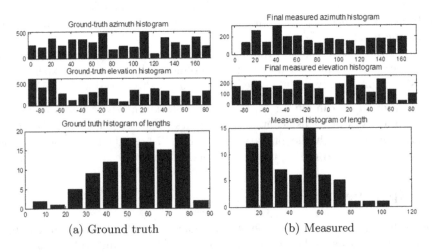

(a) Ground truth   (b) Measured

Fig. 5. Simulation 2. Isotropic bundle of random fibres with uniform distribution of orientations.

The bivariate histograms of orientations: (φ, θ) show a fair match for all simulations. The histogram of lengths are slightly biased towards shorter lengths, showing that not all fibres could have been correctly reconstructed. A few individuals with a longer length show that there has also been overreconstruction. This occurs in case of co-parallel, touching fibres.

4   Conclusion

The proposed segmentation technique is implemented in an optimal way, using hierarchical waiting lists storing the fragments of fibres, ordered in the decreasing order of the mutual angle. This ensures the best geometrical result, and the lowest computation complexity.

The filters applied to the skeleton satisfy properties required from filters, increasingness and idempotence. Moreover, both filters are anti-extensive.

The analyzed objects are reduced to their 1-D representation - a skeleton. This brings two important advantages: There's no trade off to find between locality and accuracy as in various previous works based on oriented filtering. Secondly, working with skeletons represents a considerable data reduction. Indeed, we reduce 3-D objects to 1-D, which brings a considerable reduction of the computation time.

Execution time -   The tests have been done on an AMD Opteron 2.6GHz, 16GB RAM, running MS Windows Server 2000, for volumic images 100^3 containing fibres.

Homotopic thinning, coded in C++, and implemented by using priority queues is optimal in terms of computation complexity. The thinning takes about 250 ms. The reconstruction of the fibres (with preliminary filtering), all coded in Python, takes $\approx$ 8-9 s.

The graph has been coded in XML. The time spend on its construction from the skeleton is negligeable (hundreds of ms) compared to the rest of the algorithm. The execution time of the filtering and reconstruction is proportional to the number of fibres in the volume.

Processing real data containing $2000 \times 2000 \times 1100$ samples at once is not feasible. It was done in blocks of 100^3 to 120^3 points, without overlap. The execution time was between 9 and 12 hours.

Perspective -   Obtaining the graph-based description of the fibre network allows us to evaluate other material parameters such as the local fibre density, local isotropy, boundaries between zones from different moulding pipes, etc. Coupling data from numerous samples allows to create maps of these properties. The validation of the results on real images by using maps of local orientation is an undergoing process.

The next step will be to provide 3D maps of orientations, estimated from the orientation tensor of fibres in a neighborhood of each point, as in [4]. The orientation parameters are weighted by the length of fibres, and are not sensitive to the fragmentation of fibres resulting from fibre breaks or from an incomplete reconnection at the end of the reconstruction process.

Acknowledgment. The authors are grateful to Aedinca for its support to this research project.

References

1. Soille, P., Talbot, H.: Image structure orientation using mathematical morphology. In: Conference on Pattern Recognition, June 29-July 2, vol. 2, pp. 1467–1476 (1998)
2. Soille, P., Talbot, H.: Directional morphological filtering. IEEE PAMI 23, 1313–1329 (2001)
3. Urbach, E.R., Wilkinson, M.H.F.: Efficient 2-D Grayscale Morphological Transformations With Arbitrary Flat Structuring Elements. IEEE TIP 17(1), 1–8 (2008)
4. Jeulin, D., Moreaud, M.: Segmentation of 2D and 3D textures from estimates of the local orientation. Image Analysis and Stereology 27, 183–192 (2008)
5. Chen, Y.-S., Wen-Hsing, H.: An interpretive model of line continuation in human visual perception. Pattern Recogn. 22(5), 619–639 (1989)
6. van Vliet, L., Verbeek, P.: Segmentation of overlapping objects. In: van Vliet, L., Young, I. (eds.) Abstracts of the ASCI Imaging Workshop (October 1995)
7. Chen, J., Sato, Y., Tamura, S.: Orientation space filtering for multiple orientation line segmentation. IEEE Trans. PAMI 22 (2000)
8. van Ginkel, M., van Vliet, L., Verbeek, P.: Applications of image analysis in orientation space. In: Vossepoel, A.M. (ed.) Fourth Quinquennial Review 1996-2001 Dutch Society for Pattern Recognition and Image Processing, NVPHBV, Delft, pp. 355–370 (2001)
9. Perona, P., Malik, J.: Scale-space and edge detection using anisotropic diffusion. IEEE Trans. PAMI 12(7), 629–639 (1990)
10. Frangi, A., Niessen, W., Koen, L., Viergever, M.: Multiscale vessel enhancement filtering. In: Wells, W.M., Colchester, A.C.F., Delp, S.L. (eds.) MICCAI 1998. LNCS, vol. 1496, pp. 130–137. Springer, Heidelberg (1998)
11. Stein, A.M., Vader, D.A., Jawerth, L.M., Weitz, D.A., Sander, L.M.: An algorithm for extracting the network geometry of three-dimensional collagen gels. Journal of Microscopy 232(3), 463–475 (2008)
12. Dokladal, P.: Segmentation of Grey-Scaled Images: A Topological Approach. PhD thesis, Marne-la-Valle University (January 2000)
13. Dokladal, P., Lohou, C., Perroton, L., Bertrand, G.: Liver blood vessels extraction by a 3-d topological approach. In: Taylor, C., Colchester, A. (eds.) MICCAI 1999. LNCS, vol. 1679, pp. 98–105. Springer, Heidelberg (1999)
14. Bertrand, G.: Simple points, topological numbers and geodesic neighborhoods in cubic grids. Pattern Recogn. Lett. 15(10), 1003–1011 (1994)
15. Malandain, G., Bertrand, G.: Fast characterization of 3D simple points. In: 11th ICPR, vol. III, pp. 232–235 (1992)
16. Vincent, L.: Minimal path algorithms for the robust detection of linear features in images. In: ISMM 1998, Mathematical Morphology and its Applications to Image and Signal processing, Amsterdam. Computational Imaging and Vision, vol. 12, pp. 331–338. Kluwer, Dordrecht (1998)

Spatially-Variant Morpho-Hessian Filter: Efficient Implementation and Application

Olena Tankyevych[1], Hugues Talbot[1], Petr Dokladál[2], and Nicolas Passat[3]

[1] Université Paris-Est, Département d'Informatique Gaspard-Monge,
Équipe A3SI, ESIEE Paris, Cité Descartes, BP 99, F-93162 Noisy-le-Grand Cedex,
France
{tankyevo,talboth}@esiee.fr
[2] Centre de Morphologie Mathématique, Mines-Paristech, 35 rue St Honoré, F-77300
Fontainebeau, France
petr.dokladal@mines-paristech.fr
[3] Université de Strasbourg, LSIIT UMR 7005, Strasbourg, France
passat@unistra.fr

Abstract. Elongated objects are more difficult to filter than more isotropic ones because they locally comprise fewer pixels. For thin linear
objects, this problem is compounded because there is only a restricted
set of directions that can be used for filtering, and finding this local direction is not a simple problem. In addition, disconnections can easily
appear due to noise. In this paper we tackle both issues by combining
a linear filter for direction finding and a morphological one for filtering.
More specifically, we use the eigen-analysis of the Hessian for detecting
thin, linear objects, and a spatially-variant opening or closing for their enhancement and reconnection. We discuss the theory of spatially-variant
morphological filters and present an efficient algorithm. The resulting
spatially-variant morphological filter is shown to successfully enhance
directions in 2D and 3D examples illustrated with a brain blood vessel
segmentation problem.

Keywords: Adaptive morphology, spatially-variant morphology, vascular imaging, vesselness, Hessian-based filtering, directional filtering.

1 Introduction

In this paper, we define thin objects in image as semantically consistent objects
that exhibit at least one dimension much smaller than the others. We focus
particularly on elongated thin objects, locally curve- or line-like, e.g. fibres, hair,
strings, or blood vessels.

It is generally difficult to filter such objects. Indeed, classical filters in the literature assume extended objects (median, averaging, linear convolutions, morphological filters with standard structuring elements) [1,2]. Those that do not
make this assumption still filter only in areas of low gradient in a region of
interest [3,4,5]. Elongated objects may in fact present no part with a suitable
low gradient, due to both noise and object edges, as in Figure 1. For elongated

M.H.F. Wilkinson and J.B.T.M. Roerdink (Eds.): ISMM 2009, LNCS 5720, pp. 137–148, 2009.
© Springer-Verlag Berlin Heidelberg 2009

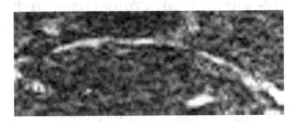

Fig. 1. A thin, noisy brain blood vessel. There is no low-gradient zone in this object, disconnections are due to noise.

objects, in the literature, it is often assumed that one dimension is long, everywhere tangent to the object. Filtering can be applied along this direction without necessarily destroying the object entirely. Within the morphology framework, one classically use families of segments as structuring elements (SEs), selecting the one best matching each object [6]. This, however, requires directional sampling, which may be prohibitive in 3D. Alternatively, path-based families of structuring elements can be used instead with no sampling required, but with similar complexities and running times [7,8]. Finally, in [9], curvature evaluation was also used in 2D in conjunction with mathematical morphology operators.

Linear (in the mathematical sense) filtering methods have also been proposed. In the case of the scale-space framework, which should be suitable to this problem, edge and ridge detection methods were proposed [10,11] utilising the Hessian or the structure tensor [12,13]. As an extension of these methods, anisotropic diffusion is often proposed for filtering, using the tensorial information for diffusing only within the object. In practice diffusion may fail if the object is very thin [14].

Elongated objects have a tendency to be sensitive to noise, often implying disconnection along the object, as is readily apparent in Figure 1. Segmentation of this kind of objects is often reliant on connection being maintained throughout the object. Diffusion is often not well suited for this task [15,14] as it will become isotropic outside the oriented objects.

In order to achieve reconnection, disconnected sets can be matched [16], but this is an ill-posed problem. Minimal path methods have been proposed to achieve reconnection [17,18], but the problem of specifying endpoints can be difficult to solve.

Recently, two similar 2D approaches were proposed combining the analysis of orientation from either the Hessian or the structure tensor, and using spatially-variant (SV) morphology to bridge the disconnection gaps [19,20]. This is a productive approach, as long as the resulting filter is indeed a morphological one, and directions are well estimated.

Spatially-variant mathematical morphology (SVMM) was introduced in [21] for binary sets using structuring functions. In [22] it was used for adaptive filtering on grey-level and colour images. An efficient algorithm for variable rectangles was proposed in [23], finally, SVMM was studied from the theoretical point of view in [24,25,26].

In this article, in the first section we expose a simple version of the theory of SVMM together with an efficient algorithm. In the second section we briefly introduce multi-scale Hessian analysis for the study of local directions. In the third section we present applications using both concepts combined.

2   Spatially-Variant Morphological Filtering

The theory of SVMM and corresponding algorithms are formulated with the purpose of filtering an image differently at various positions. In the case of elongated objects, we wish to discover their local orientation and filter them locally along this direction, for instance, with an oriented segment.

This presents a challenge. Traditionally one could use a supremum of openings or an infimum of closings with a family of segments or paths as SEs, but the range of this kind of filters is limited. For instance, there is no known way to produce in this manner a morphological equivalent to an edge-enhancing, orientation-driven, inverse diffusion filter as in [12].

In this section we present the theory upon which, the filter presented in section 4 relies on.

2.1   Adjunction and Spatially-Variant Morphology

First, we note that it is not useful to establish a distinction between binary and grey-level SVMM. Indeed, SVMM can be described on an arbitrary lattice in which SEs are available. In general, one can construct a simple SV morphological operator by computing a max or min filter using an SE that is not the same everywhere in the image. For instance, one can use parametrized disks or oriented segments, and vary respectively the diameter or the orientation according to a scalar field. Erosions and dilations with these SEs pose no problem by themselves. However, for filtering, openings and closings are the more interesting operators, but *adjunct* respective dilations and erosions are not trivially computed, as we shall see shortly.

Note that in general even though one can perform a kind of pseudo-opening or closing operation in each separate SE (by combining a max and min filter in the same window), the result of such an operator is not guaranteed to be idempotent due to the change of homothecy or rotation, and would furthermore be dependent on the order of operations. Consequently, it will not be a morphological filter [2].

In Figure 2, the concept of adjunction is illustrated in the case of SE-based operators. In Figure 2(a) we have the translation-invariant (TI) case of a 2×1 horizontal line segment. In the general case, the *transpose* of a SE B, noted $\check{B}$ is

$$\check{B}(x) = \{y \mid x \in B(y)\}, \tag{1}$$

where x and y are points, and $B(y)$ is the potentially spatially-variant structuring element originating at point y. For a TI operator as in Figure 2(a), we observe that the general definition collapses to the usual one, i.e, $\check{B} = -B$, the symmetric of B with respect to the origin. However, for the SV case this is not true. In

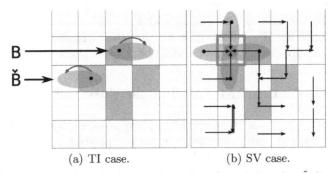

(a) TI case.   (b) SV case.

Fig. 2. Adjunct structuring elements with explicit formulation for $\check{B}$ for the TI case but not the SV one (as for the highlighted pixel)

Figure 2(b), we use the same SE, but with varying orientation (arrows denote directions). In Fig. 2(b), for the indicated pixel, the transpose of the SE is not a segment. Computing openings and closings with SV erosions and dilations requires the transpose to be computed everywhere, which can be computationally expensive requiring an exhaustive search. This becomes prohibitive for large, nD data ($n \geq 2$). In the next section we present a way to decrease the computation complexity.

2.2   Adjunct by Conditional Propagation

In the following, we describe the SV dilation δ and its adjunct erosion ε both using spatially-variant SEs .

Let $\mathcal{L}$ be a family of functions, or images, defined as a mapping from the support D to the set of values V. Let δ, ε: $\mathcal{L} \to \mathcal{L}$ be a dilation and an erosion of $f \in \mathcal{L}$ given by:

$$[\delta_B(f)](x) = [\bigvee_{b \in B(x)} f_b](x) \qquad (2)$$

$$[\varepsilon_{\check{B}}(f)](x) = [\bigwedge_{b \in \check{B}(x)} f_b](x), \qquad (3)$$

where f_b denotes the translation of f by b computed as $f_b(x) = f(x - b)$. B stands for the structuring element. In standard mathematical morphology, B is translation invariant, defined as $B \subset D$.

In the context of SVMM, B is often denoted as *structuring function* and is defined as $B : D \to \mathcal{P}(D)$, $\mathcal{P}$ being the collection of all subsets of D. As above, $B(x)$ denotes the structuring element B originating at point x.

The transpose of B, denoted $\check{B}$, is defined in Eq. 1 and used in Eq. 3. This formulation of $\check{B}$ can be computationally costly. It can a priori be of arbitrary extent depending on the B family, which becomes problematic for forming filters based on adjunctions of dilations and erosions in order to make a closing φ_B or an opening γ_B :

$$\varphi_B(f) = \varepsilon_{\check{B}}(\delta_B(f)); \qquad \gamma_B(f) = \delta_B(\varepsilon_{\check{B}}(f)). \qquad (4)$$

Note that there also exists a complementary, SE-based SV erosion $\varepsilon_B(f) = -\delta_B(-f)$ and its adjunct $\delta_{\check{B}}(f)$, and that in general ε_B and $\varepsilon_{\check{B}}$ differ.

In the following we propose an implementation of the adjunct operator identical in complexity to the normal one.

Implementation. The inf/sup-of-functions in Eqs. 2 and 3 are usually computed sequentially in raster scan order according to this scheme:

$$[\delta_B(f)](x) = \max_{b \in B(x)} f(x - b) \qquad (5)$$

$$[\varepsilon_{\check{B}}(f)](x) = \min_{b \in \check{B}(x)} f(x - b) = \min_{y \mid x \in B(y)} f(x - y) \qquad (6)$$

The dilation of Eq. 5 is computed in $\mathcal{O}(MN)$, where $N = \mathrm{Card}(D)$, and $M = \mathrm{Card}(B(x))$. In the following erosion, from Eq. 6, given some x, the set $\{y \mid x \in B(y)\}$ is a priori unknown and is computed in $\mathcal{O}(N^2)$ by exhaustive search. However, relaxing the sequential order of computing, the adjunct erosion of Eq. 6 can be computed more efficiently, as shown below.

Theorem 1 (Conditional propagation for the adjunct erosion). *Assume* $\varepsilon_{\check{B}}(f)$, *i.e. the output, originally set everywhere equal to f. One can sequentially read the input f at each point x. Considering the structuring element $B(x)$ of origin x, for all elements y of $B(x)$, we update the output value $[\varepsilon_{\check{B}}(f)](y)$ by taking the min operator between the current input value at x and the current output value at y.*

Proof. It is easy to show that both ways of computing the adjunct are equivalent. As we scan the input image at x and update the value in the output image at y, we are indeed computing a min operator between all the origins of $B(x)$ such that $B(x)$ intersects y. So, once the whole image has been scanned, if $B(x)$ intersects y, then x is in $\check{B}(y)$ and vice-versa.

This theorem is illustrated in Figure 3. In the Fig. 3(a) the original image is illustrated with the directions at each pixel as arrows. The structuring element B is originating at the selected pixel x for a SE-based dilation δ_B. The result of this dilation is shown in the Fig. 3(b). In the same figure the adjunct structuring elements $\check{B}$ are pointing at the selected pixel y for the efficient adjunct erosion. This operation is performed with the minimal value updated everywhere along $\check{B}$ as it spans the image, the result can be observed in the Fig. 3(c). This theorem leads to Algorithm 1 based on a conditional propagation of the value at the origin of each $B(x)$. In this algorithm, the for loop (lines 4-5) computes sequentially for all $x \in D$, the dilation $\delta_B(f)$ implemented by definition of Eq. 5. Every $\delta_B(f)$ is then stored in d (previously initialized (line 3)). Notice that there is no need to store the entire image $\delta_B(f)$. The erosion is implemented by a conditional propagation (lines 6-7).

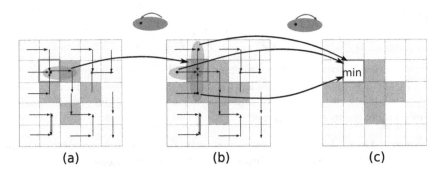

Fig. 3. Efficient SV closing (arrows at each pixel denote directions, thick long arrows indicate the operation on the selected pixel)

The result $\varphi_B(f)$ is available as soon as the raster scan of the input image ends (lines 2-7). The overall computational complexity is $\mathcal{O}(\sum_{x\in D}\mathrm{Card}(B(x)))$. If B is of constant size, this reduces to $\mathcal{O}(MN)$, where $N = \mathrm{Card}(D)$, and $M = \mathrm{Card}(B)$. The SV opening can be computed analogously.

Notes. (1) In the illustrated algorithm the closing starts with the SE-based dilation, but it would be possible to start with the complementary, adjunct dilation $\delta_{\check{B}}$, followed by the SE-based erosion ε_B. The two results differ in general. (2) Forming an opening starting with the adjunct erosion requires intermediary storage. (3) SE can be of any shape, not just a line segment.

In the following section we propose a way to derive an orientation tensor field.

Algorithm 1. SPATIALLY-VARIANT CLOSING

Input: image f, structuring function B
Result: closing $\varphi_B(f)$

Data: d - scalar

1 $\varphi_B(f) \leftarrow \infty$; *initialization*
2 **forall** $x \in \mathrm{support}(f)$ **do**
3 $\quad$ $d \leftarrow 0$; *initialization*
4 $\quad$ **forall** $b \in B(x)$ **do**
5 $\quad\quad$ $\lfloor$ $d \leftarrow \max(d, f(x-b))$; —— *dilation* $(\delta_B(f))$ ——
6 $\quad$ **forall** $b \in B(x)$ **do**
7 $\quad\quad$ $\lfloor$ $\varphi_B(f(x-b)) \leftarrow \min(\varphi_B(f(x-b)), d)$; —— *erosion* $(\varepsilon_{\check{B}}(f))$ ——

3   Multiscale Hessian Analysis

For a real, twice derivable function $L(x_1, \ldots, x_i, \ldots, x_n)$ on a support D, the Hessian operator is the order-2 tensor of the second derivatives

$$H = \left(\frac{\partial^2 L}{\partial x_i \partial x_j} \right)_{1 \leq i,j \leq n}.$$

If the second derivatives are continuous, H is symmetric semi-definite. In computer vision, the derivative operator ∂ is not well-defined because of digitization, and so H is only estimated, at a certain scale σ, via a convolution of L by a Gaussian of parameter σ and a discrete derivative operator [1]. Eigen-analysis of H can reveal local structure information. In this paper we focus on the case $n = 3$, with the three eigenvalues, $|\lambda_1|, |\lambda_2|, |\lambda_3|$, of H in increasing order, and (e_1, e_2, e_3) as their corresponding eigenvectors.

3.1   Line Detection Using the Hessian in 3D

Here we mostly follow the formulations by [11,13]. Tubular objects in 3D are indicated by features which exhibit low curvature along the main axis of the object, and strong curvature in the perpendicular plane to this axis. In Hessian eigenspace, this translates to $|\lambda_3| \geq |\lambda_2| \geq |\lambda_1|$, and the main axis is indicated by e_1 (in the case of a bright vessel). Consequently, $V = \lambda_3 - \lambda_2$ is a simple tubular indicator, which is of high value in tubular structures. Frangi [13] has a more sophisticated model for computing a so-called "vesselness measure" involving multiscale analysis to estimate the probability of a voxel to belong to a blood vessel:

$$\nu(x, \sigma) = \begin{cases} 0 & \text{if } \lambda_2 > 0 \text{ or } \lambda_3 > 0, \\ \left(1 - e^{\frac{-R_A^2}{2\alpha^2}}\right) \cdot e^{-\frac{R_B^2}{2\beta^2}} \cdot \left(1 - e^{\frac{-S^2}{2\gamma^2}}\right) & \text{otherwise,} \end{cases} \tag{7}$$

with

$$R_A = \frac{|\lambda_2|}{|\lambda_3|},$$
$$R_B = \frac{|\lambda_1|}{\sqrt{|\lambda_2 \lambda_3|}}, \tag{8}$$
$$S = \|H_\sigma\| = \sqrt{\textstyle\sum_j \lambda_j^2},$$

in which R_A differentiates between plate and line like objects, R_B describes blob-like ones, and S accounts for the intensity difference between these objects and background. Parameters α, β and γ influence the weight of the according objects. The final vesselness result is produced by its best response at different scales σ for each voxel x.

Blood vessels are not strictly made of tubular structures, but also of junctions, more complex, spongious areas, and stenosis, forming double conic shapes. In general, junctions and also thin objects that are not tube-like, e.g. membranes, surfaces or double cones, can also be characterized to some degree with the Hessian analysis but ultimately will require more sophisticated analysis. There are works that have proposed Hessian-based formulations of these kind of objects [27,28]. In the application part of this article, we only consider the tubular object detection problem.

4   Application

In this section we describe an application where combined Hessian and morphological analysis is useful. More specifically, we propose to improve Frangi's vesselness measure by morphological SV closing, allowing to reduce the noise and vessel disconnections.

4.1   Algorithm

The steps of the algorithm are illustrated in the Figure 4. In the first step of the algorithm, the Hessian analysis is performed at each pixel on multiple scales of the image. The vesselness as in Eq. 7 is calculated across these scales with the maximum final response for each voxel. This indicates which voxels belong to a tubular feature but fails to connect vessels portions when the noise level is too high or when vessels are too narrow. However, we observe that the orientation field is relatively robust within an appropriate choice of scales (see more discussion in Section 4.2).

In the second step, we use a family $B(\mathbf{e_1}(x))$ of centered segments as structuring elements oriented in the direction of $\mathbf{e_1}$, and of fixed length for the morphological closing operation with the aim of vessel reconnection. The SE-based dilation is followed by the adjunct erosion $\varepsilon_{\check{B}}(\delta_B)$ as in section 2.2. In order to propagate the objects in the space, the dense orientation field is necessary. In our case, this is required only as far as the dilation can reach. At first, the dilation is performed only on the directional image. Then, the actual intensity image is dilated according to the directions obtained in the previous dilation. This is followed by the adjunct erosion operation. This ensures an idempotent result of the closing operation, which guarantees that the resulting filter obeys all morphological rules.

Following the morpho-Hessian closing - which performs like an edge-enhancing inverse diffusion filter in some ways - noise levels are generally increased. In order to filter out the noise not connected with vessels and reconnect the vessels, we perform a gray-level reconstruction [29] using the initial vesselness image as marker, and the result of the morphological closing as mask. For the simple example from the Fig. 4, the result of the grayscale reconstruction is similar to the closing.

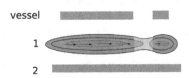

Fig. 4. Illustration of the directional closing. The original, noisy, broken vessel is at the top. Vessel directions are estimated from the Hessian (first step). A SV closing is performed in the main direction of the vessel (second step). A gray-scale reconstruction from the vesselness is also performed, the result is similar to the closing for this example.

(a) Volume rendering of small arteries of the brain (in red) in MRI-TSA modality.

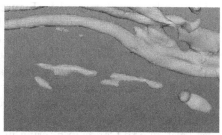

(b) Surface rendering of the vesselness response.

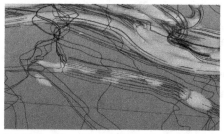

(c) Vector flow lines of the principal eigenvectors oriented in the main direction of the vessel.

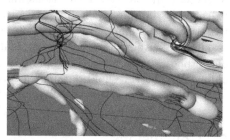

(d) Reconstructed vessel by spatially-variant closing.

Fig. 5. Application to the segmentation of brain vessels

4.2 Results and Discussion

The described algorithm was implemented with the ITK library using the available implementations of vesselness.

The image volume is produced with MRI and is of size $256 \times 256 \times 256$ pixels. The time processing of the Hessian multiscale method is 2 minutes and of the morphological closing is 1 minute 20 seconds on a single core of a 2.5GHz Intel Quad-Core Xeon processor.

The considered vesselness parameters have been set to: $\alpha = \beta = 0.5$ and $\gamma = 5$, as proposed in [13]. The SE is symmetric according to the origin projecting the opposite direction of the current pixel, its total length is fixed at 7 voxels. It would be also possible to modify the length of B according to λ_1 (this might be more costly at larger scales), but experiments indicate that this is not critical.

Commonly, the lower scales are more sensitive to noise. Furthermore, in [30] it has been reported that the vesselness response curve differs not more than 3 percent between $\sigma = 1$ and $\sigma = 2\sqrt{2}$ using a set of discrete values $\sigma_i \in 1, \sqrt{2}, 2, 2\sqrt{2}$ and that is applicable for most applications. For this filtering method, we used the scales between $\sigma = 1$ and $\sigma = 2\sqrt{2}$ with three logarithmic natural incrementation steps.

The results are illustrated in Figure 5 with an application to the segmentation of brain vessels obtained with MRI angiography modality. The original and filtered images are surface-rendered with the same threshold. The volume-rendered image (Figure 5(a)) shows that the image contains a high level of noise and disconnections that prevent using a low threshold for detecting all vessels. In Figure 5(b), it can be observed that the vesselness filters out the noise, but does not manage to reconnect the noise-corrupted vessels. However, the orientation of the first eigenvector (Figure 5(c)) remains relatively unchanged by the noise level, due to Gaussian filtering in the Hessian estimation (the turbulence-like effect of the vector flow is an artifact of a visualisation program). Thanks to these, principal vessel directions and in the contrast to the vesselness filter, morpho-Hessian filter (Figure 5(d)), accomplishes reconnections of the smaller vessel parts and eliminates the noise, up to some extent.

The gray-level reconstruction eliminates all non-connected noise of the result of morpho-Hessian filter. The final segmentation is achieved by top-hat thresholding [1]. The threshold parameter is chosen heuristically and is the same for all methods while an automated segmentation method is under development.

5   Conclusions

In this article we have presented a concise theory of spatially-variant morphology with the main result being an efficient algorithm in $\mathcal{O}(MN)$ for spatially-variant openings and closings, where N is the size of the image, and M ($M \ll N$) is the size of the structuring element. While SVMM is not in itself novel, we believe that the presentation of the algorithm in this article is simple and enlightening.

As an application, we have presented a morpho-Hessian filter for 3D image filtering, in particular, enhancement and reconnection of thin, tubular objects. Directions in 3D were obtained by eigen-analysis of the Hessian, and reconnection was achieved by SV morphology. Hessian analysis is fast, multiscale and relatively robust for object direction detection, whereas orientation analysis using known MM methods would require directional sampling, which, in turn, would be prohibitive in 3D. Conversely, anisotropic diffusion is inefficient, requiring many iterations and featuring convergence issues, whereas a closing or opening converges in one iteration.

The results obtained by applying this filter on real 3D MRI-TSA data, compared to those acquired with the vesselness function [13], underline the aptitude of the suggested routine for reconnection of the smaller vessels.

Our proposed filter also requires very few parameters, namely the range of scales in Hessian analysis and the length of the closing/opening segment for reconnection.

Overall, this approach may be seen as a productive combination of linear and non-linear techniques. Future work includes studying alternatives to the Hessian tensor for improved noise robustness and direction detection. The handling of scales with regard to the topology of the vascular tree could also be considered in further works. And, as proposed in [23], the size of the spatially-variant structuring elements could be resized according to the eigenvalues of the Hessian matrix. Lastly, a more complete application involving large-scale reconstruction of the brain vascular system, its semi- or completely automatic segmentation and classification of arteries and veins are our ultimate goals.

Acknowledgments

We wish to thank the following medical personnel from Colmar Hospital: A. Tournade, M. Musacchio, H. Oesterlé, M. Lagneau, who provided medical expertise and data. This work was partially supported by grant SURF ANR-05-BLAN-0071, reference NT05-2_45825.

References

1. Gonzalez, R., Woods, R.: Digital Image Processing, 3rd edn. Prentice-Hall, Englewood Cliffs (2007)
2. Heijmans, H.: Morphological filters for dummies. In: Maragos, P., Schafer, R.W., Butt, M.A. (eds.) Mathematical Morphology and its Applications to Image and Signal Processing, proceedings for ISMM 1996, pp. 127–137. Kluwer Acad., Dordrecht (1996)
3. Perona, P., Malik, J.: Scale-space and edge detection using anisotropic diffusion. IEEE Transactions on Pattern Analysis and Machine Intelligence 12(7), 629–639 (1990)
4. Smith, S.M., Brady, J.M.: Susan a new approach to low level image processing. Int. J. Comput. Vision 23(1), 45–78 (1997)
5. Tomasi, C., Manduchi, R.: Bilateral filtering for gray and color images. In: Sixth International Conference on Computer Vision, 1998, pp. 839–846 (1998)
6. Soille, P., Talbot, H.: Directional morphological filtering. IEEE Transactions on Pattern Analysis and Machine Intelligence 23(11), 1313–1329 (2001)
7. Heijmans, H., Buckley, M., Talbot, H.: Path openings and closings. Journal of Mathematical Imaging and Vision 22, 107–119 (2005)
8. Talbot, H., Appleton, B.: Efficient complete and incomplete paths openings and closings. Image and Vision Computing 25(4), 416–425 (2007)
9. Zana, F., Klein, J.C.: Segmentation of vessel-like patterns using mathematical morphology and curvature evaluation. IEEE Transactions on Image Processing 10(7), 1010–1019 (2001)
10. Lindeberg, T.: Feature detection with automatic scale selection. International Journal of Computer Vision 30(2), 79–116 (1998)
11. Danielsson, P.E., Lin, Q.: Efficient detection of second-degree variations in 2D and 3D images. Journal of Visual Communication and Image Representation 12, 255–305 (2001)
12. Deguchi, K., Izumitani, T., Hontani, H.: Detection and enhancement of line structures in an image by anisotropic diffusion. Pattern Recognition Letters 23(12), 1399–1405 (2002)
13. Frangi, A., Niessen, W., Vincken, K., Viergever, M.: Multiscale vessel enhancement filtering. In: Wells, W.M., Colchester, A.C.F., Delp, S.L. (eds.) MICCAI 1998. LNCS, vol. 1496, pp. 130–137. Springer, Heidelberg (1998)

14. Manniesing, R., Viergever, M., Niessen, W.: Vessel enhancing diffusion: A scale space representation of vessel structures. Medical Image Analysis 10(6), 815–825 (2006)
15. Perona, P.: Orientation diffusions. IEEE Trans. Image Processing 7, 457–467 (1998)
16. Lee, T., Talbot, H.: A fast method for detecting and matching linear features in images. In: Proc. DICTA, Brisbane, Australia, December 1995, pp. 649–654 (1995)
17. Vincent, L.: Minimal path algorithms for the robust detection of linear features in images. In: Mathematical Morphology and its Applications to Image and Signal processing, Amsterdam. Computational Imaging and Vision, Proceedings for ISMM 1998, vol. 12, pp. 331–338. Kluwer, Dordrecht (1998)
18. Cohen, L., Deschamps, T.: Segmentation of 3D tubular objects with adaptive front propagation and minimal tree extraction for 3D medical imaging. Computer Methods in Biomechanics and Biomedical Engineering 10(4), 289–305 (2007)
19. Verdú-Monedero, R., Angulo, J.: Spatially-variant directional mathematical morphology operators based on a diffused average squared gradient field. In: Blanc-Talon, J., Bourennane, S., Philips, W., Popescu, D., Scheunders, P. (eds.) ACIVS 2008. LNCS, vol. 5259, pp. 542–553. Springer, Heidelberg (2008)
20. Tankyevych, O., Talbot, H., Dokladal, P.: Curvilinear morpho-Hessian filter. In: IEEE ISBI, pp. 1011–1014 (2008)
21. Serra, J.: Image analysis and mathematical morphology. Academic Press, London (1982)
22. Lerallut, R., Decencière, E., Meyer, F.: Image filtering using morphological amoebas. Image and Vision Computing 25(4), 395–404 (2007)
23. Dokládal, P., Dokládalová, E.: Grey-scale morphology with spatially-variant rectangles in linear time. In: Blanc-Talon, J., Bourennane, S., Philips, W., Popescu, D., Scheunders, P. (eds.) ACIVS 2008. LNCS, vol. 5259, pp. 674–685. Springer, Heidelberg (2008)
24. Charif-Chefchaouni, M., Schonfeld, D.: Spatially-variant mathematical morphology. In: IEEE International Conference on Image Processing, 1994. Proceedings. ICIP 1994, November 1994, vol. 2, pp. 555–559 (1994)
25. Bouaynaya, N., Charif-Chefchaouni, M., Schonfeld, D.: Theoretical foundations of spatially-variant mathematical morphology part i: Binary images. IEEE Trans. Pattern Anal. Mach. Intell. 30(5), 823–836 (2008)
26. Bouaynaya, N., Schonfeld, D.: Theoretical foundations of spatially-variant mathematical morphology part ii: Gray-level images. IEEE Trans. Pattern Anal. Mach. Intell. 30(5), 837–850 (2008)
27. Agam, G., Armato, S.G., Wu, C.: Vessel tree reconstruction in thoracic CT scans with application to nodule detection. IEEE Transactions on Medical Imaging 24(4), 486–499 (2005)
28. Antiga, L.: Generalizing vesselness with respect to dimensionality and shape. In: ISC/NA-MIC Workshop on Open Science at MICCAI (August 2007)
29. Vincent, L.: Morphological grayscale reconstruction in image analysis: Applications and efficient algorithms. IEEE Transactions on Image Processing 2(2), 176–201 (1993)
30. Sato, Y., Westin, C.F., Bhalerao, A., Nakajima, S., Shiraga, S., Tamura, S., Kikinis, R.: Tissue classification based on 3D local intensity structures for volume rendering. IEEE Transactions on Visualization and Computer Graphics 6(2), 160–180 (2000)

Some Morphological Operators in Graph Spaces

Jean Cousty, Laurent Najman, and Jean Serra

Université Paris-Est, Laboratoire d'Informatique Gaspard-Monge, Équipe A3SI,
ESIEE
{j.cousty,l.najman,j.serra}@esiee.fr

Abstract. We study some basic morphological operators acting on the
lattice of all subgraphs of a (non-weighted) graph $\mathbb{G}$. To this end, we
consider two dual adjunctions between the edge set and the vertex set
of $\mathbb{G}$. This allows us (i) to recover the classical notion of a dilation/erosion
of a subset of the vertices of $\mathbb{G}$ and (ii) to extend it to subgraphs of $\mathbb{G}$.
Afterward, we propose several new erosions, dilations, granulometries
and alternate filters acting (i) on the subsets of the edge and vertex set
of $\mathbb{G}$ and (ii) on the subgraphs of $\mathbb{G}$.

1 Introduction

From a formal point of view, digital image processing historically consists of
analyzing the transformations that act on the subsets of $\mathbb{Z}^2$ (the sets of pixels in
a binary image) and the transformations that act on the maps from $\mathbb{Z}^2$ to $\mathbb{N}$ (the
images themselves). In such a perspective, mathematical morphology provides a
set of filtering and segmenting tools that are very useful in applications.

On the other hand, there is a growing interest for considering digital objects
not only composed of points but also composed of elements lying between them
and carrying structural information about how the points are glued together (see
[1,2] for recent examples). The simplest of these representations are the (non-
weighted) graphs. The domain of an image is considered as a graph whose vertex
set is made of the pixels and whose edge set is given by an adjacency relation on
these pixels. In this context, it becomes relevant to consider the transformations
acting on the set of all subgraphs and not only those acting on the set of all
subsets of pixels.

When dealing with a graph $\mathbb{G}$, we often need (see *e.g.* ([1,3,4,5]) to consider
the graph induced by a subset S of vertices of $\mathbb{G}$. To this end, we associate with S
the largest subset of edges of $\mathbb{G}$ such that the obtained pair is a graph. In other
cases, we have to consider a graph induced by a subset of the edges of $\mathbb{G}$.

Motivated by classifying and understanding these operations and their combinations, we propose a systematic study of the basic operators which are used
to derive a set of edges from a set of vertices and a set of vertices from a set of
edges. It turns out that these operators are dilations and erosions. They allow
us (i) to recover the classical notion of a dilation/erosion of a subset of vertices
and (ii) to extend it to subgraphs (Section 3). Then, we propose several new
erosions, dilations, granulometries and alternate sequential filters acting (i) on

M.H.F. Wilkinson and J.B.T.M. Roerdink (Eds.): ISMM 2009, LNCS 5720, pp. 149–160, 2009.

the subsets of edges and on the subsets of vertices and (ii) on the subgraphs. We emphasize that, contrarily to most of the previous work on morphology in graphs (such as [6,7,8,9]), the main operators of this paper input and output graphs.

The proofs of the properties presented in this paper will be given in a future extended version [10].

2   Lattice of Graphs

We define a *graph* as a pair $X = (X^\bullet, X^\times)$ where $X^\bullet$ is a set and $X^\times$ is composed of unordered pairs of distinct elements in $X^\bullet$, *i.e.*, $X^\times$ is a subset of $\{\{x, y\} \subseteq X^\bullet \mid x \neq y\}$. Each element of $X^\bullet$ is called a *vertex or a point (of X)*, and each element of $X^\times$ is called an *edge (of X)*. In the sequel, to simplify the notations, $e_{x,y}$ stands for the edge $\{x, y\} \in X^\times$.

Let X and Y be two graphs. If $Y^\bullet \subseteq X^\bullet$ and $Y^\times \subseteq X^\times$, then X and Y are ordered and we write $Y \sqsubseteq X$. If $Y \sqsubseteq X$, we say that Y is a *subgraph* of X, or that Y is *smaller* than X and that X is *greater* than Y.

Important remark. Hereafter, the workspace is a graph $\mathbb{G} = (\mathbb{G}^\bullet, \mathbb{G}^\times)$ and we consider the sets $\mathcal{G}^\bullet$, $\mathcal{G}^\times$ and $\mathcal{G}$ of respectively all subsets of $\mathbb{G}^\bullet$, all subsets of $\mathbb{G}^\times$ and all subgraphs of $\mathbb{G}$.

Let $\mathcal{S}_0, \mathcal{S}_1 \subseteq \mathcal{G}$ be the sets of respectively the graphs made of a single vertex and the graphs made of a pair of vertices linked by an edge, *i.e.*, $\mathcal{S}_0 = \{(\{x\}, \emptyset) \mid x \in \mathbb{G}^\bullet\}$ and $\mathcal{S}_1 = \{(\{x, y\}, \{e_{x,y}\}) \mid e_{x,y} \in \mathbb{G}^\times\}$. We set $\mathcal{S} = \mathcal{S}_0 \cup \mathcal{S}_1$. Any graph $X \in \mathcal{G}$, is *generated* by the family $\mathcal{F} = \{X_1, \dots, X_\ell\}$ of all elements in $\mathcal{S}$ smaller than X: $X = (\bigcup_{i \in [1,\ell]} X_i^\bullet, \bigcup_{i \in [1,\ell]} X_i^\times)$; we say that the elements of $\mathcal{F}$ are the *generators* of X. Conversely, any family $\mathcal{F}$ of elements in $\mathcal{S}$ generates an element of $\mathcal{G}$. Hence, $\mathcal{S}$ *(sup-) generates* $\mathcal{G}$.

Clearly, the ordering $\sqsubseteq$ on graphs amount to say that $Y \sqsubseteq X$ when all generators of Y are also generators of X. Therefore, ordering $\sqsubseteq$ provides a *lattice* structure on the set $\mathcal{G}$. Indeed, the largest graph smaller than a family $\mathcal{F} = \{X_1, \dots, X_\ell\}$ of elements in $\mathcal{G}$ is the graph generated by the generators common to all X_i, $i \in [1, \ell]$; this *infimum* is denoted by $\sqcap \mathcal{F}$. Similarly, the *supremum* $\sqcup \mathcal{F}$ is generated by the union of the families of generators of all X_i, $i \in [1, \ell]$.

If $X^\bullet \subseteq \mathbb{G}^\bullet$ (resp. $Y^\times \subseteq \mathbb{G}^\times$), we denote by $\overline{X^\bullet}$ (resp. $\overline{Y^\times}$) the *complementary set* of $X^\bullet$ *(resp. $Y^\times$) in $\mathbb{G}^\bullet$ (resp. $\mathbb{G}^\times$)*, that is $\overline{X^\bullet} = \mathbb{G}^\bullet \setminus X^\bullet$ (resp. $\overline{Y^\times} = \mathbb{G}^\times \setminus Y^\times$). Observe that, if X is a subgraph of $\mathbb{G}$, then, except in some degenerated cases, the pair $(\overline{X^\bullet}, \overline{X^\times})$ is not a graph.

Property 1. *The set $\mathcal{G}$ of the subgraphs of $\mathbb{G}$ form a complete lattice, sup-generated by the set $\mathcal{S} = \mathcal{S}_0 \cup \mathcal{S}_1$, but not complemented. The supremum and the infimum of any family $\mathcal{F} = \{X_1, \dots X_\ell\}$ of elements in $\mathcal{G}$ are given by respectively $\sqcap \mathcal{F} = (\bigcap_{i \in [1,\ell]} X_i^\bullet, \bigcap_{i \in [1,\ell]} X_i^\times)$ and $\sqcup \mathcal{F} = (\bigcup_{i \in [1,\ell]} X_i^\bullet, \bigcup_{i \in [1,\ell]} X_i^\times)$.*

3   Dilations and Erosions

In the graph $\mathbb{G}$, we can consider sets of points as well as sets of edges. Therefore, it is convenient to consider operators to go from one kind of sets to the other one. In this section, we investigate such operators and we study their morphological properties. Then, based on these operators, we propose several dilations and erosions acting on the lattice of all subgraphs of $\mathbb{G}$.

Let $X^\bullet$ be a subset of $\mathbb{G}^\bullet$, we denote by $\mathcal{G}_{X^\bullet}$ the set of all subgraphs of $\mathbb{G}$ whose vertex set is $X^\bullet$. Let $Y^\times$ be a subset of $\mathbb{G}^\times$. We denote by $\mathcal{G}_{Y^\times}$ the set of all subgraphs of $\mathbb{G}$ whose edge set is $Y^\times$.

Definition 2 (edge-vertex correspondences). *We define the operators $\delta^\bullet$, $\epsilon^\bullet$ from $\mathcal{G}^\times$ into $\mathcal{G}^\bullet$ and the operators $\epsilon^\times, \delta^\times$ from $\mathcal{G}^\bullet$ into $\mathcal{G}^\times$ as follows:*

	$\mathcal{G}^\times \to \mathcal{G}^\bullet$	$\mathcal{G}^\bullet \to \mathcal{G}^\times$
Provide the object with a graph structure	$X^\times \to \delta^\bullet(X^\times)$ such that $(\delta^\bullet(X^\times), X^\times) = \sqcap \mathcal{G}_{X^\times}$	$X^\bullet \to \epsilon^\times(X^\bullet)$ such that $(X^\bullet, \epsilon^\times(X^\bullet)) = \sqcup \mathcal{G}_{X^\bullet}$
Provide its complement with a graph structure	$X^\times \to \epsilon^\bullet(X^\times)$ such that $(\epsilon^\bullet(X^\times), \overline{X^\times}) = \sqcap \mathcal{G}_{\overline{X^\times}}$	$X^\bullet \to \delta^\times(X^\bullet)$ such that $(\overline{X^\bullet}, \delta^\times(X^\bullet)) = \sqcup \mathcal{G}_{\overline{X^\bullet}}$

In other words, if $X^\bullet \subseteq \mathbb{G}^\bullet$ and $Y^\times \subseteq \mathbb{G}^\times$, $(\delta^\bullet(Y^\times), Y^\times)$ is the smallest subgraph of $\mathbb{G}$ whose edge set is $Y^\times$, $(X^\bullet, \epsilon^\times(X^\bullet))$ is the largest subgraph of $\mathbb{G}$ whose vertex set is $X^\bullet$, $(\epsilon^\bullet(Y^\times), \overline{Y^\times})$ is the smallest subgraph of $\mathbb{G}$ whose edge set is $\overline{Y^\times}$, and $(\overline{X^\bullet}, \delta^\times(X^\bullet))$ is the largest subgraph of $\mathbb{G}$ whose vertex set is $\overline{X^\bullet}$.

These operators are illustrated in Figs. 1a-f. The following property locally characterize them. This property leads in particular to simple linear-time algorithms (with respect to the cardinality of $\mathbb{G}^\bullet$ and $\mathbb{G}^\times$) to compute $\delta^\bullet(X^\times)$, $\epsilon^\times(Y^\bullet)$, $\epsilon^\bullet(X^\times)$ and $\epsilon^\bullet(X^\times)$ without explicitly considering the families $\mathcal{G}_{X^\times}$, $\mathcal{G}_{X^\bullet}$, $\mathcal{G}_{\overline{X^\times}}$ and $\mathcal{G}_{\overline{X^\bullet}}$.

Property 3. *For any $X^\times \subseteq \mathbb{G}^\times$ and $Y^\bullet \subseteq \mathbb{G}^\bullet$:*

1. $\delta^\bullet : \mathbb{G}^\times \to \mathbb{G}^\bullet$ *is such that* $\delta^\bullet(X^\times) = \{x \in \mathbb{G}^\bullet \mid \exists e_{x,y} \in X^\times\}$;
2. $\epsilon^\times : \mathbb{G}^\bullet \to \mathbb{G}^\times$ *is such that* $\epsilon^\times(Y^\bullet) = \{e_{x,y} \in \mathbb{G}^\times \mid x \in Y^\bullet \text{ and } y \in Y^\bullet\}$;
3. $\epsilon^\bullet : \mathbb{G}^\times \to \mathbb{G}^\bullet$ *is such that* $\epsilon^\bullet(X^\times) = \{x \in \mathbb{G}^\bullet \mid \forall e_{x,y} \in \mathbb{G}^\times, e_{x,y} \in X^\times\}$;
4. $\delta^\times : \mathbb{G}^\bullet \to \mathbb{G}^\times$ *is such that* $\delta^\times(Y^\bullet) = \{e_{x,y} \in \mathbb{G}^\times \mid \text{ either } x \in Y^\bullet \text{ or } y \in Y^\bullet\}$.

In other words, $\delta^\bullet(X^\times)$ is the set of all vertices which belong to an edge of $X^\times$, $\epsilon^\times(Y^\bullet)$ is the set of all edges whose two extremities are in $Y^\bullet$, $\epsilon^\bullet(X^\times)$ is the set of all vertices which do not belong to any edge of $\overline{X^\times}$, and $\delta^\times(Y^\bullet)$ is the set of all edges which have at least one extremity in $Y^\bullet$.

From this characterization, we can recognize the general graph version of some operators introduced by Meyer and Angulo [8] (see also [9]) for the hexagonal grid. However, unlike Property 3, the important theorem of structure 9 does not appear in [8] or [9], neither Theorem 12 nor Property 14.

Before further analyzing the operators defined above, let us briefly recall some algebraic tools which are fundamental in mathematical morphology [11].

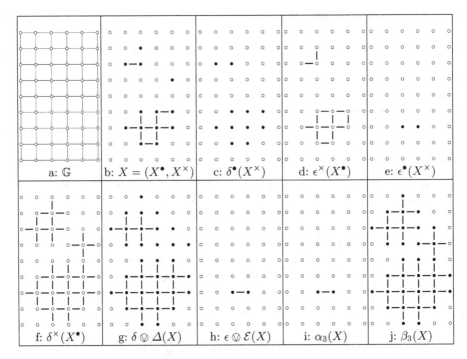

Fig. 1. Dilations and erosions

Given two lattices $\mathcal{L}_1$ and $\mathcal{L}_2$, an operator $\delta : \mathcal{L}_1 \to \mathcal{L}_2$ is called a *dilation* when it preserves the supremum (*i.e.* $\forall \mathcal{E} \subseteq \mathcal{L}_1, \delta(\vee_1 \mathcal{E}) = \vee_2 \{\delta(X) \mid X \in \mathcal{E}\}$, where $\vee_1$ is the supremum in $\mathcal{L}_1$ and $\vee_2$ the supremum in $\mathcal{L}_2$). Similarly, an operator which preserves the infimum is called an *erosion*.

Two operators $\epsilon : \mathcal{L}_1 \to \mathcal{L}_2$ and $\delta : \mathcal{L}_2 \to \mathcal{L}_1$ form an *adjunction* (ϵ, δ) when for any X in $\mathcal{L}_2$ and any Y in $\mathcal{L}_1$, we have $\delta(X) \leq_1 Y \Leftrightarrow X \leq_2 \epsilon(Y)$, where $\leq_1$ and $\leq_2$ denote the order relations on respectively $\mathcal{L}_1$ and $\mathcal{L}_2$. Given two operators ϵ and δ, if the pair (ϵ, δ) is an adjunction, then ϵ is an erosion and δ is a dilation.

Given two complemented lattices $\mathcal{L}_1$ and $\mathcal{L}_2$, two operators α and β from $\mathcal{L}_1$ into $\mathcal{L}_2$ are *dual (with respect to the complement) of each other* when, for any $X \in \mathcal{L}_1$, we have $\beta(X) = \overline{\alpha(\overline{X})}$. If α and β are dual of each other, then β is an erosion whenever α is a dilation.

Property 4 (dilation, erosion, adjunction, duality)

1. Both $(\epsilon^\times, \delta^\bullet)$ and $(\epsilon^\bullet, \delta^\times)$ are adjunctions.
2. Operators $\epsilon^\times$ and $\delta^\times$ (resp. $\epsilon^\bullet$ and $\delta^\bullet$) are dual of each other.
3. Operators $\delta^\bullet$ and $\delta^\times$ are dilations.
4. Operators $\epsilon^\bullet$ and $\epsilon^\times$ are erosions.

Let us compose these dilations and erosions to act on $\mathcal{G}^\bullet$ and $\mathcal{G}^\times$.

Definition 5 (vertex-dilation, vertex-erosion). *We define δ and ϵ that act on $\mathcal{G}^\bullet$ (i.e., $\mathcal{G}^\bullet \to \mathcal{G}^\bullet$) by $\delta = \delta^\bullet \circ \delta^\times$ and $\epsilon = \epsilon^\bullet \circ \epsilon^\times$.*

As compositions of respectively dilations and erosions, δ and ϵ are respectively a dilation and an erosion. Moreover, by composition of adjunctions and dual operators, δ and ϵ are dual and (ϵ, δ) is an adjunction.

In fact, it can be shown that δ and ϵ correspond exactly to the usual notions of an erosion and of a dilation of a set of vertices in a graph [6]. It means, in particular that, when $\mathbb{G}^\bullet$ is a subset of the grid points $\mathbb{Z}^d$ and when the edge set $\mathbb{G}^\times$ is obtained from a symmetrical structuring element, then the operators defined above are equivalent to the usual binary dilation and erosion by the considered structuring element. For instance, in Fig. 1, $\mathbb{G}^\bullet$ is a rectangular subset of $\mathbb{Z}^2$ and $\mathbb{G}^\times$ corresponds to the basic "cross" structuring element. It can be verified that the vertex sets in Fig. 1g and h, obtained by applying δ and ϵ to $X^\bullet$ (Fig. 1b), are the dilation and the erosion by a "cross" structuring element of $X^\bullet$.

We now consider a dual/adjunct pair of dilation and erosion acting on $\mathcal{G}^\times$.

Definition 6 (edge-dilation, edge-erosion). *We define Δ and $\mathcal{E}$ that act on $\mathcal{G}^\times$ by $\Delta = \delta^\times \circ \delta^\bullet$ and $\mathcal{E} = \epsilon^\times \circ \epsilon^\bullet$.*

Definition 7. *We define the operators $\delta \oslash \Delta$ and $\epsilon \oslash \mathcal{E}$ by respectively $(\delta(X^\bullet), \Delta X^\times)$ and $(\epsilon(X^\bullet), \mathcal{E}(X^\times))$, for any $X \in \mathcal{G}$.*

For instance, Figs. 1f and 1g present the results obtained by applying the operator $\delta \oslash \Delta$ and the operator $\epsilon \oslash \mathcal{E}$ to the subgraph X (Fig. 1b) of $\mathbb{G}$ (Fig. 1a).

Lemma 8. *The family $\mathcal{G}$ is closed under the operators $\delta \oslash \Delta$ and $\epsilon \oslash \mathcal{E}$. More precisely, for any subgraph X of $\mathbb{G}$, both $\delta \oslash \Delta(X)$ and $\epsilon \oslash \mathcal{E}(X)$ are subgraphs of $\mathbb{G}$.*

Theorem 9 (graph-dilation, graph-erosion). *The operators $\delta \oslash \Delta$ and $\epsilon \oslash \mathcal{E}$ are respectively a dilation and an erosion acting on the lattice $(\mathcal{G}, \sqsubseteq)$. Furthermore, $(\epsilon \oslash \mathcal{E}, \delta \oslash \Delta)$ is an adjunction.*

Note that since lattice $\mathcal{G}$ is sup-generated by set $\mathcal{S}$, it suffices to know the dilation of the graphs in $\mathcal{S}$ for characterizing the dilation of the graphs in $\mathcal{G}$.

Compared to classical morphological operators on sets, the dilations and erosions introduced in this section furthermore convey some connectivity properties different than the ones which can be deduced from classical dilations and erosions. Observe, for instance, in Fig. 1g, that some 4-adjacent vertices of $\delta(X^\bullet)$ are not linked by an edge in the graph $\delta \oslash \Delta(X)$. These properties can be useful in further processing involving for instance connected operators [12,13,14,15].

Thanks to the operators presented in Definition 2, other intersecting adjunctions (hence dilations/erosions) can be defined on $\mathcal{G}$:

1. (α_1, β_1) such that $\forall X \in \mathcal{G}$, $\alpha_1(X) = (\mathbb{G}^\bullet, X^\times)$ and $\beta_1(X) = (\delta^\bullet(X^\times), X^\times)$;
2. (α_2, β_2) such that $\forall X \in \mathcal{G}$, $\alpha_2(X) = (X^\bullet, \epsilon^\times(X^\bullet))$ and $\beta_2(X) = (X^\bullet, \emptyset)$;
3. (α_3, β_3) such that $\forall X \in \mathcal{G}$, $\alpha_3(X) = (\epsilon^\bullet(X^\times), \epsilon^\times \circ \epsilon^\bullet(X^\times))$ and $\beta_3(X) = (\delta^\bullet \circ \delta^\times(X^\bullet), \delta^\times(X^\bullet))$.

The adjunction (α_3, β_3) is illustrated in Fig. 1i and 1j. Note also that, using usual graph terminologies, β_1 (resp. α_2) can be defined as the operator which associates to a graph the graph induced by its edge set (resp. vertex set).

4   Filters

In mathematical morphology, a *filter* is an operator α acting on a lattice $\mathcal{L}$, which is increasing (*i.e.* $\forall X, Y \in \mathcal{L}$, $\alpha(X) \le \alpha(Y)$ whenever $X \le Y$) and idempotent (*i.e.* $\forall X \in \mathcal{L}$, $\alpha(\alpha(X)) = \alpha(X)$). A filter α on $\mathcal{L}$ which is extensive (*i.e.* $\forall X \in \mathcal{L}$, $X \le \alpha(X)$) is called a *closing on* $\mathcal{L}$ whereas a filter α on $\mathcal{L}$ which is anti-extensive (*i.e.* $\forall X \in \mathcal{L}$, $\alpha(X) \le X$) is called an *opening on* $\mathcal{L}$. It is known that composing the two operators of an adjunction yields an opening or a closing depending on the order in which the operators are composed [11]. In this section, the operators of Section 3 are composed to obtain filters on $\mathcal{G}^\bullet$, $\mathcal{G}^\times$ and $\mathcal{G}$.

Definition 10 (opening, closing)

1. We define γ_1 and ϕ_1, that act on $\mathcal{G}^\bullet$, by $\gamma_1 = \delta \circ \epsilon$ and $\phi_1 = \epsilon \circ \delta$.
2. We define Γ_1 and Φ_1, that act on $\mathcal{G}^\times$, by $\Gamma_1 = \Delta \circ \mathcal{E}$ and $\Phi_1 = \mathcal{E} \circ \Delta$.
3. We define the operators $\gamma \oslash \Gamma_1$ and $\phi \oslash \Phi_1$ by respectively $\gamma \oslash \Gamma_1(X) = (\gamma_1(X^\bullet), \Gamma_1(X^\times))$ and $\phi \oslash \Phi_1(X) = (\phi_1(X^\bullet), \Phi_1(X^\times))$ for any $X \in \mathcal{G}$.

Figs. 2b and 2f present the result of $\gamma \oslash \Gamma_1$ and $\phi \oslash \Phi_1$ for respectively the subgraph of Fig. 2a and the one of Fig. 2e.

In fact, by composing $\delta^\bullet$ with $\epsilon^\times$ and $\delta^\times$ with $\epsilon^\bullet$, we obtain smaller filters.

Definition 11 (half-opening, half-closing)

1. We define $\gamma_{1/2}$ and $\phi_{1/2}$, that act on $\mathcal{G}^\bullet$, by $\gamma_{1/2} = \delta^\bullet \circ \epsilon^\times$ and $\phi_{1/2} = \epsilon^\bullet \circ \delta^\times$.
2. We define $\Gamma_{1/2}$ and $\Phi_{1/2}$, that act on $\mathcal{G}^\times$ by $\Gamma_{1/2} = \delta^\times \circ \epsilon^\bullet$ and $\Phi_{1/2} = \epsilon^\times \circ \delta^\bullet$.
3. We define the operators $\gamma \oslash \Gamma_{1/2}$ and $\phi \oslash \Phi_{1/2}$ by respectively $\gamma \oslash \Gamma_{1/2}(X) = (\gamma_{1/2}(X^\bullet), \Gamma_{1/2}(X^\times))$ and $\phi \oslash \Phi_{1/2}(X) = (\phi_{1/2}(X^\bullet), \Phi_{1/2}(X^\times))$, for any $X \in \mathcal{G}$.

Thanks to Property 3, the operators defined above can be locally characterized. Let $X^\bullet \subseteq \mathbb{G}^\bullet$ and $Y^\times \subseteq \mathbb{G}^\times$, we have:

$$\gamma_{1/2}(X^\bullet) = \{x \in X^\bullet \mid \exists e_{x,y} \in \mathbb{G}^\times \text{ with } y \in X^\bullet\}$$
$$= X^\bullet \setminus \{x \in X^\bullet \mid \forall e_{x,y} \in \mathbb{G}^\times, y \notin X^\bullet\}$$
$$\Gamma_{1/2}(Y^\times) = \{u \in \mathbb{G}^\times \mid \exists x \in u \text{ with } \{e_{x,y} \in \mathbb{G}^\times\} \subseteq Y\}$$
$$= Y^\times \setminus \{u \in Y^\times \mid \forall x \in u, \exists e_{x,y} \in \mathbb{G}^\times \text{ with } e_{x,y} \notin Y^\times\}$$
$$\phi_{1/2}(X^\bullet) = \{x \in \mathbb{G}^\bullet \mid \text{ either } x \in X^\bullet \text{ or } \forall e_{x,y} \in \mathbb{G}^\times, y \in X^\bullet\}$$
$$= X^\bullet \cup \{x \in \overline{X^\bullet} \mid \forall e_{x,y} \in \mathbb{G}^\times, y \in X^\bullet\}$$
$$\Phi_{1/2}(Y^\times) = \{e_{x,y} \in \mathbb{G}^\times \mid \exists e_{x,z} \in Y^\times \text{ and } \exists e_{y,w} \in Y^\times\}$$
$$= Y \cup \{e_{x,y} \in \overline{Y^\times} \mid x \in \delta^\bullet(Y^\times) \text{ and } y \in \delta^\bullet(Y^\times)\}.$$

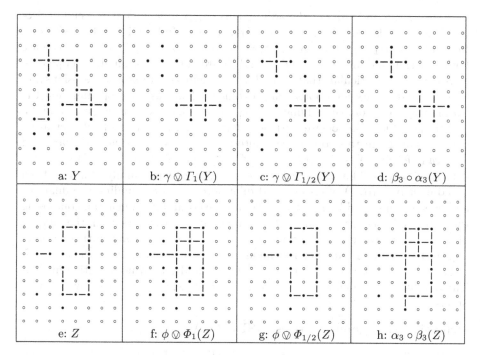

Fig. 2. Openings and closings ($\mathbb{G}$ is induced by the 4-adjacency relation)

Informally speaking, $\gamma_{1/2}$ removes from $Y^\bullet$ its isolated vertices whereas $\Gamma_{1/2}$ removes from $Y^\times$ the edges which do not contain a vertex completely covered by edges in $Y^\times$. It may be furthermore seen that $\gamma_{1/2}$ (resp. $\Gamma_{1/2}$) is the dual of $\phi_{1/2}$ (resp. $\Phi_{1/2}$). Thus, $\phi_{1/2}$ adds to $Y^\bullet$ the vertices of $\overline{X^\bullet}$ completely surrounded by elements of $Y^\bullet$ whereas $\Phi_{1/2}$ adds to $Y^\times$ the edges of $\overline{Y^\times}$ whose two extremities belong to at least one edge in $Y^\times$ (see for instance Fig. 2).

The family $\mathcal{G}$ is closed under the operators presented in Definition 10.3 since they are obtained by composition of operators also satisfying this property (Lemma 8). Furthermore, it can be deduced from the local characterization of the operators $\gamma_{1/2}$, $\Gamma_{1/2}$, $\phi_{1/2}$ and $\Phi_{1/2}$ that the family $\mathcal{G}$ is also closed under the operators of Definition 11.3. Hence, thanks to the properties of adjunctions recalled in the introduction of this section, the following theorem can be established.

Theorem 12 (graph-openings, graph-closings)

1. *The operators $\gamma_{1/2}$ and, γ_1 (resp. $\Gamma_{1/2}$ and Γ_1) are openings on $\mathcal{G}^\bullet$ (resp. $\mathcal{G}^\times$) and $\phi_{1/2}$, and Φ_1 (resp. $\Phi_{1/2}$ and ϕ_1) are closings on $\mathcal{G}^\bullet$ (resp. $\mathcal{G}^\times$).*
2. *The family $\mathcal{G}$ is closed under $\gamma \otimes \Gamma_{1/2}$, $\phi \otimes \Phi_{1/2}$, $\gamma \otimes \Gamma_1$, and $\phi \otimes \Phi_1$.*
3. *The operators $\gamma \otimes \Gamma_{1/2}$ and $\gamma \otimes \Gamma_1$ are openings on $\mathcal{G}$ and $\phi \otimes \Phi_{1/2}$ and $\phi \otimes \Phi_1$ are closings on $\mathcal{G}$.*

Composing the operators of the adjunctions (α_i, β_i), defined at the end of Section 3, also yields remarkable openings and closings. Indeed, it can be easily seen that: $\alpha_1 \circ \beta_1 = \alpha_1$, $\alpha_2 \circ \beta_2 = \alpha_2$, $\beta_1 \circ \alpha_1 = \beta_1$ and $\beta_2 \circ \alpha_2 = \beta_2$. Thus α_1 and α_2 are both a closing and an erosion and β_1 and β_2 are both a dilation and an opening. This means, in particular, that α_1 and α_2 are idempotent extensive erosions and that β_1 and β_2 are idempotent anti-extensive dilations. The opening and the closing resulting from the adjunction (α_3, β_3) are illustrated in Figs. 2d and 2h.

It is possible to associate with any lattice $\mathcal{L}$, the lattice of all increasing operators on $\mathcal{L}$. In this context, two filters φ_1 and φ_2 on the lattice $\mathcal{L}$ are said *ordered* if, for any $X \in \mathcal{L}$, $\varphi_1(X) \leq \varphi_2(X)$ or if, for any $X \in \mathcal{L}$, $\varphi_2(X) \leq \varphi_1(X)$. A usual way to build a hierarchy of filters (*i.e.* an ordered family of filters) from an adjunct pair (α, β) of erosion and dilation consists of building the dilations and erosions obtained by iterating several times α and β. In general, composing these iterated versions of α and β leads to hierarchies of filters when the number of iterations increases. In the remaining of the section, we follow this classical approach to build granulometries and alternate sequential filters in the lattice $\mathcal{G}$.

Let α be an operator acting on a lattice $\mathcal{L}$ and i be a nonnegative integer. The operator α^i is defined by the identity on $\mathcal{L}$ when $i = 0$ and by $\alpha \circ \alpha^{i-1}$ otherwise.

Definition 13 (granulometries, ASF). *Let* $\lambda \in \mathbb{N}$.

1. *We define* $\gamma \oslash \Gamma_{\lambda/2}$ *(resp.* $\phi \oslash \Phi_{\lambda/2}$*) by* $\gamma \oslash \Gamma_{\lambda/2} = (\delta \oslash \Delta)^i \circ (\gamma \oslash \Gamma_{1/2})^j \circ (\epsilon \oslash \mathcal{E})^i$ *(resp.* $\phi \oslash \Phi_{\lambda/2} = (\epsilon \oslash \mathcal{E})^i \circ (\phi \oslash \Phi_{1/2})^j \circ (\delta \oslash \Delta)^i$*), where i and j are respectively the quotient and the remainder in the integer division of λ by 2.*
2. *We define the operator* $ASF_{\lambda/2}$ *by the identity on graphs when $\lambda = 0$ and by* $ASF_{\lambda/2} = \gamma \oslash \Gamma_{\lambda/2} \circ \phi \oslash \Phi_{\lambda/2} \circ ASF_{(\lambda-1)/2}$ *otherwise.*

Note that it is possible to define a second family of operators similar to $ASF = \{ASF_{\lambda/2} \mid \lambda \in \mathbb{N}\}$ by replacing in Definition 13.2 the sequence of primitives $\gamma \oslash \Gamma_{\lambda/2} \circ \phi \oslash \Phi_{\lambda/2}$ by the sequence $\phi \oslash \Phi_{\lambda/2} \circ \gamma \oslash \Gamma_{\lambda/2}$. The following proposition (Property 14.2), which establishes that ASF is a family of alternate sequential filters, also holds true for this second family.

Property 14

1. *The families* $\{\gamma \oslash \Gamma_{\lambda/2} \mid \lambda \in \mathbb{N}\}$ *and* $\{\phi \oslash \Phi_{\lambda/2} \mid \lambda \in \mathbb{N}\}$ *are granulometries:*
   - *for any $\lambda \in \mathbb{N}$, $\gamma \oslash \Gamma_{\lambda/2}$ (resp. $\phi \oslash \Phi_{\lambda/2}$) is an opening (resp. a closing) on $\mathcal{G}$;*
   - *for any two elements $\lambda, \mu \in \mathbb{N}$ such that $\lambda \leq \mu$, we have $\gamma \oslash \Gamma_{\mu/2}(X) \sqsubseteq \gamma \oslash \Gamma_{\lambda/2}(X)$ and $\phi \oslash \Phi_{\lambda/2}(X) \sqsubseteq \phi \oslash \Phi_{\mu/2}(X)$ for any $X \in \mathcal{G}$.*
2. *The family* $\{ASF_{\lambda/2} \mid \lambda \in \mathbb{N}\}$ *is a family of alternate sequential filters:*
   - *for any two elements $\lambda, \mu \in \mathbb{N}$, $\lambda \geq \mu$ implies that $ASF_{\lambda/2} \circ ASF_{\mu/2} = ASF_{\lambda/2}$.*

Given a graph $X \in \mathcal{G}$, it must be noticed that the vertex set (resp. edge set) of $\gamma \oslash \Gamma_{\lambda/2}(X)$ and $\phi \oslash \Phi_{\lambda/2}(X)$ depends only on the vertex set (resp. edge

(a) original binary image   (b) noisy binary image

(c) classical ASF   (d) graph ASF

(e) classical ASF of double size   (f) classical ASF (double resolution)

Fig. 3. ASF illustration [see text]

set) of X. Thus, Definition 13 also induces granulometries and alternate sequential filters on $\mathcal{G}^{\bullet}$ and $\mathcal{G}^{\times}$. If we consider the case where λ is even, we deduce from the observation stated after Definition 5 that, when $\mathbb{G}^{\bullet}$ is a subset of the grid points $\mathbb{Z}^d$ and when $\mathbb{G}^{\times}$ is obtained from a symmetrical structuring element, then the vertex parts of $\gamma \otimes \Gamma_{\lambda/2}$ and $\phi \otimes \Phi_{\lambda/2}$ correspond to the usual opening and closing of size $\lambda/2$. Hence, we can see that, in the case of a set of points, the proposed framework completes the granulometries and filters which are classically used in applications by considering the odd values of λ.

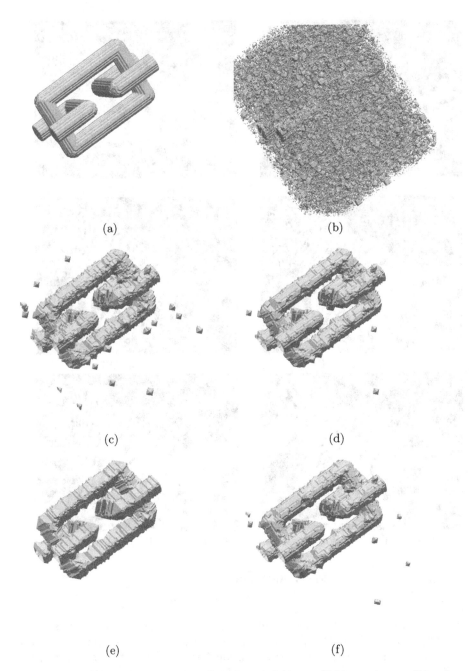

Fig. 4. Same as Fig. 3 but in 3D. Rendering of: (a) original binary image, (b) noisy binary image, (c) result of classical ASF, (d) result of graph ASF, (e) result of classical ASF of double size and (f) result of classical ASF (double resolution) [see text].

In order to illustrate the proposed framework, let us analyze the effect of our filters on the binary image of Fig. 3b obtained by adding random impulse noise of different size and shape to the digital shape shown in Fig. 3a. Fig. 3c shows the results given by the "classical" ASF (using the structuring elements corresponding to the 4-adjacency relation) of size 4. Fig. 3d presents the results given by $ASF_{8/2}$ (on the graph induced by the 4-adjacency relation) which is the corresponding alternate filter in our framework. Clearly, $ASF_{8/2}$ removes more noise than the classical ASF. However, it requires more iterations since it considers the filters $\gamma \oslash \Gamma_{\lambda/2}$ and $\phi \oslash \Phi_{\lambda/2}$ for both odd and even values of λ whereas the classical ASF only considers the even values of λ. In order to compare the proposed ASF with filters using the same number of iterations, we produce two other filtered images. The first one is obtained by filtering the image with a classical ASF of size 8 (Fig. 3e). It can be seen that more noise are removed but also that less details are preserved (see in particular the head of the zebra). The second one (Fig. 3f) is obtained in three steps: 1) double the resolution of the noisy image; 2) apply to it a classical ASF of size 8; and 3) divide by two the resolution of the output image. It can be seen that this last procedure removes more noise than the classical ASF but does not perform as well as the ASF introduced in the present paper. This qualitative assessment is confirmed by a quantitative study that will be published in a future extended version of this paper [10]. Fig. 4 provides a similar illustration for the case of a 3-dimensional synthetic binary object.

5   Conclusion

This paper investigates the lattice of all subgraphs of a graph and provides it with morphological operators. In particular, we propose new filters which input and output graphs. We show the interest of restricting these filters to sets of vertices. Indeed, they allow us to complete some classical morphological filters used in image analysis applications.

In future work, the proposed approach will be extended to (node and edge) weighted graphs considered as stacks of graphs. We will also study morphology in simplicial (and cubical) complexes (see [2] for image operators defined in cubical complexes and [16] for examples of morphological operators in 2D simplicial complexes). These topological structures extend graphs to higher dimensions in the sense that a graph is a 1-D structure made of points and edges considered as 0D and 1D elements. The proposed approach extends to general complexes by considering additional generators. In 2D, for instance, a third generator for the elementary triangles (or squares) is required.

References

1. Cousty, J., Bertrand, G., Najman, L., Couprie, M.: Watershed Cuts: Minimum Spanning Forests and the Drop of Water Principle. IEEE Trans. Pattern Analysis and Machine Intelligence 31(8), 1362–1374 (2009)

2. Couprie, M., Bertrand, G.: New characterizations of simple points in 2d, 3d, and 4d discrete spaces. IEEE Trans. Pattern Analysis and Machine Intelligence 31(4), 637–648 (2009)
3. Cousty, J., Bertrand, G., Najman, L., Couprie, M.: Watershed cuts: Thinnings, shortest-path forests and topological watersheds. IEEE Trans. Pattern Analysis and Machine Intelligence (to appear, 2009)
4. Cousty, J., Najman, L., Serra, J.: Raising in watershed lattices. In: International Conference on Image Processing, pp. 2196–2199. IEEE, Los Alamitos (2008)
5. Najman, L.: Ultrametric watersheds. In: Wilkinson, M.H.F., Roerdink, J.B.T.M. (eds.) ISMM 2009. LNCS, vol. 5720, pp. 181–192. Springer, Heidelberg (2009)
6. Vincent, L.: Graphs and mathematical morphology. Sig. Proc. 16, 365–388 (1989)
7. Heijmans, H., Vincent, L.: Graph morphology in image analysis. In: Dougherty, E. (ed.) Mathematical Morphology in Image Processing, pp. 171–203. Marcel-Dekker, New York (1992)
8. Meyer, F., Angulo, J.: Micro-viscous morphological operators. In: Mathematical Morphology and its Application to Signal and Image Processing (ISMM 2007), pp. 165–176 (2007)
9. Meyer, F., Lerallut, R.: Morphological operators for flooding, leveling and filtering images using graphs. In: Escolano, F., Vento, M. (eds.) GbRPR 2007. LNCS, vol. 4538, pp. 158–167. Springer, Heidelberg (2007)
10. Cousty, J., Najman, L., Serra, J.: Morphological operators in graph spaces. In preparation, see also Technical Report IGM 2009-08 (2009), http://www-igm.univ-mlv.fr/LabInfo/rapportsInternes/2009/08.pdf
11. Ronse, C., Serra, J.: Fondements algébriques de la morphologie. In: Morphologie mathématique 1 approche déterministe, Hermes, pp. 49–96 (2008)
12. Salembier, P., Serra, J.: Flat zones filtering, connected operators, and filters by reconstruction. IEEE Trans. on Image Processing 4(8), 1153–1160 (1995)
13. Ronse, C.: Set-theoretical algebraic approaches to connectivity in continuous or digital spaces. Journal of Mathematical Imaging and Vision 8(1), 41–58 (1998)
14. Braga-Neto, U., Goutsias, J.: Connectivity on complete lattices: new results. Computer Vision and Image Understanding 85(1), 22–53 (2002)
15. Ouzounis, G.K., Wilkinson, M.H.: Mask-based second-generation connectivity and attribute filters. IEEE Trans. Pattern Analysis and Machine Intelligence 29(6), 990–1004 (2007)
16. Loménie, N., Stamon, G.: Morphological mesh filtering and α-objects. Pattern Recognition Letters 29(10), 1571–1579 (2008)

Morphology on Graphs and Minimum Spanning Trees

Fernand Meyer and Jean Stawiaski

MINES Paristech, CMM- Centre de morphologie mathématique, Mathématiques et Systèmes,
35 rue Saint Honoré - 77305 Fontainebleau cedex, France
{fernand.meyer,jean.stawiaksi}@mines-paristech.fr
http://cmm.ensmp.fr

Abstract. This paper revisits the construction of watershed and water-fall hierarchies through a thorough analysis of Boruvka's algorithms for constructing minimum spanning trees of edge weighted graphs. In the case where the watershed of a node weighted graph is to be constructed, we propose a distribution of weights on the edges, so that the waterfall extraction on the edge weighted graph becomes equivalent with the watershed extraction on the node weighted graph.

Keywords: Waterfall, minimum spanning tree, adjunctions on graphs.

1 Introduction

Graphs are the fundamental structures used to represent images and partitions. Different graph theoretical approaches are well suited to describe morphological segmentation: 1) detection of shortest paths or cycles for various types of distances ; 2) extraction of minimum spanning trees and forests (watershed hierarchies and segmentation with markers) ; 3) maximal flows and minimal cuts. Some links between these optimal structures have already been studied [1,2,3].

From these previous studies, it appears that minimum spanning trees and forests play a central role for watershed, waterfall and hierarchical segmentations. In this paper a thorough examination of Boruvka's algorithm, known as the first algorithm created for constructing minimum spanning trees, give us a deeper insight into the links between waterfalls and watersheds. We first define a non conventional morphology based on adjunctions between nodes and edges of weighted graphs. From these new operators, we show that waterfalls can be directly obtained from a specific morphological opening. Morphological segmentations can be well described by flooding a topographic surface. We highlight how the different steps of Boruvka's algorithm can be used to build watersheds and waterfalls on edge weighted graphs. We then address the problem of the non-uniqueness of minimal spanning trees arising when graphs have plateaux and edges having the same weight. We especially give several strategies to by-pass this problem. Finally, we present an extension of this methodology in case of node weighted graphs.

M.H.F. Wilkinson and J.B.T.M. Roerdink (Eds.): ISMM 2009, LNCS 5720, pp. 161–170, 2009.
© Springer-Verlag Berlin Heidelberg 2009

2 Adjunctions on Graphs and Interpretations in Terms of Flooding

A *non oriented graph* $G = [N, E]$ is a collection of a set N whose elements are called vertices or nodes and of a set E whose elements $u \in E$ are pairs of vertices called edges. The *weights* $[e, n]$ of the graph G are represented as two functions h and k, respectively for the edges and the nodes. h_{ij} is the weight of the edge (i, j) and k_i the weight of the node i. The same graph may have various distributions of weights. If there is no ambiguity, in the case where only one distribution of weights is considered, e will also represent the distribution of weights of the edges and n the distribution of weights of the nodes.

2.1 Various Adjunctions on Graphs

Classical morphology on graphs involves operations between nodes (i.e. operations on pixels) or operations between edges. We define in this section new types of adjunctions between both nodes and edges. These operators permit us to interpret flooding in terms of basic morphological operations.

Definition 1. *We define here several operators involving both edges and nodes of a weighted graph.*

- *an erosion* $[\varepsilon_{en} n]_{ij} = n_i \wedge n_j$ *and its adjunct dilation* $[\delta_{ne} e]_i = \bigvee_{(k \ neighbors \ of \ i)} e_{ik}$.

- *a dilation* $[\delta_{en} n]_{ij} = n_i \vee n_j$ *and its adjunct erosion* $[\varepsilon_{ne} e]_i = \bigwedge_{(k \ neighbors \ of \ i)} e_{ik}$.

Proposition 1. *The operators defined above are pairwise adjunct or dual operators:*
- ε_{ne} *and* δ_{en} *are adjunct operators.*
- ε_{en} *and* δ_{ne} *are adjunct operators.*
- ε_{ne} *and* δ_{ne} *are dual operators from the edge weight function* e *to the node weight function* n.
- ε_{en} *and* δ_{en} *are dual operators from the node weight function* e *to the edge weight function* n.

As ε_{ne} and δ_{en} are adjunct operators, the operator $\varphi_n = \varepsilon_{ne} \delta_{en}$ is a closing on n and $\gamma_e = \delta_{en} \varepsilon_{ne}$ is an opening on e.

Similarly as ε_{en} and δ_{ne} are adjunct operators, the operator $\varphi_e = \varepsilon_{en} \delta_{ne}$ is a closing on e and $\gamma_n = \delta_{ne} \varepsilon_{en}$ is an opening on n.

The operators $\varepsilon_E = \varepsilon_{en} \varepsilon_{ne}$ and $\delta_E = \delta_{en} \delta_{ne}$ are respectively the elementary erosion and the elementary dilation on a neighborhood graph operating on neighboring edges : $\overline{E} \to \overline{E}$, since $(\varepsilon_E, \delta_E)$ also form an adjunction. Likewise the operators $\varepsilon_N = \varepsilon_{ne} \varepsilon_{en}$ and $\delta_N = \delta_{ne} \delta_{en}$ are the usual elementary erosions and dilations on a neighborhood graph operating on nodes: $\overline{N} \to \overline{N}$. As such $(\varepsilon_N, \delta_N)$ also form an adjunction.

2.2 Interpretation of γ_e in Terms of Flooding

Let G represent a neighboring graph of catchment basins, such as a region adjacency graph of a watershed segmentation, then the operators defined above have a physical interpretation. $[\varepsilon nee]_i$ represents the lowest pass point between the catchment basin i and its neighbors ; its altitude or weight is the overflood level of basin i. A flooding starting in basin i would flood into the neighboring basins through this pass point. So $\gamma_e(i,j) = \delta_{en}\varepsilon_{ne}(i,j)$ represents the highest overflood level of the basins i and j. γ_e will thus be invariant on all edges which are the overflood of one of the neighboring basins or in terms of graphs the lowest edge of one of their extremities. This property is used later to show how waterfalls can be constructed locally using invariants of the opening γ_e.

3 Morphological Segmentation on Edge Weighted Graphs

Edge weighted graphs are useful in segmentation since they carry a compact representation of hierarchical segmentation : cutting all edges with a weight $> \lambda$ transforms the graph into a forest, where each tree spans a region of the segmentation. For increasing threshold values, less and less edges are cut and the associated segmentation becomes coarser. The minimum spanning tree (MST) of the edge weighted graph (among all spanning trees) is particularly useful in this context, since by cutting the edges with a weight $> \lambda$ yields the same partition as by cutting the edges above the same threshold on the initial graph.

The MST is ubiquitous in morphological segmentation issues. Segmenting with markers leads to constructing a minimum spanning forest [4], where each tree is rooted in a marker. Serge Beucher and Béatrice Marcotegui [5] have shown that the waterfall hierarchy may be obtained by constructing the watershed on the MST associated to the regional minima [6]. Jean Cousty et al [7], studying the watershed defined on edge weighted graphs, have shown the equivalence between watershed and minimum spanning forests associated to the regional minima of the graph.

3.1 Waterfalls

Waterfall hierarchies have been introduced by Serge Beucher in his thesis [8,9]. The waterfall hierarchy describes the intrication of the catchment basins and the nested structures of a topographic surface. The waterfall hierarchy may be obtained by flooding. The lowest and finest level of the hierarchy corresponds to the set of all catchment basins. The first level of hierarchy is obtained by taking the catchment basins of the topographic surface after it has been submitted to the following flooding : each catchment basin is filled up to its lowest pass point. The process is then repeated for this new topographic surface and produces the successive levels of the hierarchy. The process ends when a topographic surface is created containing only one catchment basin.

The waterfall hierarchy seen through Boruvka's algorithm. A careful examination of Boruvka's algorithm for constructing MSTs give us a deeper insight in the relations between waterfalls, watersheds on graphs through invariants for the morphological opening γ_e. For the following analysis, we suppose that each node has only one lowest edge. We analyze later the general case. The waterfall hierarchy becomes apparent while constructing the MST using Boruvka's algorithm. The following different events can be distinguished by constructing iteratively a spanning tree of a graph G from an empty set of edges and nodes.

1. Two isolated nodes i and j are connected by a new edge. Since two adjacent edges have necessarily different weights, this edge is surrounded by higher edges in the graph G and hence is a regional minimum edge. Edge u is the lowest edge adjacent to i and j. Hence, as seen earlier, edge u is invariant by the opening $\gamma_e(u) = \delta_{en}\varepsilon_{ne}(u)$ and represents the lowest overflood level of the basins i and j.

2. The edge connects an isolated node i with a subtree of T. Clearly u is the lowest edge adjacent to i, otherwise node i would have been incorporated in a subtree earlier. Again edge u is invariant by the opening $\gamma_e(u) = \delta_{en}\varepsilon_{ne}(u)$.

3. The edge u connects two nodes i and j already belonging to subtrees of T. The nodes i and j have been joined to their respective subtrees through edges with lower weights than u. Hence u is not the lowest edge of the nodes i and j. For this reason u is not invariant by the opening $\gamma_e(u) = \delta_{en}\varepsilon_{ne}(u)$.

During events (1) and (2) the edges incorporated in the MST are invariant by the opening γ_e whereas events of type (3) never use edges invariant by γ_e. On the other hand, each node has one and only one lowest edge, through which it will be assigned to the MST during an event of type (1) or (2). Especially, if we only consider edges invariant by γ_e without also adding the edges of type (3), we obtain a spanning forest, and this forest is minimal as we will see in the next section. In fact this minimum spanning forest yields the same partition as the princeps waterfall algorithms proposed by Serge Beucher, in the case where G represents a region adjacency graphs, where each node represents a catchment basins and the edges are weighted with the altitude of the pass point between adjacent basins. Each basin is flooded up to the level of its lowest pass point ; like that there is no regional minimum anymore in the basins and each of them is absorbed by a neighboring basin. The edge along which this absorption takes place is precisely the lowest edge of one of its neighboring basin, that is an edge invariant by opening γ_e, belonging to the spanning forest constructed above. An example of waterfall segmentation is illustrated figure 1.

3.2   Minimum Spanning Forest

Let us now show that the spanning forest obtained by using Boruvka's algorithm is indeed minimal. If we chose an arbitrary adjacent node m_u for each minimal edge of the graph G, we obtain a family of nodes (m_u), one for each regional minimum. Adding a dummy node o linked to each m_u through an edge with valuation

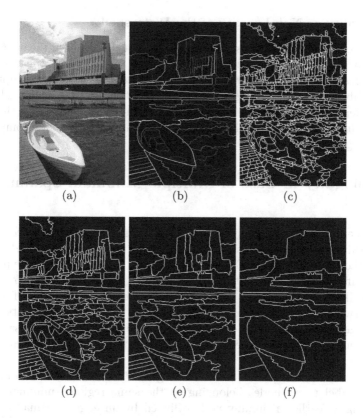

(a)   (b)   (c)

(d)   (e)   (f)

Fig. 1. (a) Original image. (b) Probabilistic gradient computed from the stochastic watershed transform [10]. (c) Watershed of the probabilistic gradient. A region adjacency graph is built from this partition, edges are weighted by the lowest passpoint between two regions. (d) Results of Boruvka's algorithm, only edges of type (1) and (2) are kept. (e-f) Second and third level of waterfall hierarchy using Boruvka's algorithm on the region adjacency graph of the partition of the previous level of the hierarchy.

-1, we get a new graph G'. Let us now construct the MST of G' starting with the dummy node o. The first steps of the algorithm visit all dummy edges with negative weights adjacent to the dummy node o. The next steps visit the edges in the same order as Boruvka's algorithm with one major difference : events of type (3) are never met, since from the beginning, there is only one tree.

After the construction is completed, we suppress the dummy node and the dummy edges and get a minimum spanning forest. This forest is identical with the forest obtained by considering only events of type (1) and (2) in Boruvka's algorithm, showing that considering only the edges invariant by opening γ_e produces indeed a minimum spanning forest. We find here by another mean the result of Serge Beucher and Beatriz Marcotegui [5], constructing the waterfall by a watershed algorithm on the MST of the neighborhood graph. This shows also that the waterfall partition of level 1 is the result of segmenting the graph G with the family of markers (m_u).

3.3   Locality or Non Locality of the Watershed ?

The edges of the spanning forest may be obtained as invariant of the opening γ_e, which is a purely local operation. Extracting the individual trees can be done via any classical labeling algorithm of connected components in a graph. It is even possible to extract a single tree, without extracting the others. This is a major difference with the watershed defined on the nodes, defined as the SKIZ of the minima for the topographic distance and constructed through competitive flooding algorithms.

4   Ambiguity in Case of Multiple Minimum Spanning Trees

In the general case, a node may have several lowest edges. Adjacent edges with the same altitude may form plateaus of any size. At each step of Boruvka's algorithm for constructing the MST, several equivalent choices of edges are possible, growing different trees. A multiplicity of spanning trees coexist, having all the same distribution of edges and producing the same nested partitions if one cuts the edges with a weight above some threshold. In particular the events of type (1), an edge connecting two nodes not connected earlier, do not necessarily produce regional minima, as this edge may belong to the inside of a plateau, which is not a regional minimum.

For this reason, one has to detect the regional minima beforehand and give the same label to all nodes belonging to the same regional minimum. After introducing as earlier a dummy node o linked by an edge of valuation -1 to each regional minimum, one may then apply Prim's algorithm for constructing the MST. Edges are considered with increasing values ; at each stage of the algorithm, only the edges adjacent to the already constructed part of the tree are considered. Nevertheless, this construction leaves a large freedom of choices and a great number of spanning trees are possible. Some of these trees do not seem desirable, as they cut the plateaus in an unfair way ; one may wish sharing the plateaus among neighboring basins by cutting them at equal distance to their lower borders.

4.1   A Hierarchical Queue Implementation for the Watershed of Edge Weighted Graphs

A fair sharing of plateaus may be reached if one resorts to the classical hierarchical queue structure: the edges are processed in increasing order, but edges with the same weight are processed according to increasing distances to the lower borders in plateaus. The structure of hierarchical queues introduces naturally the flooding order in the processing. However we loose the nice feature met above to be able to extract any tree from the watershed forest without necessarily constructing the neighboring trees which constitute its limits. In order to be able to achieve this type of extraction, one has to introduce a lexicographic order among the edges as we will see just below.

4.2   A Lower Complete Edge Graph

Our aim in this section is to complete the order relation between adjacent edges by an additional, lexicographic order relation, such that each non minimal edge has at least one neighboring edge with a lower weight. In the graph G, the edges without lower neighbor are the edges in the regional minima and the edges inside non minimal plateaus. However, each non minimal plateau has itself lower neighbors. The idea for resorbing the plateaus is to compute a geodesic distance function within the plateau towards the lower border. This distance is defined as follows : an edge u will be assigned a distance n, if n is the smallest index such $\varepsilon_E^{(n)}(u) < \varepsilon_E^{(n-1)}(u)$. This distance function may be classically obtained through a queue implementation. After this completion, each edge, except the edges inside the regional minima will have two valuations : the initial valuation $w(u)$ and the distance function $\pi(u)$ to a lower border ; the lexicographic order relation being: $u > v \Leftrightarrow \{w(u) > w(v)\}$ or $\{w(u) = w(v)\ \ and\ \ \pi(u) > \pi(v)\}$.

The lexicographic order relation between neighboring edges amounts to introducing a polarisation between edges. It can be represented by replacing the non oriented edges by oriented arcs. Recall that we only consider the edges invariant by the opening γ_e. Each such edge (i, j) is the lowest neighbor of one of its extremities, say i. The other extremity j is then necessarily the extremity of the lower neighboring edge of (i, j). This analysis shows that there exists an implicit orientation from lower towards higher edges ; replacing the non oriented edge (i, j) by an oriented arc $\overrightarrow{(j, i)}$ with an orientation opposite to the downwards direction conveys the same information as the lexicographic distance function of the previous paragraph. Of course the edges belonging to regional minima regions remain non oriented edges, as they do not possess a downwards direction.

With the introduction of the oriented edge graph, we will be able to extract individual trees or regions associated to a particular minimum or marker without constructing the whole watershed. It has been noticed that for the partial graph associated to the edges invariant by γ_e, neighboring basins may be connected, implying that a connected subgraph may contain various regional minima. It is not the case anymore with our oriented graph. We will associate to each minimum m_i the set of node which may be reached by an oriented path having its origin in the minimum. The ordinary Dijkstra algorithm of shortest oriented paths may be used for this extraction.

In the partial non oriented graph G', there exist nodes having two lowest edges with the same weight, leading to two distinct regional minima m_1 and m_2. Such nodes, together with all nodes for which they are downstream nodes, belong to the catchment basins of each adjacent minimum m_1 and m_2. Each of the oriented subtrees associated to these minima will contain this divide zone. Suppressing the zones which are common to two distinct neighboring oriented trees creates a minimal watershed tessellation which is not a partition, as the divide zones are missing.

4.3   Application to the Waterfall Hierarchy

The level 1 of the waterfall hierarchy consists in a minimum spanning forest, where each tree is centered on the minima. This minimum spanning forest differs from a minimum spanning tree by some missing edges. Contracting each tree of the forest to one node and introducing the missing edges produces a new tree T^1. In the case where all edges of the graph were with distinct weights, the edges of this tree also have distinct weights. And the waterfall hierarchy of level 2 is again the invariant set by opening γ_e. In all other cases, tree T^1 may have edges with the same weight and one of the general constructions above has to be applied.

5   Watershed on Node Weighted Graphs

In this last section we show how the construction of waterfalls and watersheds on edge weighted graph may be applied for constructing watersheds on node weighted graphs. In the context of segmentation, most often the watershed has to be constructed on a gradient image derived from the image. One of the problem to take into consideration is the problem of scale ; for a binary image, or a mosaic images (the image is constant on each tile of a mosaic) the gradient is a thin line and may be represented faithfully on an edge graph, in which each edge would be weighted by taking the absolute difference of the weights of its extremities. Such a local gradient does however not correctly the contours of natural images, where the boundaries of the objects are more or less blurred, corrupted by noise and the contour information is spread out on a larger surface than thin lines.

In this section we present how to transform a node weighted graph into an edge weighted graph. The watershed is then constructed on this edge weighted graph with the method presented earlier, producing a forest centered on the minima, yielding the same watershed as the watershed constructed on the node graph. The advantage of this way of doing are multiple:

- extending the waterfall hierarchy to node weighted graphs,
- transforming the problem of watershed construction into a problem of MST construction,
- after completion of the graph in order to suppress the plateaus, being able to extract individual trees, without constructing the watershed with competing markers,
- being able to construct a minimal watershed and isolate the thick divide zones.

Weighting the edges in order to stress the directions of steepest descent. Given a node weighted graph, we want to weight the edges in such a way that the watershed constructed on the edge graph is identical with the watershed on the node graphs. The weight of each edge (i, j) will be computed as follows.

1. Case where $w(i) > w(j)$. Then we compute the lower gradient $\varsigma(i) = w(i) - \varepsilon_N(i)$, obtained by subtracting from the weight of i the weight of its lowest

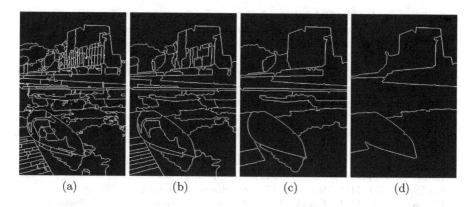

(a)   (b)   (c)   (d)

Fig. 2. (a-d) First, second, third and forth level of waterfall hierarchy using Boruvka's algorithm on the pixel adjacency graph. The waterfall was obtained from the probabilistic gradient proposed by Angulo et al. [10] illustrated in figure 1.

neighbor. The weight of edge (i,j) is then $w(i,j) = w(j) + \varsigma(i)$. In the case where j is the lowest neighbor of i, then $\varsigma(i) = w(i) - \varepsilon_N(i) = w(i) - w(j)$ and $w(i,j) = w(i)$. In all other cases, this weight will be higher than $w(i)$.

2. Case where $w(i) = w(j)$, then we take $\min(w(j) + \varsigma(i), w(i) + \varsigma(j))$. In the case where i or j have no lower neighbor, then $\varsigma(i) = \varsigma(j) = 0$ and $w(i,j) = w(i) = w(j)$, and a plateau of edges with the same values will be created.

On this edge weighted graph, we may apply the results described earlier. A waterfall segmentation of the pixel graph obtained from a gradient image is illustrated in figure 2.

6   Conclusion

We have highlighted some new properties of the watershed and the waterfall transform through a detailed analysis of Boruvka's algorithm. These properties, linked with invariants of a specific morphological opening on graphs, provide new methods and algorithms for constructing the waterfall segmentation. We have presented these properties for both edges and nodes weighted graphs. This study brings a different point of view on the importance of minimum spanning trees for watershed and waterfall based segmentation.

References

1. Allène, C., Audibert, J., Couprie, M., Cousty, J., Keriven, R.: Some links between min-cuts, optimal spanning forests and watersheds. In: Proceedings of the The 8th International Symposium on Mathematical Morphology, Rio de Janeiro, Brazil, vol. 1, pp. 253–264 (2007)

2. Audigier, R., Lotufo, R.A.: Watershed by image foresting transform, tie-zone, and theoretical relationships with other watershed definitions. In: Proceedings of the The 8th International Symposium on Mathematical Morphology, Rio de Janeiro, Brazil, vol. 1, pp. 277–288 (2007)
3. Stawiaski, J.: Mathematical morphology and graphs: Application to interactive medical image segmentation. PhD Thesis. Mines ParisTech. (2008)
4. Meyer, F.: Grey-weighted, ultrametric and lexicographical distances. In: Computational Imaging and Vision, Mathematical Morphology: 40 Years On. Proceedings of the 7th International Symposium on Mathematical Morphology, vol. 30, pp. 289–298 (2005)
5. Marcotegui, B., Beucher, S.: Fast implementation of waterfall based on graphs. In: Ronse, C., Najman, L., Decencière, E. (eds.) Mathematical Morphology: 40 Years on: Proc. 7th ISMM, Paris, pp. 177–186. Springer, Heidelberg (2005)
6. Falcao, A., Stolfi, J., Lotufo, R.: The image foresting transform: Theory, algorithms, and applications. IEEE Transactions on Pattern Analysis and Machine Intelligence 26, 19–29 (2004)
7. Cousty, J., Bertrand, G., Najman, L., Couprie, M.: Watershed cuts: minimum spanning forests, and the drop of water principle. IEEE Trans. on Pattern Analysis and Machine Intelligence (2008) (to appear)
8. Beucher, S.: Segmentation d'images et morphologie mathématique. PhD thesis, Ecole des Mines de Paris, Paris (June 1990)
9. Beucher, S.: Watershed, hierarchical segmentation and waterfall algorithm. In: Serra, J., Soille, P. (eds.) Mathematical Morphology and its applications to signal processing (Proceedings ISMM 1994), Fontainebleau, France, pp. 69–76. Kluwer Academic Publishers, Dordrecht (1994)
10. Angulo, J., Jeulin, D.: Stochastic watershed segmentation. In: Proceedings of the The 8th International Symposium on Mathematical Morphology, Rio de Janeiro, Brazil, vol. 1, pp. 265–276 (2007)

Segmentation of Complex Images Based on Component-Trees: Methodological Tools

Benoît Caldairou[1,3,4], Benoît Naegel[2], and Nicolas Passat[1]

[1] Université de Strasbourg, LSIIT, UMR CNRS 7005, France
[2] Université Nancy 1, LORIA, UMR CNRS 7503, France
[3] Université de Strasbourg, LINC, UMR CNRS 7191, France
[4] École Supérieure Chimie Physique Électronique de Lyon, France
{benoit.caldairou,passat}@unistra.fr, benoit.naegel@loria.fr

Abstract. Component-trees can be used for the design of image processing methods, and in particular segmentation ones. However, despite their ability to consider various kinds of knowledge and their tractable computation, methodological deadlocks often forbid to efficiently involve them in real applications. In this article, we explore new solutions to some of these deadlocks, and more especially those related to (*i*) complexity of the structures of interest and (*ii*) multiple knowledge handling. The usefulness of the proposed strategies is illustrated by preliminary results related to vessel segmentation from 3-D angiographic data.

Keywords: Component-trees, segmentation, attribute-filtering, grey-level images.

1 Introduction

The *component-tree* (also known as *dendrone* [1,2], *confinement tree* [3] or *max-tree* [4]) is a graph-based structure which models some characteristics of a grey-level image by considering its binary level-sets obtained from successive thresholding operations.

Initially proposed in the field of statistics, the component-tree has been (re)defined in the theoretical framework of mathematical morphology and involved, in particular, in the development of morphological operators [5,4]. Thanks to efforts devoted to its efficient computation [5,4,3,6,7] or its use in complex knowledge handling procedures [8], component-trees have been considered for the design of various kinds of grey-level image processing methods, including image filtering and segmentation [1,9,10,11,12,13], video segmentation [4], image registration [3], image compression [4], or image retrieval [14,15].

Despite the ability of component-trees to consider complex/multiple knowledge and their tractable computation, methodological deadlocks often forbid to efficiently involve them in *real* applications. In this article, we propose to explore solutions to some of these deadlocks, and more especially those related to (*i*) complexity of the (shape of) structures of interest and (*ii*) multiple knowledge handling.

M.H.F. Wilkinson and J.B.T.M. Roerdink (Eds.): ISMM 2009, LNCS 5720, pp. 171–180, 2009.

In Section 2, previous works involving component-trees in the design of segmentation methods are described, emphasising the remaining challenges to be
faced. Section 3 introduces definitions and notations required to make the article self-contained. In Section 4, some methodological considerations provide
solutions to tackle the challenges stated in Section 2. An application, described
in Section 5 for 3-D angiographic image segmentation illustrates the soundness
of the proposed framework. Section 6 summarises the contributions of this article
and points out the main perspectives.

2 Segmentation Based on Component-Trees

As mentioned above, component-trees have been considered for the development
of image segmentation methods, mainly in the field of (bio)medical imaging, and
in particular for: dermatological data [13], wood micrographs [9], cerebral MRI
[16], CT/MR angiography [17], or confocal microscopy [11].

It has to be noticed that their use is often only devoted to one specific step
of the segmentation (marker selection in [16]), or to perform filtering [17,11],
i.e. to remove "useless" parts of the processed image, leading to a superset of
an actual segmentation. Among the methods which fully use component-trees
for segmentation purpose, some can consider complex (*i.e.* multiple) knowledge
[13] or can be run without user-interaction [9], but none of them is able to
determine the correct pieces of knowledge required to perform segmentation
without guidance of the user. Moreover, such methods only deal with simple-
shape objects (circular or elliptical 2-D features in [9,13]).

This emphasises the fact that *automatic* segmentation of *complex objects*
based on the use of *multiple elements of knowledge* obviously remains an open
methodological problem in the field of component-tree-based methods, *a fortiori*
when such knowledge also needs to be automatically determined (which may be
necessary whenever the size of the parameter space becomes too large). In the
next sections, we explore some ways to deal with this difficult issue. In particular, we consider strategies enabling to decrease the potential complexity of the
structures of interest, and to determine the nodes (and thus the attributes) of
the component-trees of ground truth images, then enabling automatic learning
of correct parameters for segmentation purpose.

3 Definitions and Notations

Let $n \in \mathbb{N}^*$. In the sequel, $[a..b]$ (with $a, b \in \mathbb{Z}$) denotes the discrete interval
$[a, b] \cap \mathbb{Z}$. We set $\overline{\mathbb{Z}} = \mathbb{Z} \cup \{-\infty\}$. A discrete grey-level image can be defined as
a function $I : \mathbb{Z}^n \to \overline{\mathbb{Z}}$. The support of I is defined by $\mathrm{supp}(I) = \{x \in \mathbb{Z}^n \mid I(x) \neq -\infty\}$. We assume that for any considered image I, $\mathrm{supp}(I)$ is finite. We
will note $\mathrm{supp}(I) = E$ and $V = [a..b] \subset \mathbb{Z}$, where $a = \min\{I(x) \mid x \in E\}$ and
$b = \max\{I(x); x \in E\}$. From now on, we will assimilate an image $I : \mathbb{Z}^n \to \overline{\mathbb{Z}}$ to
its (finite) restriction $I_{|E} : E \to V$.

Let $X \subseteq E$. The connected components of X are the equivalence classes of X w.r.t. the equivalence relation on E induced by the adjacency relation chosen for $\mathbb{Z}^n$. The set of the connected components of X is noted $\mathcal{C}[X]$.

Let $v \in V$. We set $\mathcal{P}(E) = \{X; X \subseteq E\}$. Let $X_v : V^E \to \mathcal{P}(E)$ be the thresholding function defined by $X_v(I) = \{x \in E; v \leq I(x)\}$ for all $I : E \to V$.

Let $v \in V$ and $X \subseteq E$. We define the cylinder function $C_{X,v} : E \to \overline{\mathbb{Z}}$ by $C_{X,v}(x) = v$ if $x \in X$ and $-\infty$ otherwise. A discrete image $I : E \to V$ can then be expressed as $I = \bigvee_{v \in V} C_{X_v(I),v} = \bigvee_{v \in V} \bigvee_{X \in \mathcal{C}[X_v(I)]} C_{X,v}$, where $\bigvee$ is the pointwise supremum for the sets of functions.

Let $\mathcal{K} = \bigcup_{v \in V} \mathcal{C}[X_v(I)]$. The relation $\subseteq$ is a partial order on $\mathcal{K}$, and the Hasse diagram $(\mathcal{K}, L)$ of the partially ordered set $(\mathcal{K}, \subseteq)$ is a tree (*i.e.* a connected acyclic graph), the root of which is the supremum $R = \sup(\mathcal{K}, \subseteq) = E$. This rooted tree $(\mathcal{K}, L, R)$ is called the *component-tree of I*. The elements $\mathcal{K}, R$ and L are the set of the *nodes*, the *root* and the set of the *edges* of the tree, respectively.

Note that each node of $\mathcal{K}$ is a binary connected component distinct from all the other ones. However, such a connected component can be an element of $\mathcal{C}[X_v(I)]$ for several (successive) values $v \in V$. For each $X \in \mathcal{K}$, we set $m(X) = \max\{v \in V; X \in \mathcal{C}[X_v(I)]\} = \min_{x \in X}\{I(x)\}$. An image $I : E \to V$ can then be defined from its component-tree $(\mathcal{K}, L, R)$ as $I = \bigvee_{X \in \mathcal{K}} C_{X,m(X)}$.

Component-trees enable the storage - at each node - of elements of information, also called *attributes*, related to the binary connected component associated to the node. It is possible to consider any kind of quantitative/qualitative and scalar/vectorial attributes, provided they can be conveniently formalised. Pruning a component-tree $(\mathcal{K}, L, R)$ of an image I according to the attributes stored at the nodes (by removing the nodes having a non-correct attribute w.r.t. a given criterion) enables to perform filtering on I. The filtered image I_f is then defined as $I_f = \bigvee_{X \in \mathcal{K}_f} C_{X,m(X)}$ where $\mathcal{K}_f \subseteq \mathcal{K}$ is the subset of the remaining nodes after the pruning process. When performing segmentation, a binary result I_b can similarly be obtained as $I_b = \bigcup_{X \in \mathcal{K}_f} X$.

4   Methodological Concepts

In this section, we present methodological tools enabling to develop algorithms based on component-trees, and dealing with the main challenges described in Section 2. In Subsection 4.1, solutions are proposed to spatially decompose (and reconstruct) an image, thus breaking complex structures into (hopefully) simpler sub-ones. In Subsection 4.2, the way to automatically extract relevant nodes from the component-tree of a ground truth (*i.e.* a correctly segmented) image is discussed, enabling to avoid user-interaction in segmentation processes.

4.1   Image Partitioning/Reconstruction

The binary connected components at the nodes $X \in \mathcal{K}$ of a component-tree may possibly be complex and/or gather several structures of interest of the associated image. In such cases, these nodes, potentially composed of several

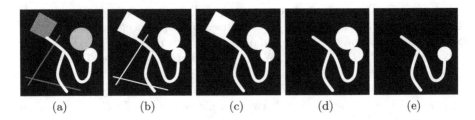

Fig. 1. (a) A grey-level image containing different semantic elements (geometric shapes). (b-e) Threshold images obtained from (a) at successive grey-levels: the obtained nodes/connected components do not enable to discriminate the visualised elements (see text).

semantic elements may be hard to detect/discriminate due to the heterogeneity of the characterising properties of these elements.

In order to illustrate this assertion, let us consider the grey-level image of Fig. 1(a), which is composed of four semantic entities: squares, disks, thin straight lines and thick curves. Here, we obtain a critical situation where only one - useless - node is available at each level of the tree, as observed in Fig. 1(b-e). Despite the existence of specific properties (elongation, straightness, compactness, etc.) for each kind of elements, their specific intensity in the image and/or their spatial organisation (connections, generation of complex shapes from simpler ones, etc.) result in a component-tree the nodes of which do not enable any characterisation.

In similar cases, the computation of attributes devoted to characterise accurate and specific properties will generally fail. It has to be noticed that such situations are not infrequent in real applications. For instance, in angiographic image analysis (see Section 5), vessels are generally organised into a unique network, thus making attributes characterising *single* tubular structures inefficient.

A solution to this general issue can consist in processing the image as a collection of smaller subimages, hence enabling to split complex structures into smaller - and hopefully easier to detect - sub-ones. A straightforward strategy based on this approach is the following one.

1. Divide $I : E \to V$ into a set of images $I^k : E^k \to V$ ($k \in [1..m]$) such that $\{E^k\}_{k=1}^m$ is a partition of E, and $I_{|E^k} = I^k$ for all $k \in [1..m]$.
2. Compute, for each $k \in [1..m]$, the component-tree of I^k and perform segmentation, then generating a binary output image $B^k \subseteq E^k$.
3. Define the segmentation result B by merging all the results B^k : $B = \bigcup_{k=1}^m B^k$.

This simple and potentially useful approach however suffers from two drawbacks: (*i*) the partition of E may split a structure of interest between several subsets E^k, thus forbidding its correct detection, and (*ii*) the size of the subsets E^k, possibly well-chosen to fit a given structure, may be non-adapted to the detection of another one.

Partitioning. A way to avoid these two drawbacks is to compute a *redundant* and *multiscale* decomposition of I, in order to fit at best the different structures of interest. The support E of I is then split by defining a set $\mathcal{E}_{\alpha,\beta} = \bigcup_{a\in\alpha}\{E_{a,\beta}^k\}_{k=1}^{m_{a,\beta}}$, such that for all $a \in \alpha$ we have

$$\forall k \in [1..m_{a,\beta}], |E_{a,\beta}^k| = |E|/a , \tag{1}$$

$$E \subseteq \bigcup_{k=1}^{m_{a,\beta}} E_{a,\beta}^k , \tag{2}$$

$$\forall x \in E, |\{X \in \{E_{a,\beta}^k\}_{k=1}^{m_{a,\beta}}; x \in X\}| = \beta , \tag{3}$$

where $\alpha \subseteq [1..|E|]$ is a set of volume ratios ("scales"), and $\beta \in \mathbb{N}^*$ is the "redundancy factor" of the pseudo-partitions $\{E_{a,\beta}^k\}_{k=1}^{m_{a,\beta}}$ at each scale $a \in \alpha$. Broadly speaking, the image support is decomposed (several times) into subsets the sizes of which are determined by Eq. (1), and for each one of these sizes, the union of these subsets has to match the whole image support[1] (Eq. (2)), while each point of this support has to belong to a given number of these subsets, this number being determined by Eq. (3). It may generally be convenient to define α as a subset of $\{2^{nk}\}_{k\geq0}$ in order to build subsets $E_{a,\beta}^k$ of $E \subset \mathbb{Z}^n$ in an "octree" fashion.

Reconstruction. Once processed, each partial image $I_{a,\beta}^k : E_{a,\beta}^k \to V$ provides a binary output $B_{a,\beta}^k \subseteq E_{a,\beta}^k$. We set $\mathcal{B}_{\alpha,\beta} = \bigcup_{a\in\alpha}\{B_{a,\beta}^k\}_{k=1}^{m_{a,\beta}}$. By opposition to the initially proposed strategy, which enables to recover $B \subseteq E$ by simply merging the subimages B^k, the one proposed above does not straightforwardly lead to a final result, since overlaps induced by both multiscale and redundancy may lead to ambiguous results for any point $x \in E$ (depending on the image $I_{a,\beta}^k$ where x is considered).

For any $x \in E$, let $\mathcal{E}_{\alpha,\beta}^x = \{E_{a,\beta}^k \in \mathcal{E}_{\alpha,\beta}; x \in E_{a,\beta}^k\}$ and $\mathcal{B}_{\alpha,\beta}^x = \{B_{a,\beta}^k \in \mathcal{B}_{\alpha,\beta}; x \in B_{a,\beta}^k\}$ (note that $0 \leq |\mathcal{B}_{\alpha,\beta}^x| \leq |\mathcal{E}_{\alpha,\beta}^x| = \beta.|\alpha|$). Final images $B_f \subseteq E$ and $I_f : E \to [0,1]$ (binary and fuzzy, respectively) can be reconstructed as follows

$$B_f = \{x \in E; \lambda \leq |\mathcal{B}_{\alpha,\beta}^x|\} \text{ for a given } \lambda \in [1, \beta.|\alpha|] , \tag{4}$$

$$I_f(x) = |\mathcal{B}_{\alpha,\beta}^x|/(\beta.|\alpha|) \text{ for all } x \in E . \tag{5}$$

It can be noticed that (i) setting $\lambda = 1$ in Eq. (4) is equivalent to define $B_f = \bigcup_{X\in\mathcal{B}_{\alpha,\beta}} X$, and (ii) B_f can be obtained by thresholding I_f at the considered value λ.

4.2 Multiple Criteria Handling

It is possible to involve arbitrarily large and heterogeneous sets of knowledge in segmentation processes by associating to each node of the component-tree vectorial attributes (containing qualitative, quantitative, structural information, etc.).

---

[1] Note that in Eq. (2) the inclusion (instead of an equality) between the two elements implies that the set $\{E_{a,\beta}^k\}_{k=1}^{m_{a,\beta}}$ is actually not a partition of E since some of the $E_{a,\beta}^k$ may be partially out of E to guarantee the same redundancy β at each point of E.

This can lead to very accurate descriptions of the structures to be segmented. However, a straightforward and undesired side effect is the difficulty to determine, among the whole (and potentially huge) parameter space Ω induced by this knowledge, the *correct subset* $\omega \subset \Omega$ characterising the structures of interest, *a fortiori* in an interactive fashion.

In such conditions it becomes fundamental to enable *automatic* determination of such characterising subsets. This can be done by using learning - and in particular classification - tools. To this end, it is necessary to find a way to put in correspondence a "ground truth" (*i.e.* correct examples of what should be segmented) and the closest result which may be obtained by the component-tree-based method.

The problem to solve may be formalised as follows. Let $I_g : E \to V$ be a ground truth image (similar to those to be further processed by the method), and $B_g \subseteq E$ be the correct segmentation of this image. Let $(\mathcal{K}, L, R)$ be the component-tree of I_g. Let $\mathcal{S} = \{\cup_{X \in C} X\}_{C \subseteq \mathcal{K}}$ be the set of all binary images which can be generated from the set of nodes $\mathcal{K}$. In general, we will - unfortunately - never have $B_g \in \mathcal{S}$. We then need to determine the "best" binary image which may be computed from $\mathcal{K}$ w.r.t. B_g. This requires to define a (pseudo)distance d on $\mathcal{P}(E)$ enabling to compare B_g and the candidate binary images of $\mathcal{S}$. In particular, the best binary image $\hat{B}$ can be defined as

$$\hat{B} = \arg \min_{B \in \mathcal{S}} \{d(B, B_g)\} . \tag{6}$$

In this context, several strategies can reasonably be considered.

- By setting $d^-(B, B_g) = |B_g \setminus B|$ if $B \subseteq B_g$ and $+\infty$ otherwise, we have $\hat{B}^- = \max_{\subseteq}\{B \in \mathcal{S}; B \subseteq B_g\}$, *i.e.* $\hat{B}^-$ is the *largest object included in B_g* which may be built from $\mathcal{K}$.
- By setting $d^+(B, B_g) = |B \setminus B_g|$ if $B \supseteq B_g$ and $+\infty$ otherwise, we have $\hat{B}^+ = \min_{\subseteq}\{B \in \mathcal{S}; B_g \subseteq B\}$, *i.e.* $\hat{B}^+$ is the *smallest object including B_g* which may be built from $\mathcal{K}$.

The first (resp. second) strategy focuses on the elimination of false positives (resp. false negatives) with the side effect of possibly authorising the preservation of false negatives (resp. false positives). It has to be noticed that these asymmetric strategies can be efficiently implemented since the set of nodes generating $\hat{B}$ can obviously be computed with a (worst case) algorithmic complexity $O(\max\{|\mathcal{K}|, |E|\})$ linear w.r.t. the number of nodes of the component-tree or the size of the image.

Some - more symmetric - strategies could also be proposed. The most straightforward one consists in setting $d^*(B, B_g) = |B_g \setminus B| + |B \setminus B_g|$. It aims at finding a "best compromise" between the amount of false positives and false negatives. In particular, we have $\hat{B}^- \subseteq \hat{B}^* \subseteq \hat{B}^+$. Also note that if $B_g \in \mathcal{S}$, we have $\hat{B}^- = \hat{B}^* = \hat{B}^+$. Surprisingly, this approach (by opposition to other symmetric ones, involving Hausdorff distances for example, which may present high algorithmic complexities) also leads to an algorithmic complexity $O(|\mathcal{K}|)$ (this claim will be proved in further works).

When a minimal set $\hat{B}$ has been extracted from $\mathcal{S}$, remains to determinate an adequate set of nodes $\hat{K} \subseteq \mathcal{K}$ associated to $\hat{B}$ (*i.e.* such that $\bigcup_{X \in \hat{K}} X = \hat{B}$). Let $\hat{C} \subseteq \mathcal{K}$ be the set defined by $\hat{C} = \{X \in \mathcal{K}; X \subseteq \hat{B}\}$ (note that the nodes of $\hat{C}$ generate a set of subtrees of the component-tree $(\mathcal{K}, L, R)$ of I_g). The set $\hat{B}$ can be generated by any set of nodes $\hat{K} \subseteq \hat{C}$ verifying $\bigcup_{X \in \hat{K}} X = \bigcup_{X \in \hat{C}} X = \hat{B}$. In order to determine such a set $\hat{K}$, two main strategies can, in particular, be considered.

- By setting $\hat{K}^+ = \hat{C}$, any node included in $\hat{B}$ is considered as a useful (*i.e.* informative) binary connected component.
- By setting $\hat{K}^- = \{X \in \hat{C}; \forall Y \in \hat{C}, X \not\subset Y\}$, only the roots of the subtrees induced by $\hat{C}$ are considered as useful binary connected components.

The first (resp. second) strategy is the one considering the largest (resp. smallest) possible set of nodes/connected components among $\hat{C}$; in particular, it can be seen as the one which focuses at most on the grey-level (resp. binary) structure of the ground truth image I_g. The choice of the strategy may then be directed by the kind (binary *vs.* grey-level) of criteria/attributes to be considered.

Once a set of nodes $\hat{K}$ has been defined from the whole set $\mathcal{K}$ (from one or possibly several ground truth image(s)), the determination of the subset of characterising knowledge $\omega \subset \Omega$ has to be performed. Let $A : \mathcal{K} \to \Omega$ be the function associating, to each node of the component-tree, its stored attribute. The determination of ω can be expressed as a classification problem consisting in partitioning Ω into two classes thanks to the samples $A(\hat{K}) = \{A(N); N \in \hat{K}\}$ (corresponding to the attributes of the structures of interest) and $A(\mathcal{K} \setminus \hat{C}) = \{A(N); N \in \mathcal{K} \setminus \hat{C}\}$. This process can, for instance, be carried out by usual classification tools (such as the Support Vector Machine (SVM) [18], which has been considered in the experiments of the next section).

5   A Case Study – Angiographic Image Segmentation

Based on the framework described above, a strategy is being developed for segmenting 3-D angiographic data (namely phase contrast magnetic resonance angiographies - PC-MRAs). We propose hereafter a preliminary and simplified description of this method, and we provide - for illustrative purpose - some obtained results. A complete description of the final method (with full validations and a larger set of involved attributes) will be found in dedicated further works.

PC-MRAs are bimodal images $(I_m, I_p) \in (V^E)^2$ where $I_p : E \to V$ is the phase (*i.e.* vascular) image while $I_m : E \to V$ is the magnitude (*i.e.* morphological) image, with $E = [0..255]^3$ and $V = [0..N] \subset \mathbb{N}$ (see Fig. 2(a,b)). The proposed method is devoted to segment phase images I_p in order to extract the vessels (and in particular to discriminate them from noise and artifacts).

In order to enable a correct segmentation of the vessels from such images, three attributes have been considered: (*i*) the second Hu's moment, (*ii*) an inertia matrix-based elongation criterion, both computed from the component-tree of

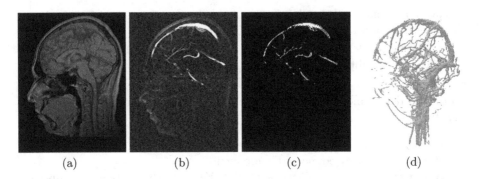

(a)   (b)   (c)   (d)

Fig. 2. (a,b) Phase contrast magnetic resonance angiography (ground truth image I_g): sagittal 2-D slices of the magnitude image I_m (a) and of the phase image I_p (b). (c,d) Ground truth segmentation B_g obtained from I_p: sagittal slice (c) and 3-D visualisation (d).

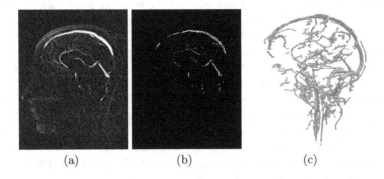

(a)   (b)   (c)

Fig. 3. (a) Phase contrast magnetic resonance angiography: sagittal 2-D slice of the phase image I_p. (b,c) Segmentation result (binary segmentation): sagittal 2-D slice (b) and 3-D visualisation (c).

I_p, and (*iii*) the (signed) distance to the brain surface, computed from both I_m (used for brain surface extraction) and the component-tree of I_p.

From these three attributes (generating a parameter space $\Omega \subset \mathbb{R}^3$), a vascular ground truth image I_g (Fig. 2(a,b)) and its segmentation B_g (Fig. 2(c,d)) have been involved in a learning process based on the computation of the best binary image $\hat{B}^-$ w.r.t. the d^- distance, and the computation of the corresponding set of nodes $\hat{K}^-$ (the choice of d^- / $\hat{B}^-$ is linked to the considered ground truth data I_g for which B_g has been slightly oversegmented by the expert, while the choice of $\hat{K}^-$ is the result of experimental considerations). An (automatic) SVM classi-fication process has then been applied on the set of binary connected components of $\hat{K}^-$ to determine an adequate set $\omega \subset \Omega$ of attribute values. PC-MRA phase images similar to I_g have then been segmented in a multiscale fashion by using the attribute values of ω. It has to be noticed that the segmentation process (and then the learning step) have been performed at several scales ($\alpha = \{1, 8, 64\}$), and with a redundancy factor $\beta = 2$. The results have been obtained by fusing the partial binary images in a binary fashion (Eq. (4) with $\lambda = 1$). An example of these results is illustrated in Fig. 3. It can been observed that, despite the presence of few false negatives, the obtained results globally present no artifacts (*i.e.* no false positive). This constitutes a satisfactory and encouraging property for the - difficult - analysis of such (non contrast-enhanced) data where vessels and artifacts present similar intensities and are often connected.

6   Conclusion

A generic framework, based on image partitioning and automatic selection of relevant structural elements from ground-truth data, has been proposed for the development of segmentation methods relying on component-trees. Methods based on this framework can automatically process complex images by use of potentially large sets of knowledge, as illustrated by an application devoted to 3-D angiographic data.

The concept of multiscale (*i.e.* spatial) decomposition has been explored. The decomposition of the image signal will also be considered in further works, leading to multiresolution approaches, permitting to enrich the proposed framework.

From an applicative point of view, a more complete version of the vessel segmentation method obtained from this framework, and introduced in Section 5 for illustrative purpose, will be described and fully validated in further works.

References

1. Hanusse, P., Guillataud, P.: Sémantique des images par analyse dendronique. In: RFIA 1991, vol. 2, pp. 577–588 (1991)
2. Chen, L., Berry, M., Hargrove, W.: Using dendronal signatures for feature extraction and retrieval. International Journal of Imaging Systems and Technology 11(4), 243–253 (2000)
3. Mattes, J., Demongeot, J.: Efficient algorithms to implement the confinement tree. In: Nyström, I., Sanniti di Baja, G., Borgefors, G. (eds.) DGCI 2000. LNCS, vol. 1953, pp. 392–405. Springer, Heidelberg (2000)
4. Salembier, P., Oliveras, A., Garrido, L.: Anti-extensive connected operators for image and sequence processing. IEEE Transactions on Image Processing 7(4), 555–570 (1998)
5. Breen, E.J., Jones, R.: Attribute openings, thinnings, and granulometries. Computer Vision and Image Understanding 64(3), 377–389 (1996)
6. Najman, L., Couprie, M.: Building the component tree in quasi-linear time. IEEE Transactions on Image Processing 15(11), 3531–3539 (2006)
7. Menotti, D., Najman, L., de Albuquerque Araújo, A.: 1D component tree in linear time and space and its application to gray-level image multithresholding. In: ISMM 2007, vol. 1, pp. 437–448. INPE (2007)
8. Urbach, E.R., Boersma, N.J., Wilkinson, M.H.F.: Vector attribute filters. In: ISMM 2005. Computational Imaging and Vision, vol. 30, pp. 95–104. Springer, Heidelberg (2005)

9. Jones, R.: Connected filtering and segmentation using component trees. Computer Vision and Image Understanding 75(3), 215–228 (1999)
10. Urbach, E.R., Wilkinson, M.H.F.: Shape-only granulometries and gray-scale shape filters. In: ISMM 2002, pp. 305–314. CSIRO Publishing (2002)
11. Ouzounis, G.K., Wilkinson, M.H.F.: Mask-based second-generation connectivity and attribute filters. IEEE Transactions on Pattern Analysis and Machine Intelligence 29(6), 990–1004 (2007)
12. Urbach, E.R., Roerdink, J.B.T.M., Wilkinson, M.H.F.: Connected shape-size pattern spectra for rotation and scale-invariant classification of gray-scale images. IEEE Transactions on Pattern Analysis and Machine Intelligence 29(2), 272–285 (2007)
13. Naegel, B., Passat, N., Boch, N., Kocher, M.: Segmentation using vector-attribute filters: methodology and application to dermatological imaging. In: ISMM 2007. INPE, vol. 1, pp. 239–250 (2007)
14. Mosorov, V.: A main stem concept for image matching. Pattern Recognition Letters 26(8), 1105–1117 (2005)
15. Alajlan, N., Kamel, M.S., Freeman, G.H.: Geometry-based image retrieval in binary image databases. IEEE Transactions on Pattern Analysis and Machine Intelligence 30(6), 1003–1013 (2008)
16. Dokládal, P., Bloch, I., Couprie, M., Ruijters, D., Urtasun, R., Garnero, L.: Topologically controlled segmentation of 3D magnetic resonance images of the head by using morphological operators. Pattern Recognition 36(10), 2463–2478 (2003)
17. Wilkinson, M.H.F., Westenberg, M.A.: Shape preserving filament enhancement filtering. In: Niessen, W.J., Viergever, M.A. (eds.) MICCAI 2001. LNCS, vol. 2208, pp. 770–777. Springer, Heidelberg (2001)
18. Vapnik, V.: Statistical Learning Theory. Wiley-Interscience, New York (1998)

Ultrametric Watersheds

Laurent Najman

Université Paris-Est, Laboratoire d'Informatique Gaspard-Monge, Equipe A3SI,
ESIEE Paris, France

Abstract. We study hierachical segmentation in the framework of edge-weighted graphs. We define ultrametric watersheds as topological watersheds null on the minima. We prove that there exists a bijection between the set of ultrametric watersheds and the set of hierarchical edge-segmentations.

Introduction

This paper is a contribution to a theory of hierarchical image segmentation in the framework of edge-weighted graphs. Image segmentation is a process of decomposing an image into regions which are homogeneous according to some criteria. Intuitively, a hierarchical segmentation represents an image at different resolution levels.

In this paper, we introduce a subclass of edge-weighted graphs that we call ultrametric watersheds. Theorem 9 states that there exists a one-to-one correspondance, also called a bijection, between the set of indexed hierarchical edge-segmentations and the set of ultrametric watersheds. In other words, to any hierarchical edge-segmentation (whatever the way the hierarchy is built), it is possible to associate a representation of that hierarchy by an ultrametric watershed. Conversely, from any ultrametric watershed, one can infer a indexed hierarchical edge-segmentation.

Following [1], we can say that, independently of its theoretical interest, such a bijection theorem is useful in practice. Any hierarchical segmentation problem is a priori heterogeneous: assign to an edge-weighted graph an indexed hierarchy. Theorem 9 allows such classification problem to become homogeneous: assign to an edge-weighted graph a particular edge-weighted graph called ultrametric watershed. Thus, Theorem 9 gives a meaning to questions like: which hierarchy is the closest to a given edge-weighted graph with respect to a given measure or distance?

The paper is organised as follow. Related works are examined in section 1. We introduce segmentation on edges in section 2, and in section 3, we adapt the topological watershed framework from the framework of graphs with discrete weights on the nodes to the one of graphs with real-valued weights on the edges. We then define (section 4) hierarchies and ultrametric distances. The last part of the paper (section 5) introduces hierarchical edge-segmentations and ultrametric watersheds, the main result being the existence of a bijection between these two sets (theorem 9).

M.H.F. Wilkinson and J.B.T.M. Roerdink (Eds.): ISMM 2009, LNCS 5720, pp. 181–192, 2009.

Apart from Theorems 2 and 3, and to the best of the author's knowledge, all the properties and theorems formally stated in this paper are new. Proofs of the various properties and theorems will be given in an extended version [2] to be published in a journal paper.

1 Related Works

1.1 Hierarchical Clustering

From its beginning in image processing, hierarchical segmentation is thought of as a particular instance of hierachical classification [3]. One of the fundamental theorems for hierarchical clustering states that there exists a one-to-one correspondance between the set of indexed hierarchical classification and a particular subset of dissimilarity measures called ultrametric distances; This theorem is generally attributed to Johnson [4], Jardine *et al.* [5] and Benzécri [3]. Since then, numerous generalisations of that bijection theorem have been proposed (see [1] for a recent review).

Our main theorem is an extension to hierarchical edge-segmentation of this fundamental hierachical clustering theorem.

1.2 Hierarchical Segmentation

There exist many methods for building a hierachical segmentation [6], which can be divided in three classes: bottom-up, top-down, and split-and-merge. A recent review of some of those approaches can be found in [7]. A useful representation of hierarchical segmentations was introduced in [8] under the name of *saliency map*. This representation has been used (under several names) by several authors, for example for visualisation purposes [9] or for comparing hierarchies [10].

In this paper, we show that any saliency map is an ultrametric watershed, and conversely.

1.3 Watersheds

For bottom-up approaches, a generic way to build a hierarchical segmentation is to start from an initial segmentation and progressively merge regions together [11]. Often, this initial segmentation is obtained through a watershed [12, 8, 13]. See [14] for a recent review of these notions in the context of mathematical morphology.

Among many others [15], topological watershed [16] is an original approach to watersheding that modifies a map (e.g., a grayscale image) while preserving the connectivity of each lower cross-section (see fig. 2). It as been proved [16,17] that this approach is the only one that preserves altitudes of the passes (named connection values in this paper) between regions of the segmentation. Pass altitudes are fundamental for hierarchical schemes [8]. On the other hand, topological watersheds may be thick. A study of the properties of different kinds of graphs with respect to the thinness of watersheds can be found in [18,19]. An interesting framework is that of edge-weighted graphs, where watersheds are naturally thin; furthermore, in that framework, a subclass of topological watersheds satisfies both the drop of water principle and a property of global optimality [20].

In this paper, we translate topological watersheds from the framework of node-weigthed-graphs to the one of edge-weighted graphs, and we identify ultrametric watersheds, a subclass of topological watersheds that is interesting for hierarchical edge-segmentation.

2   Segmentation on Edges

This paper is settled in the framework of edge-weighted graphs. Following the notations of [21], we present some basic definitions to handle such kind of graphs.

We define a *graph* as a pair $X = (V, E)$ where V is a finite set and E is composed of unordered pairs of V, *i.e.*, E is a subset of $\{\{x, y\} \subseteq V \mid x \neq y\}$. We denote by $|V|$ the cardinal of V, *i.e*, the number of elements of V. Each element of V is called a *vertex or a point (of X)*, and each element of E is called an *edge (of X)*. If $V \neq \emptyset$, we say that X is *non-empty*.

As several graphs are considered in this paper, whenever this is necessary, we denote by $V(X)$ and by $E(X)$ the vertex and edge set of a graph X.

A graph X is said *complete* if $E = V(X) \times V(X)$.

Let X be a graph. If $u = \{x, y\}$ is an edge of X, we say that x and y are *adjacent (for X)*. Let $\pi = \langle x_0, \ldots, x_\ell \rangle$ be an ordered sequence of vertices of X, π is *a path from x_0 to x_ℓ in X (or in V)* if for any $i \in [1, \ell]$, x_i is adjacent to x_{i-1}. In this case, we say that x_0 and x_ℓ are *linked for X*. We say that X *is connected* if any two vertices of X are linked for X.

Let X and Y be two graphs. If $V(Y) \subseteq V(X)$ and $E(Y) \subseteq E(X)$, we say that Y *is a subgraph of X* and we write $Y \subseteq X$. We say that Y is a *connected component of X*, or simply a *component of X*, if Y is a connected subgraph of X which is maximal for this property, *i.e.*, for any connected graph Z, $Y \subseteq Z \subseteq X$ implies $Z = Y$.

Let X be a graph, and let $S \subseteq E(X)$. The *graph induced by S* is the graph whose edge set is S and whose vertex set is made of all points which belong to an edge in S, *i.e.*, $(\{x \in V(X) \mid \exists u \in S, x \in u\}, S)$.

Important remark. *Throughout this paper $G = (V, E)$ denotes a connected graph, and the letter V (resp. E) will always refer to the vertex set (resp. the edge set) of G. We will also assume that $E \neq \emptyset$.*

Let $S \subset E$. In the following, when no confusion may occur, the graph induced by S is also denoted by S.

Typically, in applications to image segmentation, V is the set of picture elements (pixels) and E is any of the usual adjacency relations, *e.g.*, the 4- or 8-adjacency in 2D [22].

If $S \subset E$, we denote by $\overline{S}$ the *complementary set of S in E*, *i.e.*, $\overline{S} = E \setminus S$.

A set $C \subset E$ is an *(edge-)cut (of G)* if each edge of C is adjacent to two different nonempty connected components of $\overline{C}$.

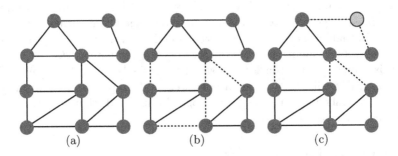

Fig. 1. Illustration of edge-segmentation and edge-cut. (a) A graph X. (b) An edge-segmentation of X; the set of dotted-lines edges is the associated edge-cut of X. (c) A subgraph of X which is not an edge-segmentation of X: the grey point is isolated.

A graph S is called an *(edge-)segmentation (of G)* if $\overline{E(S)}$ is a cut.

Any connected component of a segmentation S is called a *region (of S)*.

The previous definitions of cut and segmentation (illustrated on fig. 1) are not the usual ones. In particular, Prop. 1.(i) below states that there is no isolated point in an edge-segmentation. If we need an isolated point x, it is always possible to replace x with an edge $\{x', y'\}$. Furthemore, isolated points are often noise in an image.

It is interesting to state the definition of a segmentation from the point of view of vertices of the graph. A graph X is said to be *spanning (for V)* if $V(X) = V$. We denote by ϕ the map that associates, to any $X \subset G$, the graph $\phi(X) = \{V(X), \{\{x, y\} \in E | x \in V(X), y \in V(X)\}\}$. We observe that $\phi(X)$ is maximal among all subgraphs of G that are spanning for $V(X)$, it is thus a closing on the lattice of subgraphs of G [23]. We call ϕ the edge-closing.

Property 1. *A graph $S \subseteq G = (V, E)$ is an edge-segmentation of G if and only if*

(i) The graph induced by $E(S)$ is S;
(ii) S is spanning for V;
(iii) for any connected component X of S, $X = \phi(X)$.

3   Topological Watershed

3.1   Edge-Weighted Graphs

We denote by $\mathcal{F}$ the set of all mappings from E to $\mathbb{R}^+$ and we say that any mapping in $\mathcal{F}$ *weights* the edges of G. For any $F \in \mathcal{F}$, the pair (G, F) is called an *edge-weighted graph*. Whenever no confusion can occur, we will denote the edge-weighted graph (G, F) by F.

For applications to image segmentation, we will assume that the altitude of u, an edge between two pixels x and y, represents the dissimilarity between x and y (e.g., $F(u)$ equals the absolute difference of intensity between x and y; see [24] for a more complete discussion on different ways to set the mapping F for image segmentation). Thus, we suppose that the salient contours are located on the highest edges of (G, F).

Let $\lambda \in \mathbb{R}^+$ and $F \in \mathcal{F}$, we define $F[\lambda] = \{v \in E \mid F(v) \leq \lambda\}$. The graph (induced by) $F[\lambda]$ is called a *(cross)-section* of F. A connected component of a section $F[\lambda]$ is called a *(level λ) component* of F.

We define $\mathcal{C}(F)$ as the set composed of all the pairs $[\lambda, C]$, where $\lambda \in \mathbb{R}^+$ and C is a component of the graph $F[\lambda]$. We call *altitude of* $[\lambda, C]$ the number λ. We note that one can reconstruct F from $\mathcal{C}(F)$; more precisely, we have:

$$F(v) = \min\{\lambda \mid [\lambda, C] \in \mathcal{C}(F), v \in E(C)\}$$

For any component C of F, we set $h(C) = \min\{\lambda \mid [\lambda, C] \in \mathcal{C}(F)\}$. We define $\mathcal{C}^\star(F)$ as the set composed by all $[h(C), C]$ where C is a component of F. The set $\mathcal{C}^\star(F)$, called the *component tree* of F [25, 26], is a finite subset of $\mathcal{C}(F)$ that is widely used in practice for image filtering.

A *(regional) minimum of* F is a component X of the graph $F[\lambda]$ such that for all $\lambda_1 < \lambda$, $F[\lambda_1] \cap E(X) = \emptyset$. We remark that a minimum of F is a subgraph of G and not a subset of the points of G; we also remark that any minimum X of F is such that $|V(X)| > 1$.

We denote by $\mathcal{M}(F)$ the graph whose vertex set and edge set are, respectively, the union of the vertex sets and edge sets of all minima of F.

3.2   Topological Watersheds on Edge-Weighted Graphs

Let X be a subgraph of G. An edge $u \in \overline{E(X)}$ is said to be *W-simple (for X)* (see [16]) if X has the same number of connected components as $X + u = (V(X) \cup u, E(X) \cup \{u\})$. An edge u such that $F(u) = \lambda$ is said to be *W-destructible (for F) with lowest value* λ_0 if there exists λ_0 such that, for all λ_1, $\lambda_0 < \lambda_1 \leq \lambda$, u is W-simple for $F[\lambda_1]$ and if u is not W-simple for $F[\lambda_0]$.

A *topological watershed (on G)* is a mapping that contains no W-destructible edges.

A mapping F' is a *topological thinning (of F)* if:

– $F' = F$, or if
– there exists a mapping F'' which is a topological thinning of F and there exists an edge u W-destructible for F'' with lowest value λ such that $\forall v \neq u, F'(v) = F''(v)$ and $F'(v) = \lambda_0$, with $\lambda \leq \lambda_0 < F''(v)$.

An illustration of a topological watershed can be found in fig. 2.

The *connection value* between $x \in V$ and $y \in V$ is the number

$$F(x, y) = \min\{\lambda \mid x \in V(C), y \in V(C), [\lambda, C] \in \mathcal{C}(F)\} \tag{1}$$

In other words, $F(x, y)$ is the altitude of the lowest element $[\lambda, C]$ of $\mathcal{C}(F)$ such that x and y belong to C (rule of the least common ancestor).

Two points x and y are *separated (for F)* if $F(x, y) > \max\{\lambda_1, \lambda_2\}$, where λ_1 (resp. λ_2) is the altitude of the lowest element $[\lambda_1, c_1]$ (resp. $[\lambda_2, c_2]$) of $\mathcal{C}(F)$

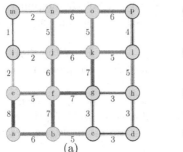
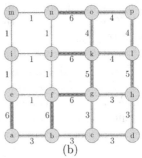

Fig. 2. Illustration of topological watershed. (a) An edge-weighted graph F. (b) A topological watershed of F. The minima of (a) are $(\{m, i\}), (\{p, l\})$, $(\{g, h\}, \{c, d\}, \{g, c\}, \{h, d\})$ and are in bold in (a).

such that $x \in c_1$ (resp. $y \in c_2$). The points x and y are λ-*separated (for F)* if they are separated and $\lambda = F(x, y)$.

The mapping F' is a *separation* of F if, whenever two points are λ-separated for F, they are λ-separated for F'.

If X and Y are two subgraphs of G, we set $F(X, Y) = \min\{F(x, y) \mid x \in X, y \in Y\}$.

Theorem 2 (Restriction to minima [16]). *Let $F' \leq F$ be two elements of $\mathcal{F}$. The mapping F' is a separation of F if and only if, for all distinct minima X and Y of $\mathcal{M}(F)$, we have $F'(X, Y) = F(X, Y)$.*

A graph X is *flat (for F)* if for all $u, v \in E(X)$, $F(u) = F(v)$. If X is flat, the *altitude* of X is the number $F(X)$ such that $F(X) = F(v)$ for any $v \in E(X)$.

We say that F' is a *strong separation* of F if F' is a separation of F and if, for each $X' \in \mathcal{M}(F')$, there exists $X \in \mathcal{M}(F)$ such that $X \subseteq X'$ and $F(X) = F(X')$.

Theorem 3 (strong separation [16]). *Let F and F' in $\mathcal{F}$ with $F' \leq F$. Then F' is a topological thinning of F if and only if F' is a strong separation of F.*

In other words, topological thinnings are the only way to obtain a watershed that preserves connection values.

In the framework of edge-weighted graphs, topological watersheds allows for a simple characterization.

Theorem 4. *A mapping F is a topological watershed if and only if:*

(i) $\mathcal{M}(F)$ is a segmentation of G;
(ii) for any edge $v = \{x, y\}$, if there exist X and Y in $\mathcal{M}(F)$, $X \neq Y$, such that $x \in V(X)$ and $y \in V(Y)$, then $F(v) = F(X, Y)$.

Note that if F is a topological watershed, then for any edge $v = \{x, y\}$ such that there exists $X \in \mathcal{M}(F)$ with $x \in V(X)$ and $y \in V(X)$, we have $F(v) = F(X)$.

4   Hierarchies and Ultrametric Distances

Let Ω be a finite set. A *hierarchy* H on Ω is a set of parts of Ω such that

(i) $\Omega \in H$
(ii) for every $\omega \in \Omega$, $\{w\} \in H$
(iii) for each pair $(h, h') \in H^2$, $h \cap h' \neq \emptyset \implies h \subset h'$ or $h' \subset h$.

An *indexed hierarchy* on Ω is a pair (H, μ), where H denotes a given hierarchy on Ω and μ is a positive function, defined on H and satisfying the following conditions:

(i) $\mu(h) = 0$ if and only if h is reduced to a singleton of Ω;
(ii) if $h \subset h'$, then $\mu(h) < \mu(h')$.

A distance d, in general, obeys the triangular inequality $d(\omega_1, \omega_2) \leq d(\omega_1, \omega_3) + d(\omega_3, \omega_2)$ where ω_1, ω_2 and ω_3 are any three points of the space. An *ultrametric distance (on Ω)* is a function d from $\Omega \times \Omega$ to $\mathbb{R}^+$ such that $d(\omega_1, \omega_2) = 0$ if and only if $\omega_1 = \omega_2$, such that $d(\omega_1, \omega_2) = d(\omega_2, \omega_1)$ and such that d obeys ultrametric inequality $d(\omega_1, \omega_2) \leq \max(d(\omega_1, \omega_3), d(\omega_2, \omega_3))$ for all $\omega_1, \omega_2, \omega_3$. The ultrametric inequality [27] is stronger than the triangular inequality.

A *partition* of Ω is a collection (Ω_i) of non-empty subsets of Ω such that any element of Ω is exactly in one of these subsets. Note that any given partition of the set Ω induces a large number of trivial ultrametric distances: $d(\omega_1, \omega_1) = 0, d(\omega_1, \omega_2) = 1$ if $\omega_1 \in \Omega_i$, $\omega_2 \in \Omega_j$, $i \neq j$, and $d(\omega_1, \omega_2) = a$ if $i = j$, $0 < a < 1$. The general connection between indexed hierarchies and ultrametric distances was proved by Benzécri [3] and Johnson [4]. This result states that there is a one-to-one correspondance between indexed hierarchies and ultrametric distances both defined on the same set. Indeed, associated with each indexed hierarchy (H, μ) on Ω is the following ultrametric distance:

$$d(\omega_1, \omega_2) = \min\{\mu(h) \mid \omega_1 \in h, \omega_2 \in h, h \in H\}. \tag{2}$$

In other words, the distance $d(\omega_1, \omega_2)$ between two elements ω_1 and ω_2 in Ω is given by the smallest element in H which contains both ω_1 and ω_2. Conversely, each ultrametric distance d is associated with one and only one indexed hierarchy.

Observe the similarity between eq. 2 and eq. 1. Indeed, connection value is an ultrametric distance on V whenever $F > 0$. More precisely, we have the following property.

Property 5. *Let $F \in \mathcal{F}$. Then $F(X, Y)$ is an ultrametric distance on $\mathcal{M}(F)$. If furthemore, $F > 0$, then $F(x, y)$ is an ultrametric distance on V.*

Let Ψ be the mapping on $\mathcal{F}$ such that for any $F \in \mathcal{F}$ the map $\Psi(F)$ and for any edge $\{x, y\} \in E$, $\Psi(F)(\{x, y\}) = F(x, y)$. It is straightforward to see that $\Psi(F) \leq F$, that $\Psi(\Psi(F)) = \Psi(F)$ and that if $F' \leq F$, $\Psi(F') \leq \Psi(F)$. Thus Ψ is an opening on the lattice $(\mathcal{F}, \leq)$ [28]. We remark that the subset of strictly positive maps that are defined on the complete graph $(V, V \times V)$ and that are open with respect to Ψ is the set of ultrametric distances on V. The mapping Ψ

is known under several names, in particular the one of subdominant ultrametric and the one of ultrametric opening. It is well known that Ψ is associated to the simplest method for hierarchical classification called single linkage clustering [5, 29], closely related to Kruskal's algorithm [30] for computing a minimum spanning tree.

Thanks to Th. 4, we observe that if F is a topological watershed, then $\Psi(F) = F$. However, an ultrametric distance d may have plateaus, and thus the weighted complete graph $(V, V \times V, d)$ is not always a topological watershed. Nevertheless, those results underline that topological watersheds are related to hierarchical classification, but not yet to hierarchical edge-segmentation; the study of such relations is the subject of the rest of the paper.

5 Hierarchical Edge-Segmentations, Saliency and Ultrametric Watersheds

Informally, a hierarchical segmentation is a hierarchy made of connected regions. However, in our framework, a segmentation is not a partition, and as the union of two disjoint connected subgraphs of G is not a connected subgraph of G, the formal definition is slightly more involved. A *hierarchical (edge-)segmentation (on G)* is an indexed hierarchy (H, μ) on the set of regions of a segmentation S of G, such that for any $h \in H$, $\phi(\cup_{X \in h} X)$ is connected (ϕ being the edge-closing defined in section 2).

For any $\lambda \geq 0$, we denote by $H[\lambda]$ the graph induced by $\{\phi(\cup_{X \in h} X) | h \in H, \mu(h) \leq \lambda\}$. The following property is an easy consequence of the definition of a hierarchical segmentation.

Property 6. *Let (H, μ) be a hierarchical segmentation. Then for any $\lambda \geq 0$, the graph $H[\lambda]$ is a segmentation of G.*

Property 5 implies that the connection value defines a hierarchy on the set of minima of F. If F is a topological watershed, then by Th. 4, $\mathcal{M}(F)$ is a segmentation of G, and thus from any topological watershed, one can infer a hierachical segmentation. However, $F[\lambda]$ is not always a segmentation: if there exists a minimum X of F such that $F(X) = \lambda_0 > 0$, for any $\lambda_1 < \lambda_0$, $F[\lambda_1]$ contains at least two connected components X_1 and X_2 such that $|V(X_1)| = |V(X_2)| = 1$. Note that the value of F on the minima of F is not related to the position of the divide nor to the associated hierarchy of minima/segmentations. This leads us to introduce the following definition.

A map $F \in \mathcal{F}$ is an *ultrametric watershed* if F is a topological watershed, and if furthemore, for any $X \in \mathcal{M}(F)$, $F(X) = 0$.

Property 7. *A map F is an ultrametric watershed if and only if for all $\lambda \geq 0$, $F[\lambda]$ is a segmentation of G.*

This property is illustrated in fig. 3.

By definition of a hierarchy, two elements of H are either disjoint or nested. If furthermore (H, μ) is a hierarchical segmentation, the graphs $E(H[\lambda])$ can be

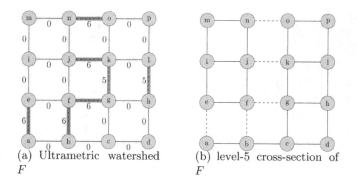

(a) Ultrametric watershed F

(b) level-5 cross-section of F

Fig. 3. An example of an ultrametric watershed F and a cross-section of F

stacked to form a map. We call *saliency map* [8] the result of such a stacking, *i.e.* a saliency map is a map F such that there exists (H, μ) a hierarchical segmentation with $F(v) = \min\{\lambda | v \in E(H[\lambda])\}$.

Property 8. *A map F is a saliency map if and only if F is an ultrametric watershed.*

A corrolary of property 8 states the equivalence between hierachical segmentations and ultrametric watersheds. The following theorem is the main result of this paper.

Theorem 9. *There exists a bijection between the set of hierachical edge-segmentations on G and the set of ultrametric watersheds on G.*

As there exists a one-to-one correspondance between the set of indexed hierarchies and the set of ultrametric distances, it is interesting to search if there exists a similar property for the set of hierarchical segmentations. Let d be the ultrametric distance associated to a hierarchical segmentation (H, μ). We call *ultrametric contour map (associated to (H, μ))* the map d_E such that:

1. for any edge $v \in E(H[0])$, then $d_E(v) = 0$;
2. for any edge $v = \{x, y\} \in \overline{E(H[0])}$, $d_E(v) = d(X, Y)$ where X (resp. Y) is the connected component of $H[0]$ that contains x (resp. y).

Property 10. *A map F is an ultrametric watershed if and only if F is the ultrametric contour map associated to a hierarchical segmentation.*

6   Conclusion

Fig. 4 is an illustration of the application of the framework developped in this paper to a classical hierarchical segmentation scheme based on attribute opening [8,25,14]. Fig. 5 shows some of the differences between applying such scheme

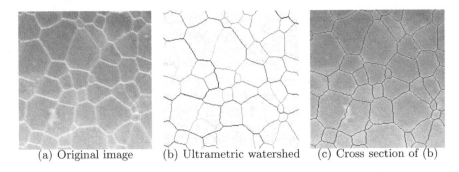

(a) Original image   (b) Ultrametric watershed   (c) Cross section of (b)

Fig. 4. Example of ultrametric watershed

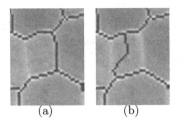

(a)   (b)

Fig. 5. Zoom on a comparison between two watersheds of a filtered version of the image 4.a. Morphological filtering tends to create large plateaus, and both watersheds (a) and (b) are possible, but only (a) is a subset of a watershed of 4.a. No hierarchical scheme will ever give a result as (b).

and applying a classical morphological segmentation scheme, *e.g.* attribute opening followed by a watershed [12]. As watershed algorithms generally place watershed lines in the middle of plateaus, the two schemes give quite different results.

It is important to note that most of the algorithms proposed in the litterature to compute saliency maps are not correct, often because they rely on wrong connection values or because they rely on thick watersheds where merging regions is difficult. Future papers will propose novel algorithms (based on the topological watershed algorithm [31]) to compute ultrametric watersheds, with proof of correctness.

On a more theoretical level, this work can be pursued in several directions.

– We will study lattices of watersheds [32] and will bring to that framework recent approaches like scale-sets [9] and other metric approaches to segmentation [10]. For example, scale-sets theory considers a rather general formulation of the partitioning problem which involves minimizing a two-term-based energy, of the form $\lambda C + D$, where D is a goodness-of-fit term and C is a regularization term, and proposes an algorithm to compute the hierarchical segmentation we obtain by varying the λ parameter. We can hope that

the topological watershed algorithm [31] can be used on a specific energy function to directly obtain the hierarchy.
- Subdominant theory (mentionned at the end of section 4) links hierachical classification and optimisation. In particular, the subdominant ultrametric d' of a dissimilarity d is the solution to the following optimisation problem for $p < \infty$:

$$\min\{\|d - d'\|_p^p \mid d' \text{ is an ultrametric distance and } d' \leq d\}$$

It is certainly of interest to search if topological watersheds can be solutions of similar optimisation problems.
- Several generalisations of hierarchical clustering have been proposed in the literature [1]. An interesting direction of research is to see how to extend in the same way the topological watershed approach, for example for allowing regions to overlap.

Acknowledgement. This work was partially supported by ANR grant SURF-NT05-2_45825

References

1. Barthélemy, J.P., Brucker, F., Osswald, C.: Combinatorial optimization and hierarchical classifications. 4OR: A Quarterly Journal of Operations Research 2(3), 179–219 (2004)
2. Najman, L.: Ultrametric watersheds: a bijection theorem for hierarchical edge-segmentation. Technical Report IGM 2009-07, Université Paris-Est, Institut Gaspard Monge (2009)
3. Benzécri, J.: L'Analyse des données: la Taxinomie, vol. 1. Dunod (1973)
4. Johnson, S.: Hierarchical clustering schemes. Psychometrika 32, 241–254 (1967)
5. Jardine, N., Sibson, R.: Mathematical taxonomy. Wiley, Chichester (1971)
6. Pavlidis, T.: Hierarchies in structural pattern recognition. Proceedings of the IEEE 67(5), 737–744 (1979)
7. Soille, P.: Constrained connectivity for hierarchical image decomposition and simplification. IEEE Trans. Pattern Anal. Mach. Intell. 30(7), 1132–1145 (2008)
8. Najman, L., Schmitt, M.: Geodesic saliency of watershed contours and hierarchical segmentation. IEEE Trans. Pattern Anal. Mach. Intell. 18(12), 1163–1173 (1996)
9. Guigues, L., Cocquerez, J.P., Men, H.L.: Scale-sets image analysis. International Journal of Computer Vision 68(3), 289–317 (2006)
10. Arbeláez, P.A., Cohen, L.D.: A metric approach to vector-valued image segmentation. International Journal of Computer Vision 69(1), 119–126 (2006)
11. Pavlidis, T.: Structural Pattern Recognition. Springer Series in Electrophysics, vol. 1, pp. 90–123. Springer, Heidelberg (1977); segmentation techniques, ch. 4–5
12. Meyer, F., Beucher, S.: Morphological segmentation. Journal of Visual Communication and Image Representation 1(1), 21–46 (1990)
13. Meyer, F.: Morphological segmentation revisited. In: Space, Structure and Randomness, pp. 315–347. Springer, Heidelberg (2005)

14. Meyer, F., Najman, L.: Segmentation, arbre de poids minimum et hiérarchies. In: Najman, L., Talbot, H. (eds.) Morphologie mathématique 1: approches déterministes, pp. 201–233. Lavoisier, Paris (2008)
15. Roerdink, J.B.T.M., Meijster, A.: The watershed transform: Definitions, algorithms and parallelization strategies. Fundamenta Informaticae 41(1-2), 187–228 (2001)
16. Bertrand, G.: On topological watersheds. J. Math. Imaging Vis. 22(2-3), 217–230 (2005)
17. Najman, L., Couprie, M., Bertrand, G.: Watersheds, mosaics and the emergence paradigm. Discrete Appl. Math. 147(2-3), 301–324 (2005)
18. Cousty, J., Bertrand, G., Couprie, M., Najman, L.: Fusion graphs: merging properties and watersheds. J. Math. Imaging Vis. 30(1), 87–104 (2008)
19. Cousty, J., Najman, L., Bertrand, G., Couprie, M.: Weighted fusion graphs: merging properties and watersheds. Discrete Appl. Math. 156(15), 3011–3027 (2008)
20. Cousty, J., Bertrand, G., Najman, L., Couprie, M.: Watershed cuts: minimum spanning forests, and the drop of water principle. IEEE Trans. Pattern Anal. Mach. Intell. (to appear, 2009)
21. Diestel, R.: Graph Theory. Graduate Texts in Mathematics. Springer, Heidelberg (1997)
22. Kong, T., Rosenfeld, A.: Digital topology: Introduction and survey. Comput. Vision Graph. Image Process. 48(3), 357–393 (1989)
23. Cousty, J., Najman, L., Serra, J.: Some morphological operators in graph spaces. In: Wilkinson, M.H.F., Roerdink, J.B.T.M. (eds.) ISMM 2009. LNCS, vol. 5720, pp. 149–160. Springer, Heidelberg (2009)
24. Cousty, J., Bertrand, G., Najman, L., Couprie, M.: Watershed cuts: thinnings, shortest-path forests and topological watersheds. IEEE Trans. Pattern Anal. Mach. Intell. (to appear, 2009)
25. Salembier, P., Oliveras, A., Garrido, L.: Anti-extensive connected operators for image and sequence processing. IEEE Trans. Image Proc. 7(4), 555–570 (1998)
26. Najman, L., Couprie, M.: Building the component tree in quasi-linear time. IEEE Trans. Image Proc. 15(11), 3531–3539 (2006)
27. Krasner, M.: Espaces ultramétrique. C.R. Acad. Sci. Paris 219, 433–435 (1944)
28. Leclerc, B.: Description combinatoire des ultramétriques. Mathématique et sciences humaines 73, 5–37 (1981)
29. Gower, J., Ross, G.: Minimum spanning tree and single linkage cluster analysis. Appl. Stats. 18, 54–64 (1969)
30. Kruskal, J.B.: On the shortest spanning subtree of a graph and the traveling salesman problem. Proc. Am. Mathemat. Soc. 7, 48–50 (1956)
31. Couprie, M., Najman, L., Bertrand, G.: Quasi-linear algorithms for the topological watershed. J. Math. Imaging Vis. 22(2-3), 231–249 (2005)
32. Cousty, J., Najman, L., Serra, J.: Raising in watershed lattices. In: 15th IEEE ICIP 2008, San Diego, USA, October 2008, pp. 2196–2199 (2008)

An Efficient Morphological Active Surface Model for Volumetric Image Segmentation

Andrei C. Jalba[1] and Jos B.T.M. Roerdink[2]

[1] Institute for Mathematics and Computing Science
Eindhoven University of Technology
P.O. Box 513, 5600 MB Eindhoven, The Netherlands
[2] Institute for Mathematics and Computing Science
University of Groningen
P.O. Box 407, 9700 AK Groningen, The Netherlands
{a.c.jalba}@tue.nl, {j.b.t.m.roerdink}@rug.nl

Abstract. Using tools from multi-scale morphology, we reformulate a region-based active-contour model using a minimum-variance criterion. Experimental results of 3D data show that our discrete model achieves similar segmentation quality as the continuous model based on the level-set framework, while being two orders of magnitude faster than optimized implementations of the original continuous model.

1 Introduction

Segmenting images using active contour models (snakes) involves evolving a curve or surface (*i.e.*, an interface), subject to constraints derived from a given input image (volume). Active contour models [1, 2, 3] found applications in medical imaging (see [4, 5] for recent surveys), geometric modeling [6], computer animation [7], texture segmentation [8] and object tracking [9, 10]. State-of-the-art active contours (see [4, 5] and references therein) are based on the level set framework [11], which enables active contours to handle complicated topologies of the underlying shapes, unlike most parametric snakes [2, 7].

Let $C(s) : [0, 1] \rightarrow \mathbb{R}^2$ be a parametrized 2D curve. Within the level set framework, C is represented implicitly via a higher-dimensional Lipschitz function ϕ, *e.g.*, $C = \{(x, y) \mid \phi(x, y) = 0\}$. The evolving curve is given by the zero level-set at time t of function $\phi(x, y, t)$. Further, evolving C in normal direction with speed F can be done solving

$$\frac{\partial \phi}{\partial t} = F \|\nabla \phi\|, \tag{1}$$

with initial condition $\phi(x, y, 0) = \phi^0(x, y)$, where $\phi^0(x, y)$ is the initial embedding (*i.e.*, the signed distance function) of C. Note that for surfaces, ϕ is defined on a subset of $\mathbb{R}^3$.

In the context of image segmentation, various formulations for the speed function F have been proposed. Traditionally F is set to some function of the gradient image [12, 13], such that the active contour stops its evolution whenever important edges in the input image are encountered. However, as shown by Siddiqi *et al.* [14], these formulations are affected by *boundary leakage* in the vicinity of blurred edges. An effective

M.H.F. Wilkinson and J.B.T.M. Roerdink (Eds.): ISMM 2009, LNCS 5720, pp. 193–204, 2009.

solution specific to gradient-based active contours is to incorporate *(global) region descriptors* into the energy functional [15,16,17]. For example, Chan and Vese [17] formulated their "active contours without edges" as a simplified version of the Mumford-Shah piecewise constant model [18], with a limited number of regions. Their energy functional includes a minimum-variance criterion of the segmented regions, see section 2. The resulting model is much more robust to noise and boundary leakage than traditional gradient-based snake models [17].

Apart from large computational requirements, some inherent problems have to be tackled by most formulations based on level sets. Most notably, the level set function ϕ, initialized to the signed distance function from the interface, has to be maintained (Lipschitz) smooth during the evolution, and thus it has to be periodically *reinitialized*. Moreover, if the speed F is only known at the locations of the zero level set, F has to be extended to all other level sets of ϕ (the so-called *velocity extension* problem).

In this paper we reformulate Chan and Vese's active contour model within the framework of mathematical morphology, relying solely on binary multi-scale morphological operators [19,20]. We show that the resulting algorithm is effective and more efficient compared to its level set counterpart.

2   Model Description

2.1   Minimum Variance Model

Let Ω be a bounded open subset of $\mathbb{R}^2$, with $\partial\Omega$ denoting its boundary. Let $I : \Omega \to \mathbb{R}$ be a given image and assume that the evolving curve C is the boundary of an open subset ω of Ω, i.e., $\omega \in \Omega$ and $C = \partial\omega$. Using the calculus of variations it can be shown that in order to minimize an energy functional

$$E_1(C) = \int_\omega f(x,y)\,dxdy, \tag{2}$$

each point of C should move under the influence of the force $\mathbf{F_1} = -f\mathbf{N}_C$, where $\mathbf{N}_C$ is the outward normal of C. Let $inside(C)$ denote the region ω and $outside(C)$ denote the region $\omega^c = \Omega \setminus \omega$. Using the result above, the optimal way to decrease

$$E_2(C) = \int_{inside(C)} f_o(x,y)\,dxdy + \int_{outside(C)} f_b(x,y)\,dxdy \tag{3}$$

is to evolve C under the force $\mathbf{F_2} = (f_b - f_o)\mathbf{N}_C$.

The active contour model of Chan and Vese minimizes the functional $E(c_1,c_2,C)$ derived from E_2 by setting $f_o = |I(x,y) - c_1|^2$ and $f_b = |I(x,y) - c_2|^2$, with c_1, c_2 some constants depending on C, and some additional regularizing terms, *i.e.*,

$$E(c_1,c_2,C) = \mu \cdot Length(C) + v \cdot Area(inside(C))$$
$$+ \lambda \int_{inside(C)} |I(x,y) - c_1|^2\,dxdy + \int_{outside(C)} |I(x,y) - c_2|^2\,dxdy, \tag{4}$$

with $\mu \geq 0$, $v \geq 0$, $\lambda > 0$ fixed parameters. Thus, the resulting force pushes C inwards when $f_b - f_o < 0$ and outwards if $f_b - f_o > 0$. In other words, the curve shrinks if the variance inside ω is larger than that of ω^c and expands otherwise.

Using the level set formalism, the steepest-descent method leads to

$$\frac{\partial \phi}{\partial t} = \left\{ \mu \cdot \kappa - v + \lambda \left[(I - c_2)^2 - (I - c_1)^2 \right] \right\} ||\nabla \phi||, \tag{5}$$

which has to be solved for ϕ, with κ the level set curvature and $c_1 = \text{average}(I)$ in $\{\phi \geq 0\}$ and $c_2 = \text{average}(I)$ in $\{\phi < 0\}$. The first term in Eq. (5) represents the curvature flow and minimizes the length of the curve, the second term represents inwards motion at constant speed and minimizes the area of the region, whereas the last term represents region competition by the minimum-variance criterion.

2.2   The Minimum-Variance Model and Discrete Multi-scale Set Morphology

We now interpret each term of Chan and Vese's minimum-variance model within the context of discrete *multi-scale morphology* [19, 20]. Thus, let us define k to be the discrete version of the continuous scale parameter t, and ε the scale step. Therefore, the link between discrete scale k and t is $t = \varepsilon k$, neglecting rounding errors.

Motion at constant speed. In this case, the speed F in Eq. (1) becomes $F_c = -v$, with $v \geq 0$. Using a forward (first-order) finite-element discretization in time, the update rule for ϕ becomes

$$\phi^{k+1} = \phi^k - \Delta t \, v \, ||\nabla \phi||. \tag{6}$$

Since ϕ should at any time be an approximate, signed distance transform, *i.e.*, $||\nabla \phi|| = 1$, the update rule above means that the graph of ϕ is translated at each iteration by $(\Delta t \, v)$ along the negative z-axis (*i.e.*, the direction of the levels). Thus, setting $S^k = inside(C^k) = \{(x, y) : \phi^k \geq 0\}$, after applying the update rule, the set becomes $S^{k+1} = S^k \ominus (\Delta t \, v) B = \{(x, y) : \phi^k \geq (\Delta t \, v)\}$, where $\ominus$ denotes set erosion and B is the unit ball induced by the Euclidean norm $||\cdot||$.

In the general case, let $B_p = \{(x, y) : ||(x, y)||_p \leq 1\}$ denote unit balls corresponding to some l_p norm $||\cdot||_p$. Let h be the spatial (grid) step size. Setting $\varepsilon = (\Delta t \, v) \leq h/4$ as required for the Courant-Friedrichs-Lewy (CFL) condition, the *multi-scale set erosion* of S^0 becomes

$$S^{\varepsilon k} = S^0 \ominus (\varepsilon k) B_p = \{(x, y) : \phi_p^0(x, y) \geq \varepsilon k\}, \tag{7}$$

with ϕ_p the distance transform with respect to the metric induced by the norm $||\cdot||_p$. Without loss of generality, let us assume that $\varepsilon = 1$, so that the scale parameter t becomes $t = k \in \mathbb{N}$. Then, for any $k \in \mathbb{N}$, an *approximate, weak solution* for the PDE describing inwards curve motion at constant speed is given by $S^k = S^0 \ominus k B_p$. Employing decomposition of the structuring element and the iteration property of set erosions, one obtains

$$S^k = S^0 \ominus B_p^k = S^0 \ominus (\overbrace{B_p \oplus B_p \oplus \cdots B_p}^{k \ times}) = ((\ldots (S^0 \ominus B_p) \ominus B_p) \ldots) \ominus B_p, \tag{8}$$

where B_p^k denotes k-fold dilation of B_p with itself. Thus, a discrete approach to curve motion at constant speed consists in iteratively eroding (or dilating for outward motion) an input set S^0 embedding the initial curve C, by a set B_p approximating the unit ball.

Region competition. The speed F in Eq. (1) is now $F_r = \lambda \left[(I - c_2)^2 - (I - c_1)^2 \right]$, with $\lambda > 0$. Following a similar reasoning as in the previous subsection, the iteration becomes

$$S^{k+1} = \left(\left(S^k \oplus B_p \right) \cap T_\varepsilon^k \right) \cup \left(\left(S^k \ominus B_p \right) \cap (T^k)_\varepsilon^c \right) \qquad (9)$$

where $T_\varepsilon^k = \{(x,y) : \varepsilon^k > 0\}$, $(T^k)_\varepsilon^c = \Omega \setminus T_\varepsilon^k$, $\varepsilon^k = \Delta t F_r^k$ and initial condition $S^0 = inside(C^0) = \{(x,y) : \phi^0 \geq 0\}$. The resulting set S^{k+1} contains points (x,y) which are either the result of a set dilation of S_k if $F_r^k(x,y) = \lambda \left[(I(x,y) - c_2^k)^2 - (I(x,y) - c_1^k)^2 \right] > 0$ or of an erosion of S_k otherwise.

Given the CFL condition and without considering the large computational overhead introduced by upwind discretizations [11] usually employed to approximate $||\nabla \phi||$ in Eq. (6) when F may change signs, computing the solution using this approach needs at least four times less iterations than the PDE-based approach (assuming $h = 1$, then $\varepsilon \leq 0.25$, so that more than four iterations are needed for $||\nabla \phi|| = 1$). This is at the same grid resolution and assuming a 3×3 discretization stencil for $||\nabla \phi||$.

Curvature flow. The Euclidean shortening flow (curvature flow) is obtained setting $F_\kappa = \mu \kappa$ in Eq. (1), where κ is the (mean) curvature.

It can be proved that iterating k times a (weighted) median filter using a window of size ε converges when $\varepsilon \to 0$, $k \to \infty$, $\varepsilon k \to t$ to the (mean) curvature flow, see for example [21, 22, 23]. Moreover it was shown [24] that an *opening-closing* filter by B_p smooths a (binary) signal similar to a median filter of (roughly) size $2 \cdot |B_p|$, where $|B_p|$ is the number of elements of B_p. Thus, an approximation of the curvature flow iteration can be obtained setting

$$S^{k+1} = MED(S^k) \approx (S^k \circ B_p) \bullet B_p, \qquad (10)$$

where $MED(\cdot)$ denotes the median, '$\circ$' denotes set opening and '$\bullet$' set closing. By the CFL condition the PDE-based approach requires for stability reasons $\varepsilon = \Delta t \mu |\kappa| \leq 0.25$. Assuming $h = 1$, $|\kappa| \leq 1$ and $\mu = 1$, it follows that the PDE-based method needs at least eight times more iterations than the discrete method based on iterated opening-closing filters, neglecting the extra computational overhead of curvature and upwind computations. However, since one opening-closing filter application requires four iterations, overall we expect that our discrete method requires at least two times less iterations than the continuous method.

2.3 Proposed Discrete Model

An obvious approach to formulate the PDE in Eq. (5), as in section 2.2, using multi-scale set morphology, would be to start from an initial state $S^0 = \{(x,y) : \phi^0 \geq 0\}$ and march through the solution S^k as follows. Assume that S^k has been solved at (discrete) time k and that we wish to compute S^{k+1}. We can solve Eq. (5) at the next iteration in three steps: start from solution $W_0 = S^k$ of the previous time step and sequentially resolve each term on the right hand size of Eq. (5) using the appropriate speed functions F, as follows:

$$W_0 \xrightarrow{F_r} W_1 \xrightarrow{F_c} W_2 \xrightarrow{F_\kappa} W_3. \qquad (11)$$

Then, the solution at $(k+1)$ is given by the last set, $S^{k+1} = W_3$. In this way, the solution using the approach in section 2.2 would be expected to require at least two times less iterations than the PDE-based approach. Of course, if a smoother solution is desired, some of the (smoothing) steps can be repeated, at the expense of increased CPU time.

Instead of pursuing this solution process, for efficiency reasons we define our approximate minimum-variance model as follows. Similar to the level set method, let $u : \Omega \to \mathbb{R}$ be the binary function

$$u(x,y) = \chi_\omega(x,y) - \chi_{\omega^c}(x,y) = \begin{cases} 1, & \text{if } (x,y) \in \omega = inside(C) \\ -1, & \text{if } (x,y) \in \omega^c = outside(C) \end{cases}, \quad (12)$$

with χ_ω the characteristic function of ω. Assume that C is embedded as the zero level set of u, i.e., $C = \{(x,y) : u(x,y) = 0\}$. If C deforms, the curve is given by the zero level-set at (discrete) time k of function u^k. Note that although zero is not a regular value of u, we can always use linear interpolation to reconstruct C within the given grid resolution, as ω and ω^c form a partition of Ω. The discrete update rule for u, describing our approximate model is

$$u^{k+1} = \text{sgn}\left(u^k * \chi_{B_p} + \text{sgn}(f^k)\left(|B_p| - 1\right)\right), \quad (13)$$

with $\text{sgn}(x) = 1$ if $x > 0$, $= -1$ otherwise, $|B_p|$ the number of elements of B_p, and '*' denoting linear convolution. The 'speed function' f^k is given by

$$f^k = \lambda\left((I - c_2^k)^2 - (I - c_1^k)^2\right) + \alpha \cdot \text{sgn}(u^k * \chi_{B_p} + \beta), \quad (14)$$

where $\lambda \geq 0$, $\alpha \in \mathbb{R}$ and $\beta \in \mathbb{Z}$.

Note that according to (13), given the definition (12), and by the duality relation of set dilations and erosions, we evolve C by thresholding the outcome of the linear convolution (similar to [25]) of the characteristic function of its interior region and that of B_p. Accordingly, curve C locally expands or shrinks as the sign of f changes. The speed function f (not limited to binary values) represents the competition of two terms: region homogeneity (minimum variance) and smoothing/regularization. Although the update rule maintains u binary, the decision function f prescribing whether to locally shrink or expand C does not suffer from this limitation. However, if one replaces $\text{sgn}(\cdot)$ with a sigmoidal function in (13) and (14), u becomes smoother as it is not restricted anymore to binary values.

Instead of using the iterative solution process of (11), the contributions of the speed terms in (14) are multiplexed in order to obtain a final decision (used in (13)) whether to locally expand or shrink the curve. Thus, using (13) instead of (11) should in principle be more efficient, see section 3. Note also that, similar to the minimum-variance model, our model extends trivially to 3D.

A whole range of smoothing/regularizing operations can be achieved varying the free parameters in the definition of f. For example, assuming that $\lambda = 0$, then setting $\alpha > 0$ and $\beta = 0$, f^k represents the output of a (binary) *median filter* applied to u^k. Thus, iterating (14), curve C is smoothed through u^k by a morphological curvature flow. If we set β to $|\beta| \in (0, |B_p| - 1)$, then other *rank filters* can be obtained. For example, a (small) negative value can be used to mimic the behaviour of the continuous minimum-variance model when $v > 0$, such that the area of ω is minimized. Setting $|\beta| = |B_p| - 1$, local *dilations* or *erosions* can be performed, see subsection 2.2. If during the iteration process, one appropriately alternates the sign of β, set *opening-closing* or *closing-opening* filters can be obtained. Note that since u in (13) is a binary symmetric function, all above morphological operations are very efficiently implemented by linear convolution. Finally, when $\lambda > 0$, a competition takes place between the data term, maintaining region homogeneity, and the smoothing terms.

The advantage of this solution process compared to the one given by (11) is that the 'compute kernel' remains the same, no matter which smoothing method is used, or whether the curve has to expand or shrink during an iteration. This is a key aspect for achieving good efficiency in a parallel implementation of the model, on shared-memory, multi-core architectures, see section 3.

3   Results

Both the PDE-based model and our discrete model were implemented on graphics processing units (GPUs), so as to take advantage of their increased computational power as compared to current CPUs. All experiments were conducted on a PC equipped with a 2.4 GHz processor, 4 GB of RAM and an NVIDIA GTX280 GPU. In order to further increase the efficiency of our model, we used the l_1 norm such that the number of texture lookups is minimal (*i.e.*, B_1 consists of only 5 elements in 2D and 7 in 3D). The PDE-based model (bound to use the l_2 norm), is discretized in 2D using a 3×3 stencil ($3 \times 3 \times 3$ in 3D) and first order upwinding. Unless explicitly mentioned, the simulations were run until *steady state* was reached, using the criterion of [26]. Accordingly, we computed the rate of change of the interface length over a fixed number of iterations. If this number was exceeded and the change in length was small, we stopped the iteration process. *No smoothing/regularization* of the input data was performed, as no gradient computations are required.

For initialization, we used disks of radius 10 in 2D and small spheres of radius 20 in 3D. These were manually placed such that at least parts of the object of interest were inside the disk/sphere; only one initialization was performed per data set.

Behaviour of the discrete model. In the first experiment we performed a brief, comparative study in 2D of the behaviour of both models, see Fig. 1, which shows the segmentation of blood vessels in angiographic data. Initialization (not shown) was done in both cases by placing a small disc inside the body of the main artery. Model parameters were varied such that the regularizing effect (without area minimization, *i.e.*, $v = 0$) of curvature flow and of its discrete counterparts became more and more pronounced. In the first three cases of Fig. 1, our model closely mimics the PDE model, when the influence of the data-dependent term is large compared to the regularization term, and when curvature flow is approximated in our model by iterated (binary) median filtering. If strong, length-minimizing regularization is desired (last column in Fig. 1), alternating opening-closing (binary) filters can be employed to achieve results similar to the PDE-based model. Although the resulting curve in the last example is certainly smoother than that delivered by our model, one should recall that we purposely used a

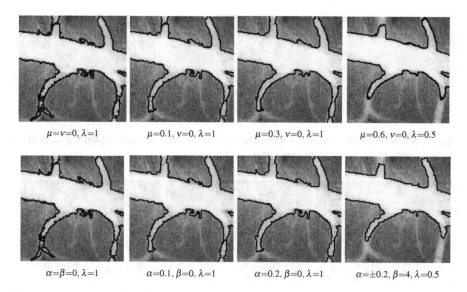

Fig. 1. Model behaviour in 2D, with different parameter settings. *First row*: PDE-based model, *second row*: discrete model.

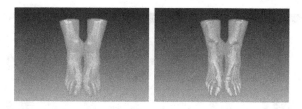

Fig. 2. Segmentation of the bone structure of the human feet in a CT scan. *Left*: volume rendering of input data, *right*: segmentation result.

crude approximation of the continuous disk (as given by the l_1 norm), thus further trading accuracy for speed, see section 3. Note that in all images shown in Fig. 1, a single object is segmented.

3D segmentation results. First, we segmented a (noise free) CT dataset, see Fig. 2. The parameters of the discrete model were set to $\alpha = 0.1$, $\beta = 0$ and $\lambda = 1$. Note that most structures were correctly segmented, although in this case the result is similar to what an iso-surface would look like. However, this is not the case for the noisy data set shown in Fig. 3.

Next, we segmented the white matter of a human brain from an MRI scan, see Fig. 4. The parameters of the model were set to $\alpha = \pm 0.1$, $\beta = 4$ and $\lambda = 1$, and opening-closing filters were employed for regularization.

Although our method is not *multi-phase*, *i.e.*, multiple objects cannot be segmented simultaneously, a sequential approach consisting in successively 'peeling' outside layers can be easily implemented using a binary mask M, see Fig. 5 for an example. While this

Fig. 3. Blood vessel extraction from a noisy data set. *Left*: volume rendering of input data, *center*: typical iso-surface, *right*: segmentation result.

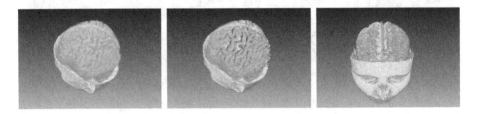

Fig. 4. Segmentation of the white matter of the human brain from an MRI scan. *Left*: volume rendering of input data, *center, right*: two views of the segmentation result.

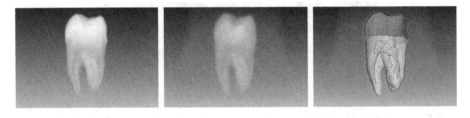

Fig. 5. Segmentation of multiple nested objects. *Left, center*: volume renderings of a tooth data set, *right*: result showing its segmented constituent parts: the enamel, dentin and the root canal.

Fig. 6. Segmentation of a large ($512 \times 380 \times 400$) data set. scan. *Left-to-right*: input volume rendering, result by the (GPU) PDE method, GPU discrete model, and a difference of the two.

approach is not as general as a multi-phase method, it can certainly be used to segment an object made of several, nested parts, as follows. Initially the binary mask M (equal in size to that of the input volume) is set to one, such that all voxels are visited when computing the region speed, F_r. Once a segmentation of the whole object is obtained (*e.g.*, the whole tooth), M is set such that voxels outside the object (set to zero in M) are skipped in further computations of F_r. Then, within the object, one can reinitialize

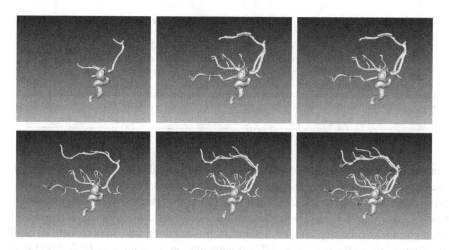

Fig. 7. Results by discrete methods. *First row*: sequential method from (11), *second row*: our final method from (13), see text.

the interface in a region showing the highest contrast compared to its surroundings (*e.g.*, the enamel). After the segmentation is obtained, those voxels of M which belong to the segmented region are set to zero, and the process is repeated for the remaining parts (regions).

We also compared the two discrete approaches based on (11) and (13) for approximating the continuous model from equation Eq. (5), see Fig. 7. The first approach consists in applying sequentially (i) region competition, (ii) area minimization, and (iii) length minimization. We modified the GPU implementation of the proposed model in (13), such that it achieves this behaviour in three iterations per time step. To allow this, the parameters of the model were set as follows:

- **region competition iteration:** $\lambda = 1$, $\alpha = \beta = 0$;
- **area-minimization iteration:** $\lambda = 0$, $\alpha = 1$, $\beta = 2$;
- **length-minimization iteration:** $\lambda = 0$, $\alpha = \pm 0.8 \ldots 0$, $\beta = 5$.

The parameters of the model in (13) were set (similarly) to $\lambda = 1$, $\alpha = \pm 0.8 \ldots 0$ and $\beta = 5$. Fig. 7 shows some of the results obtained using $\alpha = 0.8, 0.5, 0.1$. In both cases the initializations were identical and consisted of a small sphere surrounding the central aneurism. Clearly the sequential model extracts less structures than our final model, for similar parameter settings. Also, the sequential model needs about three times more iterations to converge and the order in which the above iterations are applied has a great impact on the final result. For example, if one reverses the order of iterations, the initial sphere shrinks to a point and then disappears.

Efficiency. A full-grid GPU implementation of our model and three implementations of the PDE-model were considered: full-grid running on the GPU, full-grid on the CPU and an *optimized narrow-band* CPU method [27, 28]. The CPU time required by each method to extract the main vessels from a data set of size $512 \times 380 \times 400$ (see Fig. 6)

Table 1. CPU timings of all considered methods; data-set size: $512 \times 380 \times 400$, see Fig. 6

Method		Iterations/sec. (ips)	Total iterations	Total time (min.)
GPU	Discrete, full-grid	37.2	3010	1.3
	PDE, full-grid	6.1	8210	22.8
CPU	PDE, narrow-band	$0.8 - 1.2$	9250	90.2
	PDE, full-grid	0.05	–	–

using identical initialization (a small sphere) is given in Table 1. The parameters of the discrete model were set as in the last 3D experiment above. For the PDE model, the time step was pushed close to the CFL limit (for fast convergence), while the others were: $\mu = 0.1$, $v = 0$ and $\lambda = 1$. Note that we used constant time-stepping, as opposed to recomputing the time step after each iteration, as such computation (a reduction) would introduce extra overhead for the GPU implementation. As shown in Table 1 and looking at the full-grid GPU implementations, the discrete model converges almost three times faster than the PDE model, which agrees with the expected number of iterations that should, on average, be about three times less for the discrete model, see subsection 2.3. We also see that per iteration, the discrete model is about six times faster than the PDE model. This however is not a surprise as the number of assembly GPU instructions of the compute kernel is about four times larger and also more (expensive) texture lookups have to be performed, for discretizing the continuous model. Finally, compared to the optimized narrow-band algorithm of [27, 28], our method is about two orders of magnitude faster, while delivering similar results, see Fig. 6.

4   Conclusions

We presented a discrete region-based active surface model incorporating the minimum-variance homogeneity criterion of [17]. Experiments showed that our discrete model produces similar qualitative results, compared to the continuous model based on level sets, while being almost two orders of magnitude faster. We believe that a narrow-band version of our simple, discrete model implemented on the GPU would not be more efficient than the full-grid approach, as the extra overhead involved for maintaining the narrow band around the interface will be more expensive than performing all (trivial) computations involved. Further, we would like to study the connections between both models more deeply, with respect to parameter settings and convergence rates. Moreover, we plan to develop multi-phase extensions of our discrete model.

References

1. Terzopoulos, D.: Image analysis using multigrid relaxation methods. IEEE Trans. Pattern Anal. Machine Intell. 8, 129–139 (1986)
2. Kass, M., Witkin, A., Terzopoulos, D.: Snakes: Active contour models. Int. J. Comput. Vis. 1, 321–331 (1987)

3. Terzopoulos, D., Witkin, A., Kass, M.: Constraints on deformable models: Recovering 3D shape and nonrigid motion. Artificial Intelligence 36, 91–123 (1988)
4. McInerney, T., Terzopoulos, D.: Deformable models in medical image analysis: a survey. Medical Image Analysis 1, 91–108 (1996)
5. Suri, J., Liu, K., Singh, S., Laxminarayan, S., Zeng, X., Reden, L.: Shape recovery algorithms using level sets in 2-D/3-D medical imagery: A state of the art review. IEEE Trans. on Inf. Tech. in Biomed. 6, 8–28 (2002)
6. Mallet, J.L.: Discrete smooth interpolation in geometric modelling. Comp. Aided Design 24, 178–191 (1992)
7. Terzopoulos, D., Platt, J., Barr, A., Fleischer, K.: Elastically deformable models. ACM Comput. Graph. 21, 205–214 (1987)
8. Paragios, N., Deriche, R.: Geodesic active contours for supervised texture segmentation. In: CVPR 1999, pp. 422–427 (1999)
9. Geiger, D., Gupta, A., Costa, L.A., Vlontzos, J.: Dynamic-programming for detecting, tracking, and matching deformable contours. IEEE Trans. Pattern Anal. Machine Intell. 18(5), 575 (1996)
10. Leymarie, F.F., Levine, M.D.: Tracking deformable objects in the plane using an active contour model. IEEE Trans. Pattern Anal. Machine Intell. 15(6), 617–634 (1993)
11. Osher, S., Sethian, J.A.: Fronts propagating with curvature-dependent speed: Algorithms based on Hamilton-Jacobi formulations. Journal of Computational Physics 79, 12–49 (1988)
12. Malladi, R., Sethian, J.A., Vemuri, B.C.: Shape modeling with front propagation: A level set approach. IEEE Trans. Pattern Anal. Machine Intell. 17, 158–175 (1995)
13. Caselles, V., Kimmel, R., Sapiro, G.: Geodesic active contours. In: Proc. 5th Int. Conf. Computer Vision, pp. 694–699 (1995)
14. Siddiqi, K., Lauziere, Y.B., Tannenbaum, A., Zucker, S.W.: Area and length minimizing flows for shape segmentation. IEEE Trans. Image Processing 7, 433–443 (1998)
15. Ronfard, R.: Region based strategies for active contour models. International Journal of Computer Vision 13(2) (1994)
16. Cohen, L.D.: On active contour methods and balloons. CVGIP: Image Understanding 53, 211–218 (1991)
17. Chan, T., Vese, L.: Active contours without edges. IEEE Trans. Image Processing 10, 266–277 (2001)
18. Mumford, D., Shah, J.: Optimal approximation by piecewise smooth functions and associated variational problems. Commun. Pure Applied Mathematics 42, 577–685 (1989)
19. Maragos, P.: Differential morphology and image processing. IEEE Trans. Image Processing 5(6), 922–937 (1996)
20. Brockett, R.W., Maragos, P.: Evolution equations for continuous-scale morphological filtering. IEEE Trans. Signal Processing 42(12), 3377–3386 (1994)
21. Merriman, B., Bence, J., Osher, S.: Diffusion generated motion by mean curvature. Technical report, UCLA (1992)
22. Catté, F., Dibos, F., Koepfler, G.: A morphological scheme for mean curvature motion and applications to anisotropic diffusion and motion of level sets. SIAM J. Numer. Anal. 32(6), 1895–1909 (1995)
23. Guichard, F., Morel, J.M.: Geometric partial differential equations and iterative filtering. In: ISMM 1998, Norwell, MA, USA, pp. 127–138. Kluwer Academic Publishers, Dordrecht (1998)
24. Maragos, P., Schafer, R.W.: Morphological filters. part II: Their relations to median, order-statistic, and stack filters. IEEE Trans. Acoust. Speech and Signal Processing 35, 1170–1184 (1987)
25. Koenderink, J.J., van Doorn, A.J.: Dynamic shape. Biol. Cybern. 53(6), 383–396 (1986)

26. Chaudhury, K.N., Ramakrishnan, K.R.: Stability and convergence of the level set method in computer vision. Pattern Recogn. Lett. 28(7), 884–893 (2007)
27. Peng, D., Merriman, B., Osher, S., Zhao, H., Kang, M.: A PDE-based fast local level set method. J. Comput. Phys. 155(2), 410–438 (1999)
28. Nilsson, O., Breen, D., Museth, K.: Surface reconstruction via contour metamorphosis: An Eulerian approach with Lagrangian particle tracking. IEEE Visualization, 407–414 (2005)

Ultimate Attribute Opening Segmentation with Shape Information

Jorge Hernández and Beatriz Marcotegui

Mines ParisTech
CMM- Centre de morphologie mathématique
Mathématiques et Systèmes
35 rue St Honoré 77305-Fontainebleau-Cedex, France
{hernandez,marcotegui}@cmm.ensmp.fr

Abstract. In this paper, a method for morphological segmentation using shape information is presented. This method is based on a morphological operator named ultimate attribute opening (*UAO*). Our approach considers shape information to favor the detection of specific shapes. The method is validated in the framework of two applications: façade analysis and scene-text detection. The experimental results show that our approach is more robust than the standard *UAO*.

1 Introduction

Segmentation is a fundamental problem in image analysis to distinguish between objects of interest and "the rest". It creates a partition of the image into disjoint and uniform regions, according to some features such as gray value, color, or texture [1]. An overview of morphological segmentation is presented by Meyer in [2] where a unified framework for supervised or unsupervised, multi-scale or single scale, color or grayscale and 2D or 3D images is introduced. Furthermore, a new morphological operator, named ultimate opening (*UO*) [3], has been increasingly used as a powerful segmentation method due to its various advantages (non-parametric operator, segmentation of contrasted structures, intrinsically multi-scale, etc). This morphological operator can be used for shape analysis by associating a granulometry function. Retornaz and Marcotegui have proposed and implemented ultimate attribute opening (*UAO*) [4], where attribute opening (*AO*) was introduced by Breen and Jones [5].

In this paper, we present an integrated approach for image segmentation based on *UAO* combined with shape constraints. In contrast to using only grayscale values to locate regions with the *UAO*, the proposed method uses a similarity function defined through a prior knowledge of the shapes. This similarity function is based on the characteristics of connected components *CC*s in images (shapes). Urbach *et al.* defined in [6] vector-attribute filters based on shape descriptors and a dissimilarity measure. A threshold is required in order to filter out *CC*s different from the prior one. In contrast, our method detects the most contrasted shapes similar to the prior one. Our results are shown on real applications: façade analysis and scene-text detection.

The paper is organized as follows. In Section 2, *UAO* are presented. Section 3 describes *UAO* with shape information. In Section 4, two applications are shown and the advantages of our method are illustrated. Finally, conclusions are drawn in Section 5.

M.H.F. Wilkinson and J.B.T.M. Roerdink (Eds.): ISMM 2009, LNCS 5720, pp. 205–214, 2009.

2   Ultimate Attribute Opening

Ultimate opening (UO), closing by duality, has been introduced by Beucher in [3]. This is a non-parametric method and a non-linear scale-space based on morphological numerical residues to extract (CCs). Several applications have been developed: automatic localization of text [7] and façade segmentation [8].

2.1   Ultimate Opening

The ultimate opening θ analyzes the difference between consecutive openings. This operator has two significant outputs for each pixel x from an input image I: the maximal difference between openings (Residue, $R_\theta(I)$) and the opening size, when the maximal residue is generated ($q_\theta(I)$). The equations describing the UO evolution are written as:

$$R_\theta\left(I\right) = \max\left(r_\lambda\left(I\right)\right), \forall \lambda \geq 1$$
$$\text{with } r_\lambda\left(I\right) = \gamma_\lambda\left(I\right) - \gamma_{\lambda+1}\left(I\right) \tag{1}$$
$$q_\theta\left(I\right) = \max\left(\lambda\right) + 1 : \lambda \geq 1, r_\lambda\left(I\right) = R_\theta\left(I\right) \wedge R_\theta\left(I\right) > 0$$

where, γ_λ is an opening of size λ.

2.2   Attribute Opening

A binary attribute opening, defined by Breen and Jones [5], consists in a connected opening associated with a given increasing criterion T, based on attributes. This criterion is used to keep or discard CCs. Gray attribute openings γ^T can be defined as follows:

$$\gamma^T\left(I\left(x\right)\right) = \max\left(h \mid x \in \Gamma^T\left(\mathcal{J}_h\left(I\right)\right)\right) \tag{2}$$

where, $\mathcal{J}_h\left(I\right)$ is the threshold at level h of I, and Γ^T is a binary attribute opening. If κ_{CC} is a CC attribute, λ is a scalar value and $T(CC)$ criterion is $\kappa_{CC} \geq \lambda$, an attribute opening γ^T can be denoted as γ_λ. AO and UAO are connected operators. They can only merge input flat zones but never cut them. As stated by Salembier in [9], the Max-Tree is a suitable image representation to compute connected operators.

2.3   Max-Tree

The Max-Tree, by duality Min-Tree, was introduced by Salembier [10] as a structure for computing connected operators. It is a multi-scale image representation and a hierarchical structure in which the nodes C_h^k represent k'th binary connected components of $\mathcal{J}_h\left(I\right)$. The root node corresponds to the whole image. The leaf nodes correspond to the image maxima. The links between the nodes describe the inclusion relationship of the binary connected components. Once the Max-Tree is created, several attributes can be estimated, allowing an efficient computation of attribute openings.

On Max-Tree, an attribute opening γ_λ removes a node C_h^k when its attribute $\kappa_{C_h^k}$ is smaller than a parameter λ. Fig. 1(a) illustrates a synthetic image and its corresponding Max-Tree (Fig. 2(a)). Fig. 1(d) shows the attribute opening results, with λ increasing values. The attribute κ used is height (y-extent) of C_h^k. These openings are obtained by pruning the tree at the corresponding dot-lines of Fig. 2(a).

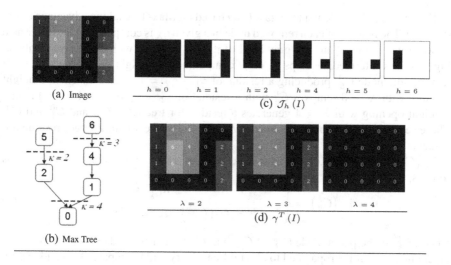

Fig. 1. (a) Original image, (b) $\mathcal{J}_h(I)$, (c) Max-Tree and (d) height openings $\gamma_T(I)$, where T : $\kappa_{C_h^k} \geq \lambda$

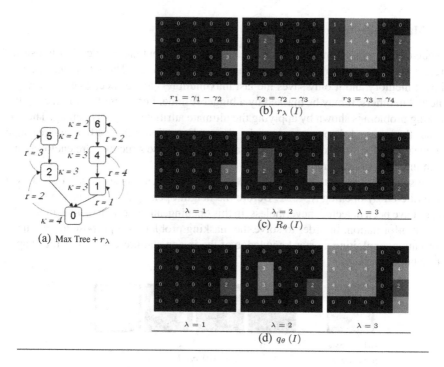

Fig. 2. (a) Max-Tree and residue $r_\lambda\left(C_h^k\right)$ and (b) height openings $R_\theta(I)$ and $q_\theta(I)$

Fabrizio in [11] implemented a fast *UAO* based on Max-Tree taking advantage of the structure. The residue of each removed node in a given γ_λ is computed by the difference between its gray level and the gray level of its first ancestor with a different attribute. Fig. 2 shows intermediate steps of *UAO* computation. A height opening with $\lambda = 2$ removes the node C_5^0, producing a residue of $\gamma_1 - \gamma_2 = 5 - 2 = 3$. Then, a height opening with $\lambda = 3$ removes C_6^0 with a residue of $\gamma_2 - \gamma_3 = 6 - 4 = 2$. Finally, a height opening with $\lambda = 4$ generates a residue for tree CCs, C_2^0, and C_1^0 and C_4^0. These residues are compared to current R_θ and only pixels with a larger residue are updated.

$r_\lambda\left(C_h^k\right)$ computation on the Max-Tree is summarized in an iterative procedure as follows:

$$
r_\lambda\left(C_h^k\right) = \begin{cases} h - h' & \kappa_{C_h^k} \neq \kappa_{C_{h'}^{k'}} \\ h - h' + r_\lambda\left(C_{h'}^{k'}\right) & \texttt{otherwise} \end{cases} \tag{3}
$$

where, $C_{h'}^{k'}$ is the parent node ($C_h^k \subset C_{h'}^{k'}$ and $h' < h$). The residue is calculated for a parent node and it is propagated to all children. Every child compares his residue with its parent, and R_θ keeps the maximum value between them. Each child becomes parent and repeats the process. q_θ is the child attribute $\kappa + 1$ when the maximal residue is generated.

2.4   Masking Problem

In spite of this non-parametric operator capacity to segment the most contrasted structures, it shows a problem of *blindness* named as "masking". The *UO* is an operator without memory and it only saves the last maximum residue. Hence, when a structure is nested in others, it may be masked by a bigger residue. For example, in Fig. 3(a), the masking problem is shown by applying the ultimate attribute (height) opening. The image has three nested shapes: a rectangular shape (dimensions 120×40), a square shape (dimensions 30×30) and a circle one (diameter 90). These structures are enclosed in a minimum gray value bounding box.

The square is the first shape found ($\lambda = 31$), and then this shape is masked by a circle filtered by opening $\lambda = 91$. Before the height opening $\lambda = 161$ is applied, two shapes have been detected; nevertheless, in this opening, an important residue masks the relevant information. In order to solve the masking problem, we propose to use shape information, exploiting a prior knowledge of the image and preserving specific shapes before being masked.

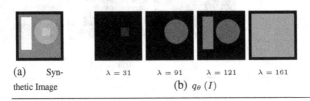

(a)   Synthetic Image $\lambda = 31$ $\lambda = 91$ $\lambda = 121$ $\lambda = 161$

(b) $q_\theta\left(I\right)$

Fig. 3. (a) Original image, **(b)** $q_\theta\left(I\right)$ of intermediate results after openings $(\lambda + 1)$

3   Ultimate Attribute Opening with Shape Information

In contrast to employing only grayscale values to locate regions with the *UAO*, the proposed method uses a similarity function based on the characteristics of connected components in images (shapes).

3.1   Shape Information

In this Section, we define our shape similarity measure. The shape definition has been widely studied in the literature. Charpiat *et al.* [12] note Ω any shape, i.e. any regular bounded subset of D, and Γ or $\partial \Omega$, its boundary, a smooth curve of $\mathbb{R}^2$. In our context, we are interested in comparing two different shapes, and their similarity measure. Many different definitions of the similarity functions ψ () between two shapes (Ω_i, Ω_j) have been proposed in the computer vision literature.

In order to compare two shapes, we propose to define a similarity function via shape descriptors κ_Ω. In our method, we propose to use the simplest shape descriptors: geometric features (height, width, etc) and their relations (fill ratio, circularity, moments, etc). We define $\psi_\kappa (\Omega_i, \Omega_j)$ by the use of Eq. 4, where τ_κ is the similarity threshold of attribute κ.

$$\psi_\kappa (\Omega_i, \Omega_j) \leftarrow \begin{cases} 1 & \kappa_{\Omega_i} = \kappa_{\Omega_j} \\ [0,1] & \left| \kappa_{\Omega_i} - \kappa_{\Omega_j} \right| < \tau_\kappa \\ 0 & \text{otherwise} \end{cases} \tag{4}$$

We have defined similarity functions with only one attribute κ. In practice, several measures are used to describe a prior shape in a same application. So, we expand the comparison between two shapes Ω_i, Ω_j to several attributes by using a simple multiplication function of similarity functions as follows: $\psi_{\forall_\kappa} = \prod_{\forall \kappa \in \Omega} \psi_\kappa (\Omega_i, \Omega_j)$.

All these possible shape attributes and similarity functions can be utilized to give an advantage over specific shapes in a segmentation process. Nevertheless, we must be careful with the selection of measures, because of the following reasons:

- **Computing Time:** Measures are computed for each tree node. To keep a reasonable computing time, we have used the simplest shape attributes, because they can straightforwardly and accurately be estimated during the Max-Tree construction.
- **Robustness of position, scale, and rotation invariant (*PSRI*):** *UO* may be intrinsically *PSRI*. However, the invariance of the proposed method also depends on the chosen measures. For example, if we choose fill ratio attribute, the new operator will only be *PSI*, besides if we choose compactness descriptor, it will be fully invariant.

3.2   Definition

We propose to consider a shape factor function $f(\Omega_i, \Omega_{ref})$ to a reference shape Ω_{ref} within the residue computation (Eq. 5). In that way, the residue of a Ω_i similar to Ω_{ref} is artificially increased. Thus, masking becomes more difficult. As the *UAO* is computed by using Max-Tree, each tree node C_h^k corresponds to a Ω_i and, $f\left(\Omega_i, \Omega_{ref}\right)$ is denoted by $f\left(C_h^k\right)$.

$$r_\lambda^\Omega \leftarrow f\left(\Omega_i, \Omega_{ref}\right) r_\lambda \qquad (5)$$

The factor function $f\left(C_h^k\right)$ is related to the similarity function $\psi_{\forall\kappa}$ of C_h^k as follows: $1 + \alpha\psi_{\forall\kappa}$. An offset of 1 is added in order to switch to standard UO when the similarity function is equal to 0 (r_λ^Ω becomes r_λ). As well, a multiplicative factor α is used to control the influence of the shape factor with respect to the gray level. Hence, $1 + \alpha$ represents the maximum value that the function may reach.

Finally, function $f_{C_h^k}$ is stored on an image $F_\theta^\Omega\left(I\right)$ when the maximal residue $\left(R_\theta(I)\right)$ is generated. With this information, we modify the original expression of UO Eq. 1 by Eq. 6.

$$\left. \begin{array}{l} R_\theta^\Omega\left(I\right) = \sup\left(r_\lambda^\Omega\left(I\right)\right) \\ F_\theta^\Omega\left(I\right) = f\left(C_h^k\right) \\ q_\theta^\Omega(I) = \max\left(\lambda\right) : \lambda \geq 1 \end{array} \right\} \begin{array}{l} R_\theta^\Omega\left(I\right) = r_\lambda^\Omega\left(I\right), \\ R_\theta^\Omega\left(I\right) > 0 \end{array} \qquad (6)$$

3.3 Example on Synthetic Image

Now, we test our approach on a synthetic image (Fig. 3(a)) to analyze it. First, we try to favor rectangular shapes. For this purpose, we use as a shape metric the fill ratio $\Upsilon_\Omega = \frac{A_\Omega}{Abbox_\Omega}$, where (A_Ω) is the shape area and $(Abbox_\Omega)$ is the bounding box area. The ratio lies in the range [0,1]; where, if the value is close to 1, it means that the shape corresponds to a rectangular polygon without rotation. We suppose that Ω_{ref} is a rectangular shape without rotation, i.e. $\Upsilon_{\Omega_{ref}} = 1$. Then, we have imposed area limit to validate the shape. We utilize a maximum area limits (90 % of image area A_I) to reject the largest regions. This factor function is translated into Eq. 7. Fig. 4(b) presents the result on a synthetic image. In this case, the masking problem is solved and the three

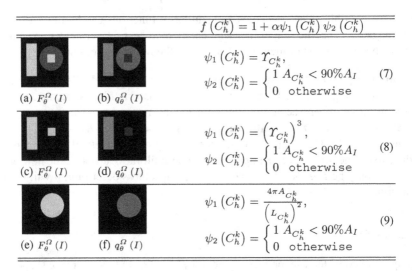

$$f\left(C_h^k\right) = 1 + \alpha\psi_1\left(C_h^k\right)\psi_2\left(C_h^k\right)$$

(a) $F_\theta^\Omega\left(I\right)$ (b) $q_\theta^\Omega\left(I\right)$

$$\psi_1\left(C_h^k\right) = \Upsilon_{C_h^k},$$
$$\psi_2\left(C_h^k\right) = \begin{cases} 1 & A_{C_h^k} < 90\%A_I \\ 0 & \text{otherwise} \end{cases} \qquad (7)$$

(c) $F_\theta^\Omega\left(I\right)$ (d) $q_\theta^\Omega\left(I\right)$

$$\psi_1\left(C_h^k\right) = \left(\Upsilon_{C_h^k}\right)^3,$$
$$\psi_2\left(C_h^k\right) = \begin{cases} 1 & A_{C_h^k} < 90\%A_I \\ 0 & \text{otherwise} \end{cases} \qquad (8)$$

(e) $F_\theta^\Omega\left(I\right)$ (f) $q_\theta^\Omega\left(I\right)$

$$\psi_1\left(C_h^k\right) = \frac{4\pi A_{C_h^k}}{\left(L_{C_h^k}\right)^2},$$
$$\psi_2\left(C_h^k\right) = \begin{cases} 1 & A_{C_h^k} < 90\%A_I \\ 0 & \text{otherwise} \end{cases} \qquad (9)$$

Fig. 4. Synthetic image segmentation (Fig. 3(a)) using UAO with shape constraints, κ is height

shapes are segmented. The importance of limits $\psi_2\left(C_h^k\right)$ is remarkable in this example, because the masking opening, size = 161, shape reaches a high factor $\Upsilon_{C_h^k} \approx 1$. Using the limits, these shapes have a factor equal to 1, thus $r_\chi^\Omega = r_\lambda$.

Even though Υ mainly favors rectangular structures, the circular shape factor is high enough to unmask it. If we want to favor only rectangular shapes, we modify $\psi_1\left(C_h^k\right)$ by a narrow function. For example, we change $\Upsilon_{C_h^k}$ in Eq. 7 by $\Upsilon_{C_h^k}$ cube value as shown in Eq. 8.

Another shape factor example is implemented to favor circular shapes. In circle detection, the most frequently used metric is circularity. The metric is the shape area (A_Ω) ratio to a circle area having the same perimeter (L_Ω): $\frac{4\pi A_\Omega}{(L_\Omega)^2}$. Eq. 9 shows factor function by using the circularity expression. Fig. 4(d) and Fig. 4(f) confirm that rectangular shapes and circular shapes are segmented, respectively.

4   Applications

The aim of the proposed method is to improve *UAO* segmentation results avoiding masking problems. In order to demonstrate the performance of our method, we illustrate two segmentation applications: façade image analysis and scene-text detection. We show all databases and result tests on the following web site: http://cmm.ensmp.fr/~hernandez/results/UOSC/testsegmentationOUSC.html.

4.1   Façade Image Analysis

Initially, we employed an ultimate attribute opening to segment façade images. For the façade structure detection, a height attribute of the *CC*s bounding box is used. Fig. 5(b) shows an example of an ultimate attribute opening using a color gradient of Fig. 5(a). In the example, all internal structures are masked in the segmentation process because the contrast between the sky and the building façade is bigger than the contrast between the wall and the windows. Most urban images contain sky information; for this reason their *UAO* segmentation is highly affected by the masking problem.

In façade images, windows and doors have particular features. Mayer and Reznik [13] describe that most windows are at least partially rectangular and the height-width ratio ($\aleph$) of a window generally lies between 0.20 and 5. With these features, we can define a similarity function for the internal structures of façades.

In the case of the synthetic image, α was selected as $\max/2$, however after several tests with façade images, the contrast between the sky and the façade is about ten times bigger than the one between the wall and the windows. For this reason, we have chosen $\alpha = 9$, i.e. $1 + \alpha = 10$. The first shape attribute is $\aleph$, we have used a dynamic function centred on $\aleph_{\Omega_{ref}} = 2.6$ and the limits have been defined on $\tau_\aleph = \pm 2.4$. The second shape attribute is Υ because we suppose that windows are partially rectangular ($\Upsilon_{\Omega_{ref}} = 1$). We have employed a square value of $\Upsilon_{C_h^k}$ as a dynamic function to penalize non-rectangular shapes or rectangular shapes with holes.

The segmentation result on the façade image is illustrated in Fig. 5. Our approach shows a better segmentation because interest structures appear (seven over nine windows/doors). Nevertheless, some structures which are neither windows nor doors become visible on the image segmentation, such as bricks (Fig. 5(d)).

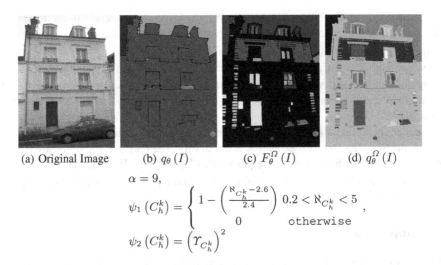

(a) Original Image   (b) $q_\theta (I)$ (c) $F_\theta^\Omega (I)$ (d) $q_\theta^\Omega (I)$

$$\alpha = 9,$$

$$\psi_1 \left(C_h^k \right) = \begin{cases} 1 - \left(\dfrac{\aleph_{C_h^k} - 2.6}{2.4} \right) & 0.2 < \aleph_{C_h^k} < 5 \\ 0 & \text{otherwise} \end{cases},$$

$$\psi_2 \left(C_h^k \right) = \left(\Upsilon_{C_h^k} \right)^2$$

Fig. 5. (a) Original image: façade, (b) $q_\theta (I)$ of *UAO*, (c)-(d) F_θ^Ω, $q_\theta (I)$ of *UAO* with shape information, κ is height. Factor function: $f \left(C_h^k \right) = 1 + \alpha \psi_1 \left(C_h^k \right) \psi_2 \left(C_h^k \right)$. Color images in the web site.

4.2 Scene-Text Detection

The text in a scene is linked to the semantic context of the image and constitutes a relevant information for content-based image indexation [7]. In several cases, the text on images is at least partially placed on a surface of different color such as: placards, posters, etc; and favors the visibility of letters. But this surface is also contrasted in comparison to its surrounding (Fig. 6(a)). When we utilize *UAO*, characters may be masked by the contrast of the signboard with its surroundings (Fig. 6(b)).

In the same way, we propose some text features to define shape information. Based on histograms of approximately 2000 analyzed characters , letter features are described as follows: 1- the range of height/width ratio mostly lies between 0.4 and 2, 2- the range of Υ falls approximately between 0.4 and 0.8 and 3- the biggest height and width of a character is $1/3$ of the image height and width respectively. We have used a similarity function analogous to the façade application. For $\aleph$ and Υ metrics, the dynamics functions are centred on $\aleph_{\Omega_{ref}} = 1.2$ and $\Upsilon_{\Omega_{ref}} = 0.6$, respectively.

In Fig. 6, the example image shows the results of the text detection using *UAO* with shape constraints. The results from this preliminary study indicate that the proposed method is superior to the classical *UAO* segmentation.

5   Discussion and Future Work

In this paper we introduce a segmentation method based on ultimate opening with shape constraints. Our approach exploits a prior knowledge to define shape information. The method can be combined with all types of similarity shape functions, thanks to the independence between the shape method computation and the classical *UAO* process.

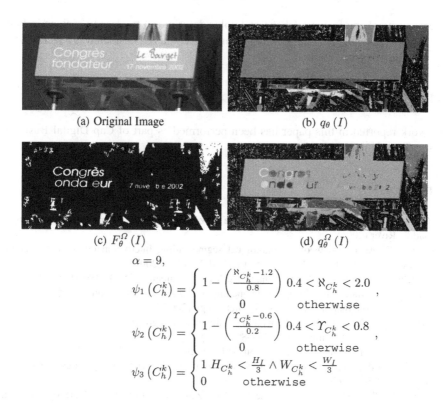

(a) Original Image   (b) $q_\theta(I)$

(c) $F_\theta^\Omega(I)$ (d) $q_\theta^\Omega(I)$

$$\alpha = 9,$$

$$\psi_1\left(C_h^k\right) = \begin{cases} 1 - \left(\frac{\aleph_{C_h^k} - 1.2}{0.8}\right) & 0.4 < \aleph_{C_h^k} < 2.0 \\ 0 & \text{otherwise} \end{cases},$$

$$\psi_2\left(C_h^k\right) = \begin{cases} 1 - \left(\frac{\Upsilon_{C_h^k} - 0.6}{0.2}\right) & 0.4 < \Upsilon_{C_h^k} < 0.8 \\ 0 & \text{otherwise} \end{cases},$$

$$\psi_3\left(C_h^k\right) = \begin{cases} 1 & H_{C_h^k} < \frac{H_I}{3} \wedge W_{C_h^k} < \frac{W_I}{3} \\ 0 & \text{otherwise} \end{cases}$$

Fig. 6. (a) Original image: placard, **(b)** $q_\theta(I)$ of *UAO*, **(c)-(d)** F_θ^Ω, $q_\theta(I)$ of *UAO* with shape information, κ is height. Factor function: $f\left(C_h^k\right) = 1 + \alpha\psi_1\left(C_h^k\right)\psi_2\left(C_h^k\right)\psi_3\left(C_h^k\right)$. Color images in the web site.

UAO provides two pieces of information, contrast ($R_\theta^\Omega(I)$) and size ($q_\theta^\Omega(I)$). The proposed method provides a third interesting piece of information: the shape factor image ($F_\theta^\Omega(I)$), that conveys a shape similarity measure with a reference shape. We store this factor when the maximal residue is generated.

The proposed method has been validated in two applications of structure extractions from façade and text images. The proposed method produces much better segmentation results than the standard *UAO*. In façade example (Fig. 5(c)), we can see that one window with shutters (down - right) is not detected, even if this shape is similar to a rectangle. The reason is its low contrast making the shape factor insufficient to avoid the masking effect. Fig. 6(c) illustrates factor images for text example. Several letters are not detected. "*m*" case is not favored ($f(\Omega_i, \Omega_{ref}) = 1$) because its $\aleph$ is outside the fixed limits. In "*f*" and "*t*" cases, the factor value is not big enough to unmask them. On the other hand, many noise *CC*s are valued with a high factor function and they are still masked thanks to their low contrast.

In the future, we will analyze in details the factor function of the segmentation algorithm in order to validate the performance on a larger databases. The detection process is the first step in computer vision problems. We intend to apply a machine learning process using shape features (shape and color descriptors) to classify regions in both applications. As well, machine learning techniques could be considered as a factor function into the presented method.

Acknowledgment

The work reported in this paper has been performed as part of Cap Digital Business Cluster Terra Numerica project.

References

1. Gonzalez, R.C., Woods, R.E.: Digital Image Processing, 3rd edn. Prentice-Hall, Inc., Upper Saddle River (2006)
2. Meyer, F.: An overview of morphological segmentation. International Journal of Pattern Recognition and Artificial Intelligence 15(7), 1089–1118 (2001)
3. Beucher, S.: Numerical residues. Image and Vision Computing 25(4), 405–415 (2007)
4. Retornaz, T., Marcotegui, B.: Ultimate opening implementation based on a flooding process. In: The 12th International Congress for Stereology (2007)
5. Breen, E.J., Jones, R.: Attribute openings, thinnings, and granulometries. Computer Vision and Image Understanding 64(3), 377–389 (1996)
6. Urbach, E.R., Boersma, N.J., Wilkinson, M.H.: Vector-attribute filters. In: ISMM 2005, Proceedings of the seventh International Symposium on Mathematical Morphology, Paris, April 18-20, pp. 95–104 (2005)
7. Retornaz, T., Marcotegui, B.: Scene text localization based on the ultimate opening. In: ISMM 2007,Proceedings of the eighth International Symposium on Mathematical Morphology, October 2007, vol. 1, pp. 177–188 (2007)
8. Hernández, J., Marcotegui, B.: Segmentation of façade images using ultimate opening. Technical report, CMM-Mines ParisTech. (2007)
9. Salembier, P., Serra, J.: Flat zones filtering, connected operators, and filters by reconstruction. IEEE Transactions on Image Processing 4, 1153–1160 (1995)
10. Salembier, P., Oliveras, A., Garrido, J.L.: Anti-extensive connected operators for image and sequence processing. IEEE Transactions on Image Processing 7, 555–570 (1998)
11. Fabrizio, J., Marcotegui, B.: Fast implementation of the ultimate opening. In: Wilkinson, M.H.F., Roerdink, J.B.T.M. (eds.) ISMM 2009. LNCS, vol. 5720, pp. 272–281. Springer, Heidelberg (2009)
12. Charpiat, G., Faugeras, O., Keriven, R.: Approximations of shape metrics and application to shape warping and empirical shape statistics. Foundations of Computational Mathematics 5(1), 1–58 (2005)
13. Mayer, H., Reznik, S.: Building façade interpretation from image sequences. In: CMRT 2005, vol. XXXVI, pp. 55–60 (2005)

Hierarchical Shape Decomposition via Level Sets

Sibel Tari

Middle East Technical University, Department of Computer Engineering,
Ankara, TR-06531
stari@metu.edu.tr

Abstract. A new tool for shape decomposition is presented. It is a function defined on the shape domain and computed using a linear system of equations. It is demonstrated that the level curves of the new function provide a hierarchical partitioning of the shape domain into visual parts, without requiring any features to be estimated. The new tool is an unconventional distance transform where the minimum distance to the union of the shape boundary and an *unknown critical curve* is computed. This curve divides the shape domain into two parts, one corresponding to the coarse scale structure and the other one corresponding to the fine scale structure.

The connection of the new function to a variety of morphological concepts (Skeleton by Influence Zone, Aslan Skeleton, and Weighted Distance Transforms) is discussed.

Keywords: level set methods, PDEs and variational methods, curve evolution, shape representation, shape decomposition.

1   Introduction

Let Ω, a connected, bounded, open domain of $\mathbf{R}^2$, represent a planar shape. Let $\partial\Omega$ denote the shape boundary. Consider the following PDE:

$$(\triangle - \alpha)\,v \;=\; -\alpha \tag{1}$$
$$\text{with } v(\mathbf{x}) = 0 \ \text{ for } \mathbf{x} = (x, y) \in \partial\Omega$$

where $\triangle$ denotes the Laplace operator and α is a constant scalar. The parameter α inversely indicates the strength of diffusion. As $\alpha \to 0$, the trivial solution which is identically zero is attained.

This equation has been proposed by Tari, Shah and Pien [1,2] as an alternative method to compute a weighted distance transform [3,4] where the local steps between neighboring points are given different costs. The level curves of v are related to the motion of fronts propagating with curvature dependent speed in the direction of the inward normal and it has been demonstrated that the gradient of v along a level curve approximates the curvature of level curves, thus, suggesting a robust method for skeleton extraction by locating the extrema of the gradient of v along the level curves. The method is illustrated in Fig. 1 using a cat silhouette given on a 256×256 lattice. The level curves of v for $\alpha = 1$

M.H.F. Wilkinson and J.B.T.M. Roerdink (Eds.): ISMM 2009, LNCS 5720, pp. 215–225, 2009.
© Springer-Verlag Berlin Heidelberg 2009

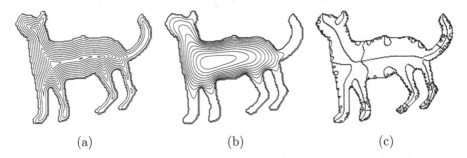

(a)   (b)   (c)

Fig. 1. The method of Tari, Shah and Pien [1,2]. (a)-(b) Level curves of v for $\alpha = 1$ and $\alpha = 1/\sqrt{128} \approx 0.88$, respectively. (a) The cat is given on a 256×256 lattice. For large α, the v function resembles the standard distance transform. As α is reduced, the level curves get smoother; and the number of extrema decreases. Observe that in (b) the v function has a single extrema located at the center. (c) Skeletons computed from the v function depicted in (b).

and $\alpha = 1/\sqrt{128} \approx 0.88$ are shown in (a) and (b), respectively. For large α, the v function resembles the standard distance transform. As α is reduced, the level curves get smoother; and the number of extrema decreases. Observe that in (b) the v function has a single extremum located at the center. In (c) the skeletons computed from the v function depicted in (b) is shown. The skeleton computation method of Tari, Shah and Pien [1,2] exploits connections among morphology, distance transforms and fronts propagating with curvature dependent speeds. (Such connections have stimulated many interesting approaches in solving shape related problems [4].) Tari, Shah and Pien approach is important in the sense that it is a first attempt to unify segmentation and local symmetry computation into a single formulation by exploiting the connection between (1) and the Mumford and Shah [5] segmentation functional (via its Ambrosio and Tortorelli [6] approximation). It naturally extends to shapes in arbitrary dimension [7].

The success of the v function as well as other diffused distance functions, can be attributed to the replacement of the Euclidean distance with a diffusion distance; as this allows wider interaction among boundary points. However, the skeletons extracted from v exhibit instability when a limb is close to a neck. The remedy proposed by Aslan and Tari [8,9,10] is to gradually increase the diffusion such that each shape is forcefully interpreted as a single blob. Despite its successful use in shape matching [9,10,11] neither the v function nor its modification [8,9,10] provide a natural decomposition into part structure. One has to rely on skeleton branches whose computation requires high enough resolution. Furthermore, after the ad-hoc modification by Aslan and Tari [8,9,10], the method cannot be applied to shapes which can not be reduced to a single blob e.g. shapes with holes or dumbbell-like shapes.

In this paper, we propose a new function w defined over the shape domain Ω whose level sets provide a parameter-free, robust, hierarchical decomposition of the shape into visual parts. This new function is computed by solving a set of linear equations which is obtained from (1) after discretization and modification steps.

The paper is organized as follows. In Section 2, the construction of the function w is explained. In Section 3, experiments are presented and discussed. Connection between the new function and a variety of morphological concepts are also discussed in Section 3. Finally, in Section 4, a summary is provided.

2   The Method

In the discrete setting, (1) takes the following form:

$$(\mathbf{L} - \alpha\mathbf{I})\mathbf{v} = -\alpha\mathbf{1} \tag{2}$$

where $\mathbf{v}$ is the discretized and vectorized v whose dimension N is the number of pixels that cover the shape domain Ω, $\mathbf{L}$ is the $N \times N$ matrix representation of the Laplace operator defined on the discrete shape domain, $\mathbf{I}$ is the $N \times N$ identity matrix, and $-\alpha\mathbf{1}$ is the constant right hand side term.

Let d_Ω be the distance transform of the domain Ω and $\mathbf{d}_\Omega$ be the vector representation of the discrete distance transform. Throughout the text, we use small letters to represent continuous functions and bold small letters to denote their discrete representations as vectors. Let $\mathbf{J}$ be the $N \times N$ matrix whose all entries are equal to one. Consider the linear equation:

$$(\mathbf{L} - \alpha\mathbf{I} - \beta\mathbf{J})\,\mathbf{w} = -\alpha\mathbf{d}_\Omega \tag{3}$$

which is obtained from (2) by

- changing the right hand side term from constant to linear;
- adding a new term $-\beta\mathbf{J}$ to the operator.

Claim: The function w (with a vectorized form $\mathbf{w}$) is a smooth and *necessarily an oscillatory* approximation of d_Ω.

To support our claim, we relate the linear system (3) to a minimization problem. First, ignoring the third term, we establish the connection between (3) and a quadratic energy, when $\beta = 0$.

Proposition: When $\beta = 0$, the solution $\mathbf{w}$ of (3) is the discretization of a smooth approximation of d_Ω with a smoothing level proportional to $1/\alpha$.

Proof: In the continuous setting, consider the following energy

$$\int\int_\Omega |\nabla w|^2 dxdy + \alpha \int\int_\Omega (w - d_\Omega)^2 dxdy \tag{4}$$

$$\text{with } w(\mathbf{x}) = 0 \text{ for } \mathbf{x} = (x, y) \in \partial\Omega$$

The above energy is minimized if the function w is close to d_Ω and if its squared gradient is small. This means that the function w is a regularized approximation of d_Ω. Setting the first variation of (4) equal to zero, to find its minimizer, yields

$$\nabla \cdot (\nabla w) - \alpha\,(w - d_\Omega) = 0 \Rightarrow \nabla \cdot (\nabla w) - \alpha w = -\alpha d_\Omega \tag{5}$$

$$\text{with } w(\mathbf{x}) = 0 \text{ for } \mathbf{x} = (x, y) \in \partial\Omega$$

Discretizing (5) using finite difference approximation gives (3) for the special choice $\beta = 0$.   **Q.E.D.**

Now, let us discuss the effect of the third term of the operator in (3). The operator $\mathbf{J}$ computes the sum of the components of the vector that it is applied to. Consequently, $\frac{1}{N}\mathbf{J}$ is a global averaging operator over the shape domain. When it is applied to a vector $\mathbf{w}$, it produces $\mu\mathbf{1}$ where μ denotes the average of the components of the vector $\mathbf{w}$. (We remark that $\frac{1}{N}\mathbf{J}$ is an idempotent operator.) If the components of $\mathbf{w}$ sum up to 0, then the application of the operator $\frac{1}{N}\mathbf{J}$ produces $\mathbf{0}$. Therefore, for $\mathbf{w} \neq \mathbf{0}$, $\mathbf{w}^T \frac{1}{N}\mathbf{J}\mathbf{w} \geq 0$.

Adding $-\frac{1}{N}\mathbf{J}$ to $(\mathbf{L} - \alpha\mathbf{I})$ may be interpreted as adding a *non-local* term to the energy minimization problem given in (4). This term in discrete form is:

$$\frac{1}{N}\left(\sum_{k=1}^{N} w(kh)\right)^2$$

where h is the spatial discretization step size. Notice that the minimum for this expression is attained when the values of the function at the nodal points add up to *zero*. Hence, the function w whose discrete form $\mathbf{w}$ is obtained from (3) should simultaneously satisfy the following three competing criteria:

1. being smooth,
2. being close to d_Ω,
3. having zero-crossings (attaining both positive and negative values inside the domain).

The relative importance of each criterion depends on α when the other parameter, β, is fixed at $\frac{1}{N}$. The only difference between $\mathbf{L}$ and $(\mathbf{L}-\alpha\mathbf{I})$ from the numerical algebra point of view is the reduction in the condition number. The role of α in adjusting the trade-off between the second and the third criteria is more critical. Intuitively, we expect that the size of the negative region to shrink as α increases. In Fig. 2, for $1-D$ case, the solutions to (3) for various choices of α are displayed. In (a), the solid blue line is the solution when $\alpha = 1/N$ (that is $\alpha = 1/31$). Each of the dashed line plots are obtained by reducing the value of α by a factor of 2 $(1/62, 1/124, 1/248)$. In (b), the solid blue line is the distance transform. The dashed line indicates the solution for $\alpha = 2/N$. The dash-dotted line represents the solution for $\alpha = 2$. Clearly, the data fidelity criterion dominates over the other two criteria, thus, the solution is quite close to the distance transform. Notice that the choice of α affects neither the location of the zero-crossings nor the location of the extrema, as long as α does not significantly exceed $1/N$. Thus, both parameters are set to $\frac{1}{N}$, giving

$$\left(\mathbf{L} - \frac{1}{N}\mathbf{I} - \frac{1}{N}\mathbf{J}\right)\mathbf{w} = -\frac{1}{N}\mathbf{d}_\Omega \qquad (6)$$

The right hand side of the equation may be multiplied with a constant scalar (e.g. N), without altering the qualitative behavior of the level curves.

The transformation facilitated by (6) takes both local and global interactions within the shape domain into account, as $\mathbf{L}$ is a local operator; $\mathbf{I}$ is a point-wise operator; $\mathbf{J}$ is a global operator.

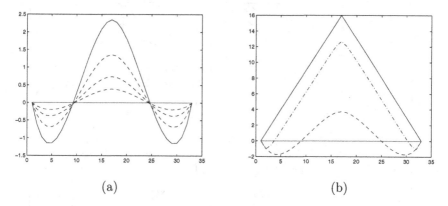

Fig. 2. The effect of α. (a) Plot of w for various choices of α not exceeding $\frac{1}{N}$. Seemingly the choice of α affects neither the location of the zero-crossings nor the location of the extrema. (b) The solid blue line is d_Ω. The dashed line is w for $\alpha = \frac{2}{N}$. The dash-dotted line is the solution for $\alpha = 2$. In this case, data fidelity criterion dominates.

3   Experimental Results and Discussion

In Fig. 3 and Fig. 4, the level curves of w for a few sample shapes are depicted. The new function attains both positive and negative values. The inner zero-level curve shown in black separates the shape domain Ω into two disjoint open sets $\{\Omega^- : (x, y) \in \Omega$ s.t $w(x, y) < 0\}$ and $\{\Omega^+ : (x, y) \in \Omega$ s.t $w(x, y) > 0\}$ with a common boundary $\partial\Omega^+$. This boundary, which is the zero-level curve, divides the shape domain into two parts capturing, respectively, the coarsest structure (in the sense of Aslan and Tari [8,10]), and the peripheral structure including limbs, protrusions, boundary texture and noise. This is the first partitioning level in the hierarchy. The size of Ω^- relative to the size of Ω^+ (i.e. $\frac{|\Omega^-|}{|\Omega^+|}$) is a measure of boundary roughness [12] which attains its smallest value for a circle. The ratio increases for a shape with significant protrusions. Let us consider the human silhouettes given in Fig. 3. The first level of partitioning with the help of zero level set separates the main body from the legs, the arms, and the head. The restriction of w to Ω^- (note that this is the lower level set at threshold 0) has five local minima, located roughly at the centers of the two legs, the two arms, and the head. Five watershed [13,14] regions within Ω^- capture the legs, the arms and the head. The restriction of w to Ω^+ (note that this is the upper level set at threshold 0) has a single maximum located at the center. The main body is a single blob without any boundary concavities indicating otherwise.

For most cases, Ω^+ is composed of a single blob. In some cases, such as the dog-bone like shapes in Fig. 4, Ω^+ is composed of multiple blobs. In (a)-(d), the restrictions of w to Ω^+ has two local maxima. These points may be interpreted as *seed* points. In the process of growth, two growing blobs may merge at the unique saddle point and continue to grow together (Fig. 4 (a) and (b)). Notice

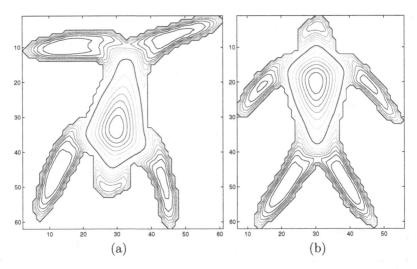

(a)   (b)

Fig. 3. The level curves of w for two human shapes. The inner black level curve is the zero-level curve. It partitions the shape into disjoint sets inside which w attains positive and negative values, respectively. This partitioning separates the coarse structure from the peripheral structure. Intuitive parts are also captured. See the text for discussion.

that $\partial \Omega^+$ has a pair of concavities indicating two blobs. For the shapes shown in Fig. 4 (c) and (d), the peripheral structure grows before the merge of the two inner blobs. In these cases, the restriction of w to Ω^- does not have a saddle point. In all cases, Ω^+ captures the coarse scale structure.

Similar to the Eccentricity transform by Kropatsch et al. [15], the new function is a peculiar distance transform. In the classical distance transform and its smooth analogues [10,3,16,2], for each point on the shape interior, the minimum distance to the shape boundary $\partial \Omega$ is computed. Thus, the value of the distance function at a point depends monotonically on the minimum distance of the point to the shape boundary $\partial \Omega$. Curiously, the value of w at a point seems to be a monotonic function of the minimum distance to either $\partial \Omega$ or $\partial \Omega^+$ (yet unknown), whichever is closer.

The new function w has an intriguing connection to the v function in [2,8,10]. Once Ω^+ and Ω^- are separated, the level curves in both subdomains are analogous to the level curves of the v function. Thus the skeleton computation method in [1,2] becomes readily applicable, as demonstrated in Fig. 5. $\partial \Omega^+$ resembles the level curve passing through the last disconnection point in [10]. Notice that $\partial \Omega^- \equiv \partial \Omega^+ \cup \partial \Omega$.

We offer a connection between the new function and a morphological concept called SKIZ (Skeleton by Influence Zone). Consider the human shapes displayed in Fig. 3. Imagine six seed points, respectively, in the centers of the body, arms, legs and head. Let a growth start from these seed points, each claiming a region of influence. These growing regions finally meet at the zero-level curve separating the coarse scale structure from the fine scale structure. We illustrate this concept

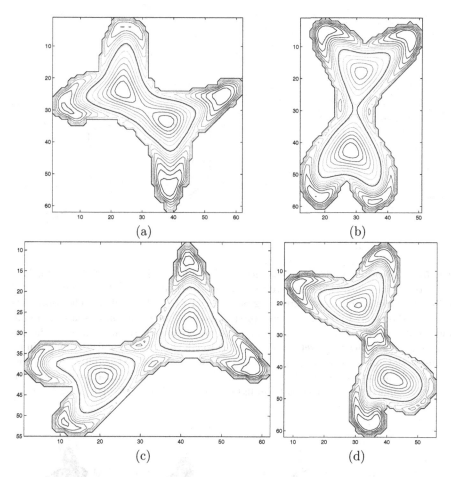

Fig. 4. The level curves of w for dog-bone like shapes. The inner black level curve is the zero-level curve. See the text for discussion.

in Fig. 6 using the dog-bone-like shape from Fig. 4(a). Each figure depicts a union of a lower level set $\{\chi_{\lambda_-} : (x,y) \in \chi_{\lambda_-} \text{ s.t. } w(x,y) < \lambda_-\}$, and an upper level set $\{\chi_{\lambda_+} : (x,y) \in \chi_{\lambda_+} \text{ s.t. } w(x,y) > \lambda_+\}$. λ_- is negative and λ_+ is positive. From left to right, λ_- increases, while λ_+ decreases. Both thresholds gradually approach to zero at the rightmost figure.

The behavior of the level sets is being explored, using the harmonic analysis point of view, in a future paper [17].

Finally, in Fig. 7, sample decomposition results are displayed. Watershed zones are extracted by applying Matlab's watershed command directly to the restriction of w to Ω^-. Parts of Ω^- are shown in bright colors. Dark blue pixels are the watershed boundaries. Ω^+ is shown in gray. Note that the decomposition is not applied to Ω^+ which in some cases such as the one in (n) is actually composed of two blobs of approximately equal size. (See also Fig.4(a)-(b).) Multiple instances

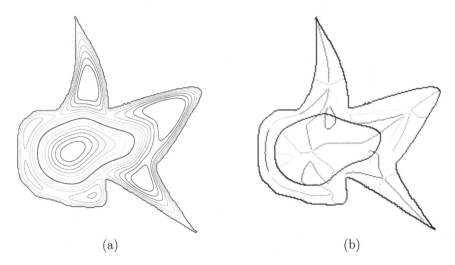

(a)   (b)

Fig. 5. A shape on a 200 × 225 lattice [9]. (a) The level curves of w. (b) Skeleton extracted from w using the skeleton extraction method of Tari, Shah and Pien [1,2] followed by the prunning and grouping procedure by C. Aslan [8,9,10]. The procedures are applied directly to w, without any pre or post procesing. The outer shape boundary and the inner zero-level curve is superimposed merely for illustration purpose. The zero level curve resembles the level curve passing through the last disconnection point in the disconnected Aslan skeleton [8,10,9]. Notice that some of the skeleton branches code the main body and some of the skeleton branches code the individual parts; consistent with the implied part structure. (The figure is prepared by E. Erdem using the codes of C. Aslan and the author.)

Fig. 6. Connection to SKIZ. Imagine four seed points located at the local minima inside Ω^-, and two seed points located at the local maxima inside Ω^+. These seeds grow and merge at the zero-level curve. Each figure depicts a union of a lower level set $\left\{ \chi_{\lambda_-} : (x,y) \in \chi_{\lambda_-} \text{ s.t. } w(x,y) < \lambda_- \right\}$, and an upper level set $\left\{ \chi_{\lambda_+} : (x,y) \in \chi_{\lambda_+} \text{ s.t. } w(x,y) > \lambda_+ \right\}$. λ_- is negative and λ_+ is positive. From left to right, λ_- increases, while λ_+ decreases. Both thresholds gradually approach to zero at the rightmost figure.

from the same shape category are used in order to demonstrate the insensitivity of the w function to visual transformation including local variations and articulations. Similar shapes are decomposed similarly and the detected parts are compatible with our intuition. There are some little discrepancies that may

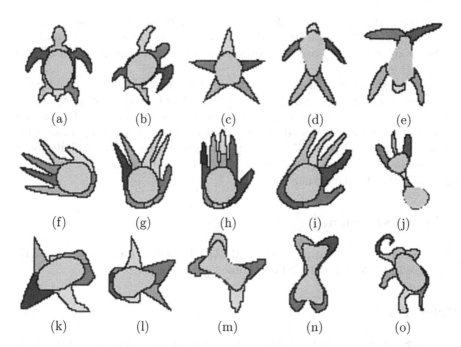

(a)   (b)   (c)   (d)   (e)

(f)   (g)   (h)   (i)   (j)

(k)   (l)   (m)   (n)   (o)

Fig. 7. Sample decomposition results. Shapes are drawn at most on a 60×60 lattice. Parts of Ω^- are shown in bright colors. Dark blue pixels are the watershed boundaries. Ω^+ is shown in gray. Parts are extracted by applying Matlab's watershed command directly to the restriction of w to Ω^-. Multiple instances from the same shape category are used in order to demonstrate the insensitivity of the w function to visual transformations including local variations and articulations. Similar shapes are decomposed similarly and the detected parts are compatible with our intuition.

seem like an instability at first look. The middle finger in (h) is decomposed into two parts. This is due to a discretization error in the original shape. There is a two pixel thick bridge between the ring and the middle fingers, forming an artificial hole in between the two parts. Despite this complication, the partitioning is quite similar to the others. We may assign a saliency to each partitioning in order to eliminate less salient partitions. For example we may use the value of the w function at the point of break as a saliency measure. Another possibility is to compare neighboring parts for a possible merge.

4   Summary and Conclusion

We presented a linear equation (6) whose solution is an unconventional distance function. The level sets of this new function provide a hierarchical decomposition of the shape domain in the form of watershed zones. The new function takes both positive and negative values inside the shape domain. The zero level curve resembles the level curve passing through the last disconnection point in the disconnected Aslan skeleton [8,9,10]; and it divides the domain into a coarse scale structure (Ω^+) and a fine scale structure (Ω^-). Inside both Ω^+ and Ω^-, the behavior of the level curves is equivalent to that of of the Tari, Shah Pien v function [1,2,7].

Features extracted from distance transform or its analogues has been previously used for part decomposition e.g. [18,19]. Unlike these works, we do not extract any features. The method does not require any parameters. In our computations, we retain the locally diffusive effect of the Laplace operator while introducing a global effect. The perceptual implications of the exploitation of the combined local/global character of shapes is a subject matter of an upcoming publication by the author [20].

Acknowledgements

The work is supported in part by TUBITAK through research grants 105E154 (May 2006-May 2008) and 108E015 (June 2008-June 2010).

References

1. Tari, S., Shah, J., Pien, H.: A computationally efficient shape analysis via level sets. In: MMBIA 1996: Proceedings of the 1996 Workshop on Mathematical Methods in Biomedical Image Analysis (MMBIA 1996), pp. 234–243 (1996)
2. Tari, S., Shah, J., Pien, H.: Extraction of shape skeletons from grayscale images. CVIU 66(2), 133–146 (1997)
3. Borgefors, G.: Distance transforms in digital images. Comput. Vision, Graphics and Image Processing 34, 344–371 (1986)
4. Maragos, P., Butt, M.A.: Curve evolution, differential morphology and distance transforms as applied to multiscale and eikonal problems. Fundamentae Informatica 41, 91–129 (2000)
5. Mumford, D., Shah, J.: Optimal approximations by piecewise smooth functions and associated variational problems. Commun. Pure Appl. Math. 42(5), 577–685 (1989)
6. Ambrosio, L., Tortorelli, V.: On the approximation of functionals depending on jumps by elliptic functionals via Γ-convergence. Commun. Pure Appl. Math. 43(8), 999–1036 (1990)
7. Tari, S., Shah, J.: Local symmetries of shapes in arbitrary dimension. In: ICCV, pp. 1123–1128 (1998)
8. Aslan, C., Tari, S.: An axis-based representation for recognition. In: ICCV, pp. 1339–1346 (2005)
9. Aslan, C.: Disconnected skeletons for shape recognition. Master's thesis, Department of Computer Engineering, Middle East Technical University (May 2005)
10. Aslan, C., Erdem, A., Erdem, E., Tari, S.: Disconnected skeleton: Shape at its absolute scale. IEEE Trans. Pattern Anal. Mach. Intell. 30(12), 2188–2203 (2008)
11. Baseski, E., Erdem, A., Tari, S.: Dissimilarity between two skeletal trees in a context. Pattern Recogn. 42(3) (2009)
12. Maragos, P.: Pattern spectrum and multiscale shape representation. IEEE Trans. Pattern Anal. Mach. Intell. 11(7), 701–716 (1989)

13. Lantejoul, C., Beucher, S.: On the use of the geodesic metric in image analysis. J. of Microscopy 121, 39–49 (1981)
14. Vincent, L., Soille, P.: Watersheds in digital spaces: An efficient algorithm based on immersion simulations. IEEE Trans. Pattern Anal. Mach. Intell. 13(6), 583–598 (1991)
15. Kropatsch, W., Ion, A., Haxhimusa, Y., Flanitzer, T.: The eccentricity transform (of a digital shape). In: Kuba, A., Nyúl, L.G., Palágyi, K. (eds.) DGCI 2006. LNCS, vol. 4245, pp. 437–448. Springer, Heidelberg (2006)
16. Gorelick, L., Galun, M., Sharon, E., Basri, R., Brandt, A.: Shape representation and classification using the poisson equation. IEEE Trans. Pattern Anal. Mach. Intell. 28(12), 1991–2005 (2006)
17. Tari, S., Burgeth, B., Tari, I.: How to use a positive definite operator for shape analysis (submitted)
18. Arcelli, C., Serino, L.: From discs to parts of visual form. Image Vision Comput. 15(1), 1–10 (1997)
19. Pan, X., Chen, Q., Liu, Z.: 3d shape recursive decomposition by poisson equation. Pattern Recognition Letters 30(1), 11–17 (2009)
20. Tari, S.: Extracting parts of 2d shapes using local and global interactions simultaneosuly. In: Chen, C. (ed.) Handbook of Pattern Recognition and Computer Vision. World Scientific, Singapore (2010)

Morphological Exploration of Shape Spaces

Jesús Angulo and Fernand Meyer

CMM-Centre de Morphologie Mathématique, Mathématiques et Systèmes, MINES
Paristech; 35, rue Saint Honoré, 77305 Fontainebleau Cedex, France
jesus.angulo@ensmp.fr, fernand.meyer@ensmp.fr

Abstract. The aim of this paper is to propose efficient tools for analysing
shape families using morphological operators. The developments include
the definition of shape statistics (mean and variance of shapes, modes of
shape variation) and the interpolation/extrapolation in shape geodesic
paths. The main required ingredients for the operators and the algorithms
here introduced are well known in mathematical morphology such as the
median set, the watershed on distance functions or the interpolation function. In addition, the projection of shapes in spaces with reduced dimensions using PCA or ISOMAP techniques permits to apply morphological
interpolation techniques in shape manifolds.

1   Introduction

Let $\mathcal{X} = \{X_1, X_2, \cdots, X_N\} = \{X_i\}_{i=1}^N$ be a family (or collection) of N shapes,
where $X_i \in \mathcal{P}(E)$ represents the set (or binary image) of the shape i, and the
support space E is a nonempty set. Typically for the digital 2D images $E \subset \mathbb{Z}^2$.
The set X_i is a compact set (and typically a closed simply connected set). The
family $\mathcal{X}$ can be considered as a random variable of shape, where X_i represents
a realization of this random variable. The family may also viewed as defined in
a shape space, where $\mathcal{X}$ is modelled as a low dimensional manifold embedded in
a higher-dimensional space. The aim of this paper is to propose efficient tools
for analysing shape families using morphological operators. This kind of analysis
includes the definition of shape statistics (mean and variance of shape, modes of
variation of shape) and the interpolation/extrapolation in shape manifolds or in
shape geodesic paths.

Statistical theory of shapes has been studied by Kendall [9], representing the
shapes as a finite number of salient points; and by Grenander [6], considering the
shapes as points on some infinite-dimensional differentiable manifold, under the
actions of Lie groups. More in relation with our study, Klassen *et al.* [10] proposed statistical shape analysis and shape interpolation by differential geometry
methods, where the shapes are represented by curvature functions. Whitaker [16]
proposed a method for image blending by progressive minimisation of a difference metric in a variational framework (i.e., a pair of coupled nonlinear PDE),
where the metric is based on computing the distance between level-set shapes
(distance function for binary images). Charpiat *et al.* [3] and Etyngier *et al.* [5]
formalised the problem by optimizing mappings based on the Hausdorff metric
and the signed distance functions.

M.H.F. Wilkinson and J.B.T.M. Roerdink (Eds.): ISMM 2009, LNCS 5720, pp. 226–237, 2009.
© Springer-Verlag Berlin Heidelberg 2009

Background notions. The main required ingredients for the operators and the algorithms introduced here are well known in mathematical morphology.

Let E be a metric space equipped with a distance $d_M : E \times E \to \mathbb{R}_+$, and let $\mathcal{K}'$ be the class of the non empty compact sets of E. The *distance of point x to set Y* is defined as $d_M(x, Y) = \inf\{d_M(x, y), y \in Y\}$, $x \in E$ and $Y \in \mathcal{K}'$. Then, the *distance function of set X* according to metric d_M is the mapping $\Delta_M X : E \to \mathbb{R}_+$, such that

$$\Delta_M X(x) = d_M(x, X^c) = \inf\{\|x, y\|_M : y \in X^c\}.$$

We also use in this study the notion of distance between two shapes. Given two sets $X, Y \in \mathcal{K}'$, the most basic mapping $\mathcal{K}' \times \mathcal{K}' \to \mathbb{R}_+$ to compare two sets is their *Euclidean distance*, i.e., $d_E(X, Y) = \sum_{\mathbf{x} \in E} 1_{\mathbf{x} \in [(X \cup Y) \setminus (X \cap Y)]}$. Classically, it is considered most useful in practice the distance associated to the *Jacquard coefficient*:

$$d_J(X, Y) = 1 - \frac{\sum_{\mathbf{x} \in E} 1_{\mathbf{x} \in (X \cap Y)}}{\sum_{\mathbf{x} \in E} 1_{\mathbf{x} \in (X \cup Y)}} = \frac{\sum_{\mathbf{x} \in E} 1_{\mathbf{x} \in X \triangle Y}}{\sum_{\mathbf{x} \in E} 1_{\mathbf{x} \in (X \cup Y)}},$$

where $X \triangle Y = [(X \cup Y) \setminus (X \cap Y)]$. Furthermore, the natural metric to compare spatial shapes is the Hausdorff distance:

$$d_H(X, Y) = \max\left\{\sup_{x \in X} d(x, Y) ; \sup_{y \in Y} d(y, X)\right\}.$$

The *Hausdorff distance* can also be expressed by means of the dilations by the balls of space E [14]:

$$d_H(X, Y) = \inf\{\lambda : X \subseteq \delta_\lambda(Y); Y \subseteq \delta_\lambda(X)\},$$

with $\delta_\lambda(X)$ being the dilation of $X \in \mathcal{K}'$ by a radius of size $\lambda \in \mathbb{R}_+$: $\delta_\lambda(X) = \cup\{B_\lambda(x), x \in X\}$, where $B_\lambda(x)$ stands for the compact ball of centre x and of radius λ.

The theory of morphological interpolation was introduced in [11,2,14]. In particular, the *interpolation distance function* [11],

$$\underset{X}{\overset{Y}{\mathrm{interp}}}(x) = \frac{d_X^Y(x)}{d_X^Y(x) + d_Y^X(x)};$$

and the *morphological median set* [2]:

$$m(X, Y) = Y \triangle \{x : \underset{X}{\overset{Y}{\mathrm{interp}}}(x) \leq 0.5\},$$

will be frequently used below. The distance $d_X^Y(x) : E \times E \to \mathbb{R}_+$ to set X in set Y is defined as [13]:

$$d_X^Y(x) = n \quad \text{if} \quad \left(\varepsilon_X^n(Y) = 0 \quad \text{and} \quad \varepsilon_X^{n-1}(Y) = 1\right),$$

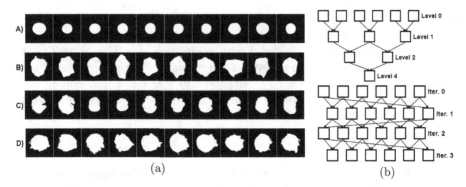

Fig. 1. (a) Four population of cells, each one representing a spatially equivalent shape family $\mathcal{X}$. (b) Two approaches for computing the mean shape using the median set: top, merger algorithm and bottom, iterative algorithm.

where $\varepsilon_X^1(Y) = X \cap \delta_1(X \cap Y)$ for points on $Y \subseteq X$, $\varepsilon_X^1(Y) = X \cup \varepsilon_1(X \cup Y)$ for points on $X \subseteq Y$; and $\varepsilon_X^n(Y) = \varepsilon_X^1 \varepsilon_X^{n-1}(Y)$.

Shape analysis makes no sense without a renormalisation of shapes. We only consider *spatially equivalent shape families:* two shapes will be considered as equivalent if there exists a rotation and a translation transforming one shape in the other. In practice, the mass center and the angle of orientation of principal axis are obtained by computing the second order inertia moments. Then the shapes are aligned and rotated to impose the same centroid and the same principal axis of variation, see in Fig. 1(a) four examples of spatially equivalent shape families. Other algorithms of centring and orientation can be considered, e.g. embedding which maximises the intersection between the sets [7], or which minimises the Hausdorff distance between the sets [14].

2   Shape Statistics

Let us start with the basics of shape statistics, the mean shape $\mu_\mathcal{X}$ and the variance of shape $\sigma_\mathcal{X}^2$ from a shape family $\mathcal{X}$. These basic statistics are needed for instance to build prototypes or shape priors in model-based image segmentation or to define primary grains for Boolean modelling and simulation. We describe and compare 3 different methods.

2.1   Computation of Mean Shapes

Approach based on the median set $\mu_\mathcal{X}^{ms}$. The morphological median set is defined only for two sets, i.e., $m(X_1, X_2)$. The extension to N sets requires consequently the combination of successive median sets. Fig. 1(b) illustrates two different cascaded median set operators to compute $\mu_\mathcal{X}^{ms}$. The merger algorithm leads sequentially to a single final shape, whereas the iterative algorithm is applied until that the cumulated distance between sets is lower than a fixed threshold (i.e., convergence to the mean set). Both algorithms depend on the ordering of the operators since the median set is not associative, i.e., $m(m(X_1, X_2), X_3) \neq m(X_1, m(X_2, X_3))$. Empirical observations show that the iterative algorithm depends less upon the ordering of the sets and converges after a few iterations, however the merger algorithm requires less computation and at the end the results are quite similar.

Extrinsic mean by thresholding the sum of distance functions $\mu_{\mathcal{X}}^{thres}$.
The sum of the distance functions of the sets in $\mathcal{X}$ has been widely used in the literature to build the theory of averaging shapes [4,1], although the associated algorithms to estimate the mean shapes are often inefficient. Inspired by the work [1], we propose to use both the inner and outer distance functions to estimate two extrinsic means, where the inner distance function of the family is $\Delta\mathcal{X}(x) = \sum_i^N \Delta X_i(x)$ and the outer distance function is $\Delta\mathcal{X}^c(x) = \sum_i^N \Delta X_i^c(x)$. The algorithm aims at computing an optimal level set in $\Delta\mathcal{X}(x)$.

Let $X^u \in \mathcal{P}(E)$ be the set obtained by thresholding the inner distance function at value u, i.e.,

$$X^u = \{x \in E : \Delta\mathcal{X}(x) \geq u\}, \quad u \in [0, \max(\Delta\mathcal{X}(x))[.$$

We define the cumulative distance of shape family $\mathcal{X}$ to set X^u by

$$D_{\Delta\mathcal{X}(x)}(u) = \sum_i^N d_M(X_i, X^u),$$

where $d_M(\cdot, \cdot)$ is the distance between the two sets. Then the inner extrinsic mean is defined as

$$\mu_{\mathcal{X}}^{inner} = \arg_u \min D_{\Delta\mathcal{X}(x)}(u),$$

this minimization problem can be solved by an exhaustive search algorithm (i.e., discretization of $\Delta\mathcal{X}(x)$ in K thresholded sets and selection of the minimum). A similar outer extrinsic mean $\mu_{\mathcal{X}}^{outer}$ can be defined from optimal thresholding on function $D_{\Delta\mathcal{X}^c(x)}(u)$. The associated extrinsic mean shape $\mu_{\mathcal{X}}^{thres}$ is then defined as the median set between $\mu_{\mathcal{X}}^{inner}$ and $\mu_{\mathcal{X}}^{outer}$. Fig. 2(a) gives an example of the various elements for an example. An important parameter of this algorithm is the distance $d_M(\cdot, \cdot)$. We have compared the performance of both the Jacquard distance and the Hausdorff distance: it appears that the obtained mean shapes are more interesting when the Jacquard distance is chosen.

Locally optimal mean by watershed of sum of squared distance functions $\mu_{\mathcal{X}}^{wshed}$. The previous approach presents two main limitations: i) the inner and outer distance functions are used separately, ii) the obtained mean shape is optimal only for a constant level set. A more original and powerful technique to exploit the sum of distance functions is based on the classical definition of the mean μ of N samples: μ is the value such that $\sum_i^N (\mu - x_i)^2$ is minimal, which leads to $\sum_i^N (\mu - x_i) = 0$ and consequently to $N\mu = \sum_i^N x_i$. In the extension

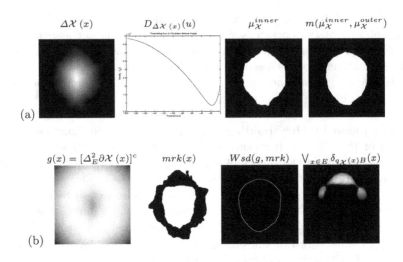

Fig. 2. (a) Computation of extrinsic mean by thresholding the sum of distance functions. (b) Computation of locally optimal mean by watershed of sum of squared distance functions. $\mathcal{X}$ is the population B) of Fig. 1(a), where the last image represents an intermediate step of the quench function reconstruction.

to the case of the shape family $\mathcal{X}$, we start by constructing the sum of distance functions to the frontier sets ∂X_i using the squared Euclidean distance, i.e.,

$$\Delta_E^2 \partial \mathcal{X}(x) = \sum_i^N \Delta_E^2 \partial X_i(x),$$

which takes simultaneously the inner/outer distance functions. The locally minimal contour of $\Delta_E^2 \partial \mathcal{X}(x)$ corresponds, by definition, to the mean shape.

This optimal contour can be easily obtained by computing the watershed line of the inverse of this distance function, i.e.,

$$\partial \mu_{\mathcal{X}}^{wshed} = Wshed([\Delta_E^2 \partial \mathcal{X}(x)]^c, mrk(x)),$$

where the marker function is $mrk(x) = \varepsilon_B(mrk^{in}(x)) \cup \varepsilon_B(mrk^{out}(x))$, with $mrk^{in} = \{\bigcap_i X_i\}$ and $mrk^{out} = \{\bigcup_i X_i\}^c$. Fig. 2(b) depicts an example of the algorithm. Replacing the L^2 norm by the L^1 norm and calculating the watershed from the inverse of $\Delta_E \partial \mathcal{X}(x)$ leads to the contour of the median shape of the family.

We have compared the three approaches $\mu_{\mathcal{X}}^{ms}$, $\mu_{\mathcal{X}}^{thres}$ and $\mu_{\mathcal{X}}^{wshed}$ for computing mean shape from the same family. The three algorithms yield very similar results. However the last method, summing the squared distance function to the contours and extracting its thalweg line (watershed of the inverse function) is by far the most efficient. Moreover, as we show below, it is also useful to compute the shape variance.

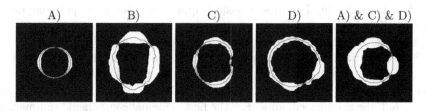

Fig. 3. Mean shapes $\partial\mu_{\mathcal{X}}^{wshed}$ (in red) and std. dev. of shape $\sigma_{\mathcal{X}}$ (in white) for the populations of cells of Fig. 1(a).

2.2 Variance of a Shape Family

Having defined a mean shape, we can now explore the computation of the variance of a shape family. In fact, it is more interesting, for the purpose of representation as an image, to obtain the standard deviation $\sigma_{\mathcal{X}}$.

If we remind that the variance of a set of points is $\sigma^2 = 1/N \sum_i^N (\mu - x_i)^2$, it is evident that the variance can be easily computed from $\Delta_E^2 \partial\mathcal{X}(x)$. More precisely, starting from the squared quench function of the family of shapes $\mathcal{X}$, which is defined as:

$$q_{\mathcal{X}}^2(x) = \begin{cases} (1/N) \cdot \Delta_E^2 \partial\mathcal{X}(x) & \text{if } x \in \partial\mu_{\mathcal{X}}^{wshed} \\ \\ 0 & \text{if } x \in [\partial\mu_{\mathcal{X}}^{wshed}]^c \end{cases}$$

then, the image representation of the standard deviation of shape is obtained by the *reconstruction* of the quench function:

$$\sigma_{\mathcal{X}} = \bigvee_{x \in E} \delta_{q_{\mathcal{X}}(x)}(x).$$

In Fig. 3 are given the mean shape and the std. dev. of shape for the populations of cells of Fig. 1(a).

The notion of shape variance can be also obtained using alternative algorithms. For instance, after computing the distance of each shape X_i to their mean $\mu_{\mathcal{X}}$, i.e., $d_{\mu_{\mathcal{X}}}^{X_i}(x)$, the variance on the frontier of the mean shape, $\partial\mu_{\mathcal{X}}$, can be approached by $1/N \sum_i^N (d_{\mu_{\mathcal{X}}}^{X_i})^2$.

3 Linear Methods for Dimensionality Reduction: Eigenshapes, Modes of Shape Variation

The computation of the mean shape (and variance of shape) has real sense only in the case of homogenous shape families since on collections of very heterogeneous shapes, the mean tends to be a circle. The application of standard techniques of multivariate data analysis can help the exploration of shape families (to determine the homogenous subfamilies) and their representation in spaces of reduced dimension. The most classical approach is the principal component analysis (PCA) [8].

The basic idea is to represent the sets as vectors: $X_i \in \mathcal{P}(E) \rightarrow \boldsymbol{x}_i \in \mathbb{R}^D$ (D is the cardinal of discrete space E), thus the shape family is now given by the following matrix of data $\mathbf{X} = [\boldsymbol{x}_1 \; \boldsymbol{x}_2 \; \cdots \; \boldsymbol{x}_N]$. The covariance matrix of centred data: $\mathbf{C_{XX}} = cov(\widehat{\mathbf{X}})$, where $\widehat{\mathbf{X}} = \mathbf{X} - \overline{\mathbf{X}}$ (if the average $\overline{\mathbf{X}}$ is not subtracted, the average will appear as the first principal component) summarizes the variability of the family, analysed by solving the following spectral problem $\mathbf{C_{XX}} \boldsymbol{w} = \lambda \boldsymbol{w}$, whose eigendecomposition leads to $[\Lambda, \mathbf{W}] = eig(\mathbf{C_{XX}})$, where Λ is the diagonal matrix of different eigenvalues λ_j and $\mathbf{W}$ is the matrix of the associated eigenvectors $\boldsymbol{w}_j$. The relative value of λ_j (i.e., the variance of shape explained by the axis j) is used to determine the number of significant dimensions K.

Fig. 4 illustrates the method. First of all, it is possible to produce an image representation of the K first shape modes $\{\boldsymbol{v}_j\}_{j=1}^K$: $\boldsymbol{v}_j = \widehat{\mathbf{X}} \boldsymbol{w}_j$. The corresponding images of the eigenvectors, $V_j(x)$, are the eigenshapes which correspond to the principal modes of shape variation (see Fig. 4(a)). In addition, the N values of each eigenvector correspond to the projection of each shape onto this vector (see Fig. 4(a)). This can be used typically for shape clustering (i.e., unsupervised classification in shape space in order to identify sub-families of shapes). Another application is the computation of an intrinsic mean shape as follows:

$$\hat{\nu}_\mathcal{X} = \arg_{k \in 1,2,\cdots,N} \min \sum_{i=1}^{N} \left(\sum_{j=1}^{K} (\boldsymbol{s}_j(i) - \boldsymbol{s}_j(k))^2 \right),$$

where $\boldsymbol{s}_j(i) = \mathbf{W}^T \boldsymbol{x}_i$, i.e., $\hat{\nu}_\mathcal{X}$ is the shape which minimises his cumulated distance to the other shapes in the PCA space (see Fig. 4(b)).

PCA has already been applied to shape analysis [7], but the exploitation of the eigenshapes has not yet considered in detail. One of the basic objectives is to decompose the eigenshapes into binary images representing the orthogonal modes of variation. As we observe in the eigenshapes images $V_j(x)$, the modes are differentiated by positive/negative structures on a reference intensity. Using the classical close-holes operator, we can decompose both phases into two different images:

$$V_{j\downarrow}(x) = [CloseHoles(V_j^c(x))]^c \; ; \; V_{j\uparrow} = [CloseHoles(V_j(x))].$$

The objective is to construct two closed binary shapes from $V_{j\downarrow}$ and $V_{j\uparrow}$, but as we can observe in Fig. 4(c), the "modes of shape variation" require an additional "average shape" $V_0(x)$, which is obtained from the image of the average: $\overline{\mathbf{X}} \rightarrow V_0(x)$. The gradient of each image $V_{j\downarrow}$ and $V_{j\uparrow}$ is combined by sum with the gradient of the average image, i.e., $g_0(x) = \delta_1(V_0)(x) - \varepsilon_1(V_0)(x)$. Hence, the two phases of mode of variation j can be now segmented with the watershed transformation as follows:

$$Wshed(\hat{g}_{j\downarrow}(x), mrk(x)) \; ; \; Wshed(\hat{g}_{j\uparrow}(x), mrk(x)),$$

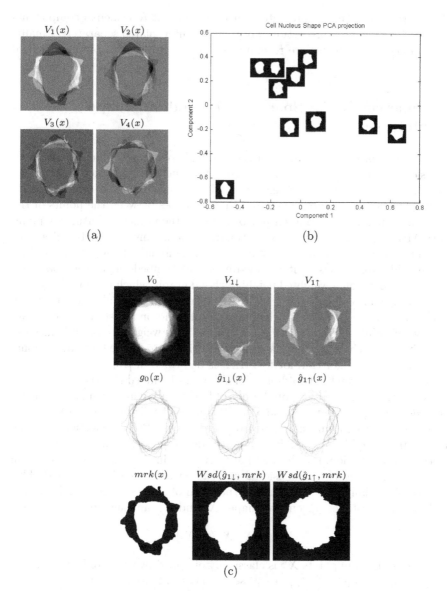

Fig. 4. Shape analysis using PCA of family $\mathcal{X}$ (population B) of Fig. 1(a)): (a) Four first eigenshapes. (b) Projection of shapes on the two first components (in red, intrinsic mean shape). (c) Morphological segmentation of modes of variation from the first eigenshape (see the text for full details).

where $\hat{g}_{j\downarrow}(x) = g_{j\downarrow}(x) + g_0(x)$ with $\hat{g}_{j\downarrow}(x) = \delta_1(V_{j\downarrow})(x) - \varepsilon_1(V_{j\downarrow})(x)$. The obtained images for the example are also given in Fig. 4(c).

Morphological interpolation does not always a good job if two shapes are too dissimilar; PCA can then be used as a useful preprocessing: the shapes X and Y

to be interpolated are first projected onto the different K principal components produced by PCA ; the projections are then are interpolated separately. Finally, the K interpolated shapes are recombined linearly in order to obtain the final shape.

4   Isometric Shape Spaces and Geodesic Shape Interpolation

PCA is based on the covariance matrix of the shape family, which corresponds to consider a dimensionality reduction using the Euclidean distance of shapes. A lot of effort has been paid in recent years to introduce nonlinear dimensionality reduction techniques compatible with other distances between the points of the space. Particularly interesting for our purposes is the isometric feature mapping (ISOMAP) [15]. It is a method for estimating the intrinsic geometry of a data manifold based on a rough positioning of the neighbours of each data point on the manifold. More precisely, it is a low-dimensional embedding method based on geodesic distances on a weighted neighbourhood graph, which is then reduced by multidimensional scaling (MDS). ISOMAP depends on being able to choose the neighbourhood size (k-nearest neighbours graph) and on a distance to compare each pair of points (weights of edges of graph). This weighted graph defines the connectivity of each data point via its nearest neighbours in the high-dimensional space. The precise algorithm for ISOMAP is described in [15].

In Fig. 5(a) is given the two-dimensional ISOMAP embedding (with the neighbourhood graph) for the four populations of Fig. 1(a). We have compared various distances to weight the graph, and again the Jacquard distance outperforms the Hausdorff distance in our examples. Compared to the PCA projection of shape families, the ISOMAP embedding allows to define geodesic paths between the shapes, and in addition, the shortest path distances in the neighbourhood graph are preserved in the two dimensional embedding recovered by ISOMAP. This property is specially useful for the interpolation of shapes in the family $\mathcal{X}$ (see Fig. 5(b)). For instance, given two shapes X_i and X_j, an Euclidean shape path

$$[X^0 = X_i, X^P = X_j]$$

of $P - 1$ intermediate points X^k is classically obtained by thresholding the interpolating function $\text{interp}_{X^0}^{X^P}(x)$ at values $\lambda = (1/P) \cdot k$, with $k = 1, 2, \cdots, P - 1$. Now, using the ISOMAP graph, we can define the geodesic shape path

$$\Pi_{P+1}(X_i, X_j) = \left(X^0 = X_i, X^1, \cdots, X^P = X_j \right),$$

which includes the Q shapes of the family $\mathcal{X}$ belonging to the path. The remaining $(P - Q - 1)$ shapes are computed by the interpolation function according to their respective geodesic distance, i.e., the number of intermediate shapes between two successive shapes X_n and $X_m \in \Pi_{P+1}(X_i, X_j)$ is

$$(P - Q - 1) \cdot (d^{geo}(X_n, X_m)/(d^{geo}(X_i, X_j)),$$

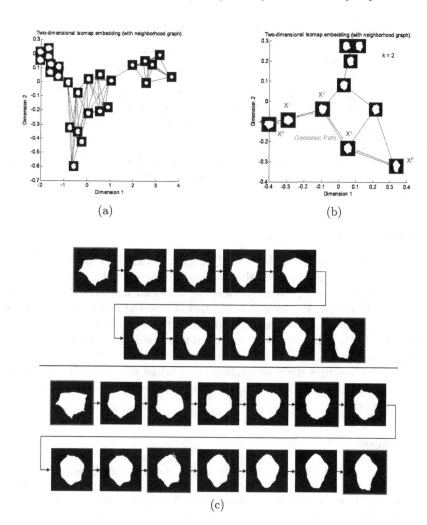

Fig. 5. (a) Projection of four populations of shapes on the two first ISOMAP dimensions (using the Jacquard distance). (b) Idem. for family $\mathcal{X}$: population B) of Fig. 1(a)). (c) Morphological shape interpolation of 8 intermediate shapes between X^0 and X^P (in red): Top, interpolation along an Euclidean shape path; bottom, interpolation along a geodesic path (including 3 shapes of the family in blue).

where $d^{geo}(X,Y)$ is the geodesic distance between the shapes X and Y. See example in Fig. 5(c).

Shape interpolation in reduced spaces has been also studied in [5], with a Delaunay triangulation of the training family of shapes in the reduced space. When a new shape Y is projected in the trained space (which is a difficult problem), the corresponding triangle determines the 3 initial sets which can be used to approach Y by a barycentric-weighted mean shape. This problem can also be solved in our framework. Let us define the weighting interpolation function between sets X and Y as

$$\operatorname*{interp}_{X}^{Y}(x;\omega_X,\omega_Y) = \frac{\omega_X d_X^Y(x)}{\omega_X d_X^Y(x) + \omega_Y d_Y^X(x)},$$

where ω_X and ω_Y are the weights (i.e., ω_X/ω_Y corresponds to the speed of propagation between sets X and Y). The gravity centre between three sets of the family X_i, X_j and X_k is obtained as $X^{i,j,k} = \{x : \operatorname{interp}_{X^{i,j}}^{X_k}(x;2,1) \le 0.5\}$, where $X^{i,j} = \{x : \operatorname{interp}_{X_i}^{X_j}(x;1,1) \le 0.5\}$. The result depends on the processing order however the differences are negligible in practical examples. For the interpolation of a not centred shape in a triangle, the coefficient for each set can be proportionally set up.

5   Conclusions

We have proposed efficient tools for analysing shape families using morphological operators. Various algorithms for the computation of mean shape and variance of shape as well as for the construction of shape priors for modes of shape variation have been introduced. We have also illustrated how the morphological interpolation can be used in shape manifolds to obtain more relevant results. The main motivation of this study was to define prototypes and shape targets for model-based morphological segmentation. More generally, the statistical analysis of shapes families requires the computation of advanced notions such as covariance and probability distribution in shape spaces. This last point will be the object of ongoing work.

References

1. Baddeley, A.J., Molchanov, I.S.: Averaging of Random Sets Based on Their Distance Functions. Journal of Mathematical Imaging and Vision 8, 79–92 (1998)
2. Beucher, S.: Sets, partitions and functions interpolations. In: Mathematical Morphology and its Applications to Image and Signal Processing (Proc. of ISMM 1998), pp. 307–314. Kluwer, Dordrecht (1998)
3. Charpiat, G., Faugeras, O., Keriven, R., Maurel, P.: Distance-based shape statistics. In: IEEE ICAPS 2006, vol. 5, pp. 925–928 (2006)
4. Delfour, M.C., Zolésio, J.-P.: Shape analysis via orientated distance funtions. J. Funct. Anal. 123, 129–201 (1994)
5. Etyngier, P., Segonne, F., Keriven, R.: Shape Priors using Manifold Learning Techniques. In: IEEE ICCV 2007, pp. 1–8 (2007)
6. Grenander, U.: General Pattern Theory. Oxford Univ. Press, Oxford (1993)
7. Horgan, G.W.: Principal Component Analysis of Random Particles. Journal of Mathematical Imaging and Vision 12, 169–175 (2000)
8. Jolliffe, I.T.: Principal Component Analysis. Springer, New York (1989)
9. Kendall, D.G.: A survey of statistical theory of shape. Statistical Science 4, 87–120 (1989)
10. Klassen, E., Srivastava, A., Mio, M., Joshi, S.H.: Analysis of planar shapes using geodesic paths on shape spaces. IEEE Trans. Pattern Anal. Mach. Intell. 26(3), 372–383 (2004)

11. Meyer, F.: Morphological interpolation method for mosaic images. In: Mathematical Morphology and its Applications to Image and Signal Processing (Proc. of ISMM 1996). Kluwer, Dordrecht (1996)

12. Meyer, F.: The levelings. In: Mathematical Morphology and its Applications to Image and Signal Processing (Proc. of ISMM 1998), pp. 199–206. Kluwer, Dordrecht (1998)

13. Meyer, F.: L'espace des formes entre 2 formes X et Y. Rapport Technique CMM-Ecole des Mines de Paris, N-/09/MM (2009)

14. Serra, J.: Hausdorff distance and interpolations. In: Mathematical Morphology and its Applications to Image and Signal Processing (Proc. of ISMM 1998), pp. 107–114. Kluwer, Dordrecht (1998)

15. Tenenbaum, J.B., de Silva, V., Langford, J.C.: A Global Geometric Framework for Nonlinear Dimensionality Reduction. Science 290, 2319–2323 (2000)

16. Whitaker, R.T.: A Level-Set Approach to Image Blending. IEEE Trans. Image Proc. 9(11), 1849–1861 (2000)

From Scalar-Valued Images to Hypercomplex Representations and Derived Total Orderings for Morphological Operators

Jesús Angulo

CMM-Centre de Morphologie Mathématique, Mathématiques et Systèmes, MINES
Paristech; 35, rue Saint Honoré, 77305 Fontainebleau Cedex, France
{jesus.angulo}@ensmp.fr

Abstract. In classical mathematical morphology for scalar images, the
natural ordering of grey levels is used to define the erosion/dilation and
the derived operators. Various operators can be sequentially applied to
the resulting images always using the same ordering. In this paper we
propose to consider the result of a prior transformation to define the
imaginary part of a complex image, where the real part is the initial image. Then, total orderings between complex numbers allow defining subsequent morphological operations between complex pixels. In this case,
the operators take into account simultaneously the information of the
initial image and the processed image. In addition, the approach can be
generalised to the hypercomplex representation (i.e., real quaternion) by
associating to each image three different operations, for instance a directional filter. Total orderings initially introduced for colour quaternions
are used to define the derived morphological transformations. Effects of
these new operators are illustrated with different examples of filtering.

1 Introduction

Let $f(\mathbf{x}) = t$ be a scalar image, $f : E \to \mathcal{T}$. In general $t \in \mathcal{T} \subset \mathbb{Z}$ or $\mathbb{R}$, but for the
sake of simplicity of our study, $\mathcal{T} = \{0, 1, \cdots, t_{\max}\}$ (e.g., $t_{\max} = 255$ for 8 bits
images) is considered as an ordered set of grey-levels; and typically, for digital 2D
images $\mathbf{x} = (x, y) \in E$ where $E \subset \mathbb{Z}^2$ is the support of the image. For 3D images
$\mathbf{x} = (x, y, z) \in E \subset \mathbb{Z}^3$. According to the natural scalar partial ordering $\leq$, $\mathcal{T}$
is a complete lattice, and then $\mathcal{F}(E, \mathcal{T})$ is a complete lattice too. Morphological
operators are naturally defined in the framework of functions $\mathcal{F}(E, \mathcal{T})$ [15,16,9].
Various operators can be sequentially applied to the resulting images always
using the same ordering $\leq$.

The aim of this paper is to construct (hyper)-complex image representations
which will be endowed with total orderings and consequently, which will lead
to complete lattices. More precisely, it is proposed to use the result of a prior
morphological transformation to define the imaginary part of a complex image,
where the real part is the initial scalar image. Then, total orderings between
complex numbers allow defining subsequent morphological operations between

M.H.F. Wilkinson and J.B.T.M. Roerdink (Eds.): ISMM 2009, LNCS 5720, pp. 238–249, 2009.

complex pixels. In this case, the operators take into account simultaneously the scalar intensities of both the initial and the transformed images. The complex scalar value brings information about the invariance of intensities with respect to a particular size and shape structure (i.e., using openings, closings or alternate sequential filters) as well as information about the local contrast of intensities (i.e., by means of top-hat transformations). In addition, the approach is then generalised to the hypercomplex representation (i.e., real quaternion) by associating to each image three different operations, for instance a series directional filters.

The motivation of these methodological developments is to obtain "regularized" morphological operators whose result depends not only on the sup/inf of the grey values, locally computed in the structuring element, but also on differential information or more regional information. The problem has been previously addressed using an inf-semilattice framework [10], working on fuzzy logic morphology [4] or introducing microviscous effects by second-order operators [13].

Geometric algebraic representations have been previously used for image modelling and processing. Classically, in one-dimensional (1D) signal processing, the analytic signal is a powerful complex-model which provides access to local amplitude and phase. The complex signal is built from a real signal by adding its Hilbert transform -which is a phase-shifted version of the signal- as an imaginary part to the signal. The approach was extended to 2D signals and images in [3] by means of the quaternionic Fourier transform. In parallel, another theory introduced in [7] to extend the analytic model in 2D is based on the application of the Riesz transform as generalised Hilbert transform, leading to the notion of monogenic signal which delivers an orthogonal decomposition into amplitude, phase and orientation. Later, the monogenic signal was studied in the framework of scale-spaces [8]. More recently, in [19], the 2D scalar-valued images are embedded into the geometric algebra of the Euclidean 4D space and then the structure are decomposed using monogenic curvature tensor. The quaternion-based representations have been also used to deal with colour image processing, such as colour Fourier transform, colour convolution and linear filters, have been studied mainly by [5,14,6], and to build colour PCA by [18]. We have recently explored also the interest of colour quaternions for extending mathematical morphology to colour images [2].

2   Complex Representation, Total Orderings and Complex Operators

Let $\psi : \mathcal{T} \to \mathcal{T}$ be a morphological operator for scalar images. We need to recall a few notions which characterise the properties of morphological operators. ψ is increasing if $\forall f, g \in \mathcal{F}(E, \mathcal{T})$, $f \leq g \Rightarrow \psi(f) \leq \psi(g)$. It is anti-extensive if $\psi(f) \leq f$ and it is extensive if $f \leq \psi(f)$. An operator is idempotent if $\psi(\psi(f)) = \psi(f)$.

The transformation ψ is applied to $f(\mathbf{x}) \in \mathcal{F}(E, \mathcal{T})$ according to the shape and size associated to the structuring element B and it is denoted as $\psi_B(f)(\mathbf{x})$.

We may now define the following ψ-complex image

$$\mathbf{f}_C(\mathbf{x}) = f(\mathbf{x}) + i\psi_B(f)(\mathbf{x}),$$

with $\mathbf{f}_C \in \mathcal{F}(E, \mathcal{T} \times i\mathcal{T})$. The data of the bivalued image are discrete complex numbers: $\mathbf{f}_C(\mathbf{x}) = \mathbf{c}_n = a_n + ib_n$, where a_n and b_n are respectively the real and the imaginary part of the complex of index n in the discrete space $\mathcal{T} \times i\mathcal{T} \subset \mathbb{C}$. Let us consider the polar representation, i.e., $\mathbf{c}_n = \rho_n \exp(i\theta_n)$, where the modulus is given by $\rho_n = |\mathbf{c}_n| = \sqrt{a_n^2 + b_n^2}$ and the phase is computed as $\theta_n = \arg(\mathbf{c}_n) = \mathrm{atan2}(b_n, a_n) = \mathrm{sign}(b_n)\,\mathrm{atan}(|b_n|/a_n)$, with $\mathrm{atan2}(\cdot) \in (-\pi, \pi]$. The phase can be mapped to $[0, 2\pi)$ by adding 2π to negative values.

Working in the polar representation, two alternative total orderings based on lexicographic cascades can be defined for complex numbers:

$$\mathbf{c}_n \leq_{\Omega_1^{\theta_0}} \mathbf{c}_m \Leftrightarrow \begin{cases} \rho_n < \rho_m \ or \\ \rho_n = \rho_m \ and \ \theta_n \preceq_{\theta_0} \theta_m \end{cases}; \ \mathbf{c}_n \leq_{\Omega_2^{\theta_0}} \mathbf{c}_m \Leftrightarrow \begin{cases} \theta_n \prec_{\theta_0} \theta_m \ or \\ \theta_n =_{\theta_0} \theta_m \ and \ \rho_n \leq \rho_m \end{cases}$$

where $\preceq_{\theta_0}$ depends on the angular difference to a reference angle θ_0 on the unit circle, i.e.,

$$\theta_n \preceq_{\theta_0} \theta_m \ \Leftrightarrow \ \begin{cases} (\theta_n \div \theta_0) > (\theta_m \div \theta_0) \ or \\ (\theta_n \div \theta_0) = (\theta_m \div \theta_0) \ and \ \theta_n \leq \theta_m \end{cases}$$

such that

$$\theta_p \div \theta_q = \begin{cases} |\,\theta_p - \theta_q\,| & if \ |\,\theta_p - \theta_q\,| \leq \pi \\ 2\pi - |\,\theta_p - \theta_q\,| & if \ |\,\theta_p - \theta_q\,| > \pi \end{cases}$$

These total orderings can be easily interpreted. In $\leq_{\Omega_1^{\theta_0}}$, priority is given to the modulus, in the sense that a complex is bigger than another if its modulus is bigger, and if both have the same modulus the bigger value is the one whose phase is closer to the reference θ_0. In case of equal phase angular distances, the last condition for a total ordering is based on closeness to the phase origin, i.e., $\theta = 0$. The ordering $\leq_{\Omega_2^{\theta_0}}$ uses the same priority conditions, but they are reversed. Furthermore, by the equivalence of norms, we can state that $\rho_n \leq \rho_m \Leftrightarrow |a_n| + |b_n| \leq |a_m| + |b_m|$.

Given now a set of pixels of the initial image $[f(\mathbf{z})]_{\mathbf{z} \in Z}$, the basic idea behind our approach is to use, for instance $\leq_{\Omega_1^{\theta_0}}$, for ordering the set Z of initial pixels. Formally, we have

$$\mathbf{f}_C(\mathbf{y}) \leq_{\Omega_1^{\theta_0}} \mathbf{f}_C(\mathbf{z}) \Rightarrow f(\mathbf{y}) \preccurlyeq_{\Omega_1^{\theta_0}} f(\mathbf{z}),$$

where the *indirect total ordering* $\preccurlyeq_{\Omega_1^{\theta_0}}$ allows to compute the supremum $\widetilde{\bigvee}_{\Omega_1^{\theta_0}}$ and the infimum $\widetilde{\bigwedge}_{\Omega_1^{\theta_0}}$ in the original scalar-valued image, i.e., $\mathbf{f}_C(\mathbf{y}) = \bigvee_{\Omega_1^{\theta_0}} [\mathbf{f}_C(\mathbf{z})] \Rightarrow f(\mathbf{y}) = \widetilde{\bigvee}_{\Omega_1^{\theta_0}} [f(\mathbf{z})]$.

We notice that the complex total orderings are only defined once the transformation ψ_B is totally defined. The next question to be studied is what kind of morphological operators are useful to build basic operators such as dilations and erosions.

Adjunction and duality by complementation. The theory of adjunctions on complete lattices has played an important role in mathematical morphology [15,16,9]. The operator ε between the complete lattice $\mathcal{T}$ and itself is an erosion if $\varepsilon\left(\bigwedge_{\Omega^{\theta_0}}[f(\mathbf{x}_k)]\right) = \bigwedge_{\Omega^{\theta_0}} \varepsilon\left([f(\mathbf{x}_k)]\right)$, $k \in I$, for every function $f \in \mathcal{F}(E, \mathcal{T})$. A similar dual definition holds for dilation δ (i.e., commutation with the supremum). The pair (ε, δ) is called an adjunction between $\mathcal{T} \to \mathcal{T}$ iff $\delta\left(f\right)(\mathbf{x}) \leq_{\Omega^{\theta_0}} g(\mathbf{x}) \Leftrightarrow f(\mathbf{x}) \leq_{\Omega^{\theta_0}} \varepsilon\left(f\right)(\mathbf{x})$. If we have an adjunction for the ordering Ω^{θ_0}, the products of (ε, δ) such as the openings and the closings can be defined in a standard way. Hence, it is important that the proposed complex erosions/dilations verifies the property of adjunction.

One of the most interesting properties of standard grey-level morphological operators is the duality by the complementation $\complement$. The complement image (or negative image) $\complement f$ is defined as the reflection of f with respect to $t_{\max}/2$; i.e., $\complement f(\mathbf{x}) = t_{\max} - f(\mathbf{x}) = f^c(\mathbf{x})$, $\forall \mathbf{x} \in E$. Let the pair (ε, δ) be an adjunction, the property of duality holds that $\varepsilon(f^c) = (\delta(f))^c \Rightarrow \varepsilon(f) = (\delta(f^c))^c$, and this is verified for any other pair of dual operators, such as the opening/closing. In practice, this property allows us to implement exclusively the dilation, and using the complement, to be able to obtain the corresponding erosion. In our case, the transformation $f \to \complement f \Rightarrow \mathbf{f}_C \to \tilde{\mathbf{f}}_C = \complement f + i\psi_B(\complement f) = \complement f + i\complement\xi_B(f)$, where $\xi_B(f) = \complement\psi_B(\complement f)$ is the dual operator of ψ_B. Note that this is different from the complement of the ψ-complex image: $\complement \mathbf{f}_C = \complement f + i\complement\psi_B(f)$.

Ordering invariance and commutation under anamorphosis. The concepts of ordering invariance and of commutation of sup and inf operators under intensity image transformations is also important in the theory of complete lattices [11,17]. More precisely, in mathematical morphology a mapping $A : \mathcal{T} \to \mathcal{T}$ which satisfies the criterion $t \leq_{\Omega} s \Leftrightarrow A(t) \leq_{\Omega} A(s) \forall t, s \in \mathcal{T}$ is called an anamorphosis. Then, we say that the ordering $\leq_{\Omega}$ is invariant under A. Any increasing morphological operator ψ commutes with any anamorphosis, i.e., $\psi_B(A(f)) = A(\psi_B(f))$. It is well known for the grey-tone case that any strictly increasing mapping A is an anamorphosis.

A typical example is the linear transformation $A(t) = K(t)$ if $0 \leq K(t) \leq t_{\max}$, $A(t) = 0$ if $K(t) < 0$ and $A(t) = t_{\max}$ if $K(t) > t_{\max}$, where $K(t) = \alpha t + \beta$, with $\alpha \geq 0$. In our case, we have for the ψ-complex image: $f \to f' = A(f) \Rightarrow \mathbf{f}_C \to \mathbf{f}'_C = A(f) + i\psi_B(A(f)) = A(f) + iA(\psi_B(f))$, i.e., both axes of complex plane are modified according to the same mapping (scaled and shifted for the example of the linear transformation). Obviously, the partial ordering according to the modulus is invariant under A. The partial ordering with respect to the phase is also invariant if θ is defined in the first quadrant. Hence, the orderings $\Omega_1^{\theta_0}$ and $\Omega_2^{\theta_0}$ commutes with anamorphosis applied on the scalar function f.

γ-complex dilation and φ-complex erosion. A morphological filter is an increasing operator that is also idempotent (i.e., erosion and dilation are not idempotent). The two basic morphological filters, the opening γ_B and the closing φ_B, seem particularly appropriate to build the ψ-complex image. Besides the idempotence, the opening (closing) is an anti-extensive (extensive) operator which removes bright structures and peaks of intensity (dark structures and valleys of intensity) that are thinner than the structuring element B, the structures larger than B preserve their intensity values.

Let us use the diagram depicted in Fig. 1(a) for illustrating how the complex values are ordered. If we consider for instance $\psi \equiv \gamma$ and $\Omega_1^{\theta_0}$, the pixels in structures invariant according to B have module values which are bigger than pixels having the same initial grey value but belonging to structures that do not match B. In the diagram, c_1 is bigger than c_3, but for c_1 and c_2 which have the same modulus, a reference θ_0 is needed. By the anti-extensivity, we have $f(\mathbf{x}) \geq \gamma_B(f)(\mathbf{x})$ and hence $0 \leq \theta \leq \pi/4$. By taking $\theta_0 = \pi/2$, we consider that, with equal modulus, a point is bigger than another if the intensities before and after the opening are more similar (i.e., more invariant). Or in other words, when the ratio $\gamma_B(f)(\mathbf{x})/f(\mathbf{x})$ is closer to 1 or θ is closer to $\pi/4$, and consequently to $\pi/2$. This implies that the opening is an appropriate transformation to define a dilation which propagate the bright intensities associated in priority to B-invariant structures.

The same analysis leads to easily justify the choice of $\psi \equiv \varphi$ and $\Omega_1^{\pi/2}$ for the complex erosion. Note that now $\varphi_B(f)(\mathbf{x})/f(\mathbf{x}) \geq 1 \Rightarrow \pi/4 \leq \theta \leq \pi/2$. By taking the reference $\theta_0 = \pi/2$, the idea of intensities invariance before and after the closing is again used, in the ordering by θ, for considering now that a point is smaller than another if both have the same modulus and the first is closer to $\theta_0 = \pi/4$ than the second (in the example of Fig. 1(a), c_5 is smaller than c_4).

Mathematically, the γ-*complex dilation* is defined by

$$\begin{cases} \mathbf{f}_C(\mathbf{x}) = f(\mathbf{x}) + i\gamma_{B_C}(f)(\mathbf{x}), \\ \delta_{\langle \gamma_{B_C}, B \rangle}(f)(\mathbf{x}) = \{f(\mathbf{y}) : \mathbf{f}_C(\mathbf{y}) = \bigvee_{\Omega_1^{\pi/2}}[\mathbf{f}_C(\mathbf{z})], \mathbf{z} \in B(\mathbf{x}))\}. \end{cases}$$

and the dual φ-*complex erosion* is formulated as follows:

$$\begin{cases} \mathbf{f}_C(\mathbf{x}) = f(\mathbf{x}) + i\varphi_{B_C}(f)(\mathbf{x}), \\ \varepsilon_{\langle \varphi_{B_C}, B \rangle}(f)(\mathbf{x}) = \{f(\mathbf{y}) : \mathbf{f}_C(\mathbf{y}) = \bigwedge_{\Omega_1^{\pi/2}}[\mathbf{f}_C(\mathbf{z})], \mathbf{z} \in B(\mathbf{x}))\}. \end{cases}$$

The complex operator requires two independent structuring elements: B_C associated to the imaginary part; and B which is properly the structuring element of the complex transformation. Obviously, B_C and B can have different size and shape.

The pair $\left(\delta_{\langle \gamma_{B_C}, B \rangle}, \varepsilon_{\langle \varepsilon_{B_C}, B \rangle} \right)$ is an *adjunction*, i.e., $\delta_{\langle \gamma_{B_C}, B \rangle}(f)(\mathbf{x}) \preccurlyeq_{\Omega_1^{\pi/2}} g(\mathbf{x})$ $\Leftrightarrow f(\mathbf{x}) \preccurlyeq_{\Omega_1^{\pi/2}} \varepsilon_{\langle \varphi_{B_C}, B \rangle}(g)(\mathbf{x})$, for any $f, g \in \mathcal{F}(E, \mathcal{T})$. The proof is as follows. We consider the values of points $\mathbf{z} \in B(\mathbf{x})$, and we have $\bigvee_{\Omega_1^{\pi/2}}[\mathbf{f}_C(\mathbf{z})] \leq_{\Omega_1^{\pi/2}} \mathbf{g}_C(\mathbf{x})$ $\Leftrightarrow \bigvee_{\Omega_1^{\pi/2}}[f(\mathbf{z}) + i\gamma_{B_C}(f)(\mathbf{z})] \leq_{\Omega_1^{\pi/2}} g(\mathbf{x}) + i\gamma_{B_C}(g)(\mathbf{x}) \Leftrightarrow f(\mathbf{x}) + i\gamma_{B_C}(f)(\mathbf{x})$ $\leq_{\Omega_1^{\pi/2}} g(\mathbf{x}) + i\gamma_{B_C}(g)(\mathbf{x}) \leq_{\Omega_1^{\pi/2}} g(\mathbf{x}) + i\varphi_{B_C}(g)(\mathbf{x})$. On the other hand, we have $f(\mathbf{x}) + i\varphi_{B_C}(f)(\mathbf{x}) \leq_{\Omega_1^{\pi/2}} \bigwedge_{\Omega_1^{\pi/2}}[g(\mathbf{z}) + i\varphi_{B_C}(g)(\mathbf{z})] \Leftrightarrow f(\mathbf{x}) + i\gamma_{B_C}(f)(\mathbf{x}) \leq_{\Omega_1^{\pi/2}}$ $f(\mathbf{x}) + i\varphi_{B_C}(f)(\mathbf{x}) \leq_{\Omega_1^{\pi/2}} g(\mathbf{x}) + i\varphi_{B_C}(g)(\mathbf{x})$. Consequently, we establish that $\bigvee_{\Omega_1^{\pi/2}}[\mathbf{f}_C(\mathbf{z})] \leq_{\Omega_1^{\pi/2}} \mathbf{g}_C(\mathbf{x}) \Leftrightarrow \mathbf{f}_C(\mathbf{x}) \leq_{\Omega_1^{\pi/2}} \bigwedge_{\Omega_1^{\pi/2}}[\mathbf{g}_C(\mathbf{z})].$

In addition, if we apply the dilation to the complemented original image, we have $\bigvee_{\Omega_1^{\pi/2}}[f^c(\mathbf{z}) + i\gamma_{B_C}(f^c)(\mathbf{z})] = \bigvee_{\Omega_1^{\pi/2}}[f^c(\mathbf{z}) + i\varphi_{B_C}^c(f)(\mathbf{z})] = \left[\bigwedge_{\Omega_1^{\pi/2}}[f^c(\mathbf{z}) + i\varphi_{B_C}^c(f)(\mathbf{z})]^c\right]^c = \left[\bigwedge_{\Omega_1^{\pi/2}}[f(\mathbf{z}) + i\varphi_{B_C}(f)(\mathbf{z})]\right]^c$, $\mathbf{z} \in B(\mathbf{x})$. Hence, we have the following classical result of *duality*: $\delta_{\langle\gamma_{B_C},B\rangle}(f) = \left[\varepsilon_{\langle\varphi_{B_C},B\rangle}(f^c)\right]^c$.

It should be remarked that the dilation is extensive according to the ordering $\Omega_1^{\pi/2}$: $f(\mathbf{x}) \preccurlyeq_{\Omega_1^{\pi/2}} \delta_{\langle\gamma_{B_C},B\rangle}(f)(\mathbf{x})$, but not necessarily according to the standard ordering: $f(\mathbf{x}) \not\leq \delta_{\langle\gamma_{B_C},B\rangle}(f)(\mathbf{x})$. If this last property is required for any reason, we can define the γ-*complex upper dilation* as:

$$\widehat{\delta}_{\langle\gamma_{B_C},B\rangle}(f)(\mathbf{x}) = \delta_{\langle\gamma_{B_C},B\rangle}(f)(\mathbf{x}) \vee f(\mathbf{x}).$$

Using the standard infimum $\wedge$, the dual definition leads to the φ-*complex lower erosion* $\widehat{\varepsilon}_{\langle\phi_{B_C},B\rangle}(f)(\mathbf{x})$, which is anti-extensive according to the grey level ordering.

As we show below, because they constitute an adjunction, the γ-complex dilation and the φ-complex erosion can be combined to construct derived γ, φ-complex operators such as gradients, openings/closing and even geodesic operators (e.g., opening by reconstruction, leveling, etc.). Instead of a morphological opening/closing for the γ-complex dilation and the φ-complex erosion, any other pair of anti-extensive extensive dual transformation can play a similar role, e.g. opening/closing by reconstruction, or thinning/thickening which are not idempotent operators. It is also possible to consider the ordering $\Omega_2^{\theta_0}$ for computing a complex dilation and erosion where the complex part is an opening or a closing respectively. However, we prefer to illustrate other possible ways for introducing the complex part.

τ^+-complex dilation and τ^--complex erosion.

Let us consider another family of complex dilation/erosion using now the residues of the opening/closing. We remind that the top-hat and the dual top-hat are respectively the residue of the opening and the closing [12], i.e., $\tau_B^+(f)(\mathbf{x}) = f(\mathbf{x}) - \gamma_B(f)(\mathbf{x})$ and $\tau_B^-(f)(\mathbf{x}) = \varphi_B(f)(\mathbf{x}) - f(\mathbf{x})$. The top-hat transformations yield positive grey-level images and are used to extract contrasted components (i.e., smaller than the structuring element used for the opening/closing) with respect to the background and removing the slow trends. The top-hat is an idempotent transformation and if $f(\mathbf{x}) \geq 0$ then $\tau^+(f)(\mathbf{x})$ is anti-extensive and $[\tau^-(f)]^c(\mathbf{x})$ is extensive.

We introduce, with the help of the top-hats, the τ^+-complex dilation as

$$\begin{cases} \mathbf{f}_C(\mathbf{x}) = f(\mathbf{x}) + i\tau_{B_C}^+(f)(\mathbf{x}), \\ \delta_{\langle\tau_{B_C}^+,B\rangle}(f)(\mathbf{x}) = \{f(\mathbf{y}) : \mathbf{f}_C(\mathbf{y}) = \bigvee_{\Omega_2^{\pi/4}}[\mathbf{f}_C(\mathbf{z})], \mathbf{z} \in B(\mathbf{x}))\}. \end{cases}$$

and the equivalent τ^--complex erosion defined by

$$\begin{cases} \mathbf{f}_C(\mathbf{x}) = f(\mathbf{x}) - i[\tau_{B_C}^-(f)]^c(\mathbf{x}), \\ \varepsilon_{\langle\tau_{B_C}^-,B\rangle}(f)(\mathbf{x}) = \{f(\mathbf{y}) : \mathbf{f}_C(\mathbf{y}) = \bigwedge_{\Omega_2^{-\pi/4}}[\mathbf{f}_C(\mathbf{z})], \mathbf{z} \in B(\mathbf{x}))\}. \end{cases}$$

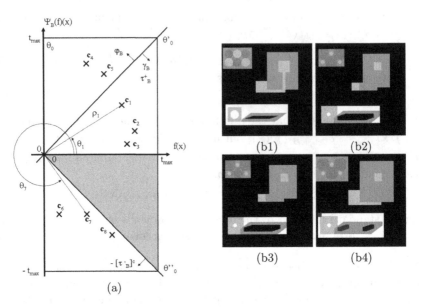

(a)

Fig. 1. (a) Complex plane for points $\in f(\mathbf{x}) + i\psi_B(f)(\mathbf{x})$. Standard erosion vs. complex erosions: (b1) original image f, (b2) erosion $\varepsilon_B(f)(\mathbf{x})$, (b3) φ-complex erosion $\varepsilon_{\langle\varphi_{B_C},B\rangle}(f)(\mathbf{x})$, (b4) τ^--complex erosion $\varepsilon_{\langle\tau^-_{B_C},B\rangle}(f)(\mathbf{x})$. For the three examples B is a square of size 10 pixels and B_C is a square of 40 pixels.

We must remark that for the τ^--erosion $\mathbf{f}_C \in \mathcal{F}(E, \mathcal{T} \times -i\mathcal{T})$. We can use again the diagram of Fig. 1(a) to interpret these operators. By using the ordering $\Omega_2^{\theta_0}$, the supremum favours the complex points closer to $\pi/4$ which correspond to those where the initial intensity is more similar to the intensity of the top-hat (the point $\mathbf{c}_1$ is bigger than the points $\mathbf{c}_2$ and $\mathbf{c}_3$); or in other words, the points belonging to structures well contrasted with respect to B_C. The intensity itself has only a secondary influence which is introduced by the modulus (second condition in the lexicographic ordering $\Omega_2^{\theta_0}$). In the case of the erosion, a point is smaller than another is the θ of the first is closer to $-\pi/4$ than the θ of the second. In the diagram, $\mathbf{c}_7$ is smaller than $\mathbf{c}_6$ and $\mathbf{c}_8$ is the smallest between the three (even if $\mathbf{c}_8$ presents the biggest initial intensity). In summary, by means of the τ^+-complex dilation and τ^--complex erosion, we obtain a mechanism of filtering based on the local contrast of structures and the expected results should be quite different of the standard dilation and erosion. Due to the fact that the modulus of the complex number at point x is strongly correlated to its initial intensity at x, the γ-complex dilation and φ-complex erosion lead to filtering effects more similar to the standard ones. This analysis is illustrated by the comparative example of erosion depicted in Fig. 1(b). The properties of the pair of operators $\left(\delta_{\langle\tau^+_{B_C},B\rangle}, \varepsilon_{\langle\tau^-_{B_C},B\rangle}\right)$ cannot be explored in detail by the limited length of the paper.

3   Generalisation to Multi-operator Cases Using Real Quaternions

We generalise the ideas introduced above by the extension to image representations based on hypercomplex numbers or real quaternions. Before that, we remind the foundations of quaternions.

Remind on quaternionic representations. A quaternion $\mathbf{q} \in \mathbb{H}$ may be represented in hypercomplex form as $\mathbf{q} = a + bi + cj + dk$, where a, b, c and d are real. A quaternion has a *real part or scalar part*, $S(\mathbf{q}) = a$, and an *imaginary part or vector part*, $V(\mathbf{q}) = bi + cj + dk$, such that the whole quaternion may be represented by the sum of its scalar and vector parts as $\mathbf{q} = S(\mathbf{q}) + V(\mathbf{q})$. A quaternion with a zero real/scalar part is called a *pure quaternion*. The addition of two quaternions, $\mathbf{q}, \mathbf{q}' \in \mathbb{H}$, is defined as follows $\mathbf{q} + \mathbf{q}' = (a+a') + (b+b')i + (c+c')j + (d+d')j$. The addition is commutative and associative. The quaternion result of the *product of two quaternions* $\mathbf{q}'' = \mathbf{qq}' = S(\mathbf{q}'') + V(\mathbf{q}'')$ can be written in terms of dot product $\cdot$ and cross product $\times$ of vectors as $S(\mathbf{q}'') = S(\mathbf{q})S(\mathbf{q}') - V(\mathbf{q}) \cdot V(\mathbf{q}')$ and $V(\mathbf{q}'') = S(\mathbf{q})V(\mathbf{q}') + S(\mathbf{q}')V(\mathbf{q}) + V(\mathbf{q}) \times V(\mathbf{q}')$. The multiplication of quaternions is not commutative, i.e., $\mathbf{qq}' \neq \mathbf{q}'\mathbf{q}$; but it is associative.

Any quaternion may be represented in polar form as $\mathbf{q} = \rho e^{\xi\theta}$, with $\rho = \sqrt{a^2 + b^2 + c^2 + d^2}$, $\xi = \frac{bi+cj+dk}{\sqrt{b^2+c^2+d^2}} = \overline{\xi}_i i + \overline{\xi}_j j + \overline{\xi}_k k$ and $\theta = \arctan\left(\frac{\sqrt{b^2+c^2+d^2}}{a}\right)$. Then a quaternion can be rewritten in a trigonometric version as $\mathbf{q} = \rho(\cos\theta + \xi \sin\theta)$. In the polar formulation, $\rho = |\mathbf{q}|$ is the modulus of $\mathbf{q}$; ξ is the pure unitary quaternion associated to $\mathbf{q}$ (by the normalisation, the pure unitary quaternion discards "intensity" information, but retains orientation information), sometimes called *eigenaxis*; and θ is the angle, sometimes called *eigenangle*, between the real part and the 3D imaginary part. It is possible to describe vector decompositions using the product of quaternions. A full quaternion $\mathbf{q}$ may be decomposed about a pure unit quaternion $\mathbf{p}^u$ [5,6]: $\mathbf{q} = \mathbf{q}_\perp + \mathbf{q}_\|$, the *parallel part* of $\mathbf{q}$ according to $\mathbf{p}^u$, also called the *projection part*, is given by $\mathbf{q}_\| = S(\mathbf{q}) + V_\|(\mathbf{q})$, and the *perpendicular part*, also named the *rejection part*, is obtained as $\mathbf{q}_\perp = V_\perp(\mathbf{q})$ where $V_\|(\mathbf{q}) = \frac{1}{2}(V(\mathbf{q}) - \mathbf{p}^u V(\mathbf{q})\mathbf{p}^u)$ and $V_\perp(\mathbf{q}) = \frac{1}{2}(V(\mathbf{q}) + \mathbf{p}^u V(\mathbf{q})\mathbf{p}^u)$.

Total orderings for quaternions. Total orderings introduced initially for colour quaternions [2] can be used also to define the derived morphological transformations. Hence, we can generalise the polar-based total orderings proposed above, now named $\leq_{\Omega_1^{qo}}$ and $\leq_{\Omega_2^{qo}}$, by including an additional condition in the cascade associated to the distance of the eigenaxis between the quaternions n and m and a reference quaternion, i.e., $\|\xi_n - \xi_0\| \geq \|\xi_m - \xi_0\|$. Note that here the reference θ_0 is the eigenangle from the reference quaternion q_0 (and ξ_0 is the eigenaxis of the reference). The total $\leq_{\Omega_3^{qo}}$ is defined by considering as first condition in the lexicographical cascade the distance to reference eigenangle. But we can also introduce another total ordering based on the $\| / \perp$ decomposition as follows:

$$\mathbf{q}_n \leq_{\Omega_4^{\mathbf{q}_0}} \mathbf{q}_m \quad \Leftrightarrow \quad \begin{cases} |\mathbf{q}_{\| \, n}| < |\mathbf{q}_{\| \, m}| \ \ or \\ |\mathbf{q}_{\| \, n}| = |\mathbf{q}_{\| \, m}| \ \ and \ \ |\mathbf{q}_{\perp \, n}| \geq |\mathbf{q}_{\perp \, m}| \end{cases}$$

The pure unitary quaternion required for the $\| \, / \perp$ decomposition is just the corresponding to the reference quaternion $\mathbf{q}_0$. The last one is only a pre-ordering, i.e., two distinct quaternions can verify the equality of the ordering. In order to have total ordering, the lexicographic cascade can be completed with a priory in the choice of the various hypercomplex components.

$(\boldsymbol{\Psi}^+, \Omega_+^{\mathbf{q}_0})$-**hypercomplex dilation and** $(\boldsymbol{\Psi}^-, \Omega_-^{\mathbf{q}_0})$-**hypercomplex erosion.** Given the four-variate transformation $\boldsymbol{\Psi} = (\psi_{B_0}^0, \psi_{B_I}^I, \psi_{B_J}^J, \psi_{B_K}^K)$, the $\boldsymbol{\Psi}$-hypercomplex image is defined as

$$\mathbf{f}_H(\mathbf{x}) = \psi_{B_0}^0(f)(\mathbf{x}) + i\psi_{B_I}^I(f)(\mathbf{x}) + j\psi_{B_J}^J(f)(\mathbf{x}) + k\psi_{B_K}^K(f)(\mathbf{x}).$$

After choosing a particular $\boldsymbol{\Psi}^+$-hypercomplex representation as well as a particular quaternionic total ordering $\Omega_+^{\mathbf{q}_0}$, one defines the $(\boldsymbol{\Psi}^+, \Omega_+^{\mathbf{q}_0})$-hypercomplex dilation as $\delta_{\langle \boldsymbol{\Psi}^+, B \rangle}(f)(\mathbf{x}) = \{f(\mathbf{y}) : \mathbf{f}_H(\mathbf{y}) = \bigvee_{\Omega_+^{\mathbf{q}_0}} [\mathbf{f}_H(\mathbf{z})], \mathbf{z} \in B(\mathbf{x}))\}$. Similarly, associated to the pair $(\boldsymbol{\Psi}^-, \Omega_-^{\mathbf{q}_0})$, a $\boldsymbol{\Psi}^-$-hypercomplex erosion is defined. We should notice that this framework generalise the complex operators introduced in previous section; e.g., $\boldsymbol{\Psi}^+ = (\mathrm{Id}, \gamma_{B_C}, 0, 0)$ and $\Omega_+^{\mathbf{q}_0} \equiv \Omega_1^{\mathbf{q}_0}$ with $\mathbf{q}_0 = 1 + i$ corresponds to the γ-complex dilation. Therefore, two kinds of degrees of freedom must be set up (i.e., the hypercomplex transformation and the quaternionic ordering, which includes the choice of the reference quaternion) to have totally stated the operator. As we show below in the examples, the four-variate transformation can be used for instance to introduce directional openings, directional top-hats or directional gradients for the imaginary part, and real counterpart being an isotropic operator which cannot be necessary the identity. This framework can be specifically exploited by working with $\Omega_4^{\mathbf{q}_0}$ and selecting the appropriate reference quaternion for the $\| \, / \perp$ decomposition. The properties of these generic four-variate hypercomplex operators should be studied in depth in ongoing research.

4 Examples of Derived Morphological Operators and Applications

The hypercomplex dilations/erosions introduced in the paper can be used to build more advanced operators such as gradients, openings/closings (and their residues), alternate sequential filters; and geodesic operators, such as the openings/closings by reconstruction or the levelings. The principle entails using always a homogeneous pair of basic operators and then applying the standard definitions for the evolved operators.

In Fig. 2 are given three comparative examples of morphological filters applied to natural images. The idea is to compare the standard result with various hypercomplex operators. In the first case we calculate a symmetric gradient (i.e., dilation minus erosion) of a very noisy image. It is observed that each

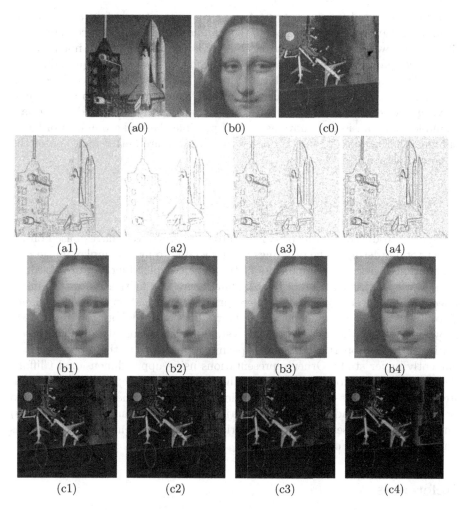

Fig. 2. First row: original images. Second row, comparison of gradient $\varrho_B(f) = \delta_B(f) - \varepsilon_B(f)$ (B is a square of 3 pixels): (a1) standard transformation, (a2) τ^+-complex dilation and τ^--complex erosion where $B_C = D_5$ is a square of size 5, (a3) $\boldsymbol{\Psi}^{+/-} = (\gamma_{B_0}/\varphi_{B_0}, \gamma_{B_I}/\varphi_{B_I}, \gamma_{B_J}/\varphi_{B_J}, 0)$, $\Omega_3^{\mathbf{q_0}}$ with $\mathbf{q_0} = i + j$, where $B_0 = D_5$, $B_I = L_5^x$ is an horizontal line of length 5 and $B_J = L_5^y$ is an vertical line of length 5, (a4) $\boldsymbol{\Psi}^{+/-} = \left(\tau_{D_5}^+/ - [\tau_{D_5}^-]^c, \tau_{L_5^x}^+/ - [\tau_{L_5^x}^-]^c, \tau_{L_5^y}^+/ - [\tau_{L_5^y}^-]^c, 0\right)$, $\Omega_3^{\mathbf{q_0}}$ with $\mathbf{q_0} = i + j$. Third row, comparison of $\gamma_B\varphi_B(f)$ (B is a square of 3 pixels): (b1) standard transformation, (b2) based on γ-complex dilation and φ-complex erosion with $B_C = D_{10}$, (b3) $\boldsymbol{\Psi}^{+/-} = \left(\gamma_{D_{10}}/\varphi_{D_{10}}, \gamma_{L_{10}^x}/\varphi_{L_{10}^x}, \gamma_{L_{10}^y}/\varphi_{L_{10}^y}, 0\right)$, $\Omega_3^{\mathbf{q_0}}$ with $\mathbf{q_0} = i + j$, (b4) idem. with $\mathbf{q_0} = i$. Four row, comparison of top-hat τ_B^+ (B is a square of 25 pixels): (c1) standard transformation, (c2) based on γ-complex dilation and φ-complex erosion with $B_C = D_{10}$, (c3) $\boldsymbol{\Psi}^{+/-} = \left(\gamma_{D_{10}}/\varphi_{D_{10}}, \gamma_{L_{10}^x}/\varphi_{L_{10}^x}, \gamma_{L_{10}^y}/\varphi_{L_{10}^y}, 0\right)$, $\Omega_3^{\mathbf{q_0}}$ with $\mathbf{q_0} = i + j$, (c4) $\boldsymbol{\Psi}^{+/-} = \left(\gamma_{D_{10}}/\varphi_{D_{10}}, \varrho_{L_{10}^x}/ - \varrho_{L_{10}^x}, \varrho_{L_{10}^y}/ - \varrho_{L_{10}^y}, 0\right)$, $\Omega_3^{\mathbf{q_0}}$ with $\mathbf{q_0} = i + j$.

hypercomplex gradient presents particular characteristics but in any case the results are regularised with respect to the standard gradient. The second example illustrates how an alternate open/close filter is used to simplify a noisy image. We notice again that the behaviour of hypercomplex operators is quite different of standard ones, and in particular, we observe the way to introduce directional effects on the filters by decomposing the quaternions according to a particular privileged direction. The effects of a top-hat are finally compared in the last example. The aim is to remove as well as possible the background, in order to enhance the aeroplanes. As we observe, the results are better for some of the hypercomplex operators.

5   Conclusions and Perspectives

We have introduced morphological operators for grey-level images based on indirect total orderings. The orderings are associated to hypercomplex image representations where the components of the hypercomplex function are obtained from a prior transformation of the original image. The motivation was to introduce in the basic erosion/dilation operators some information on size invariance or on relative contrast of structures. The results obtained from the initial tests showed their potential applicative interest. However, a more detailed characterisation of their properties and some specific applications of these operators are currently under study. Other representations using upper dimensional Clifford Algebras [1] can be foreseen in order to have a more generic framework not limited to four-variables image representations. In addition, the approach can also be extended to already natural multivariate images (i.e., multispectral images) and, in this last case, it seems appropriate to envisage tensor representations and associated total orderings.

References

1. Ablamowicz, R., Sobczyk, G.: Lectures on Clifford (Geometric) Algebras and Applications. Birkhäuser, Basel (2004)
2. Angulo, J.: Quaternion colour representations and derived total orderings for morphological operators. In: Proc. of the CGIV 2008, pp. 417–422 (2008)
3. Bülow, T., Sommer, G.: Hypercomplex Signals - A Novel Extension of the Analytic Signal to the Multidimensional Case. IEEE Trans. Signal Proc. 49(11), 2844–2852 (2001)
4. Deng, T.Q., Heijmans, H.J.A.M.: Gray-scale Morphology Based on Fuzzy Logic. J. Math. Imaging Vision 16(2), 155–171 (2002)
5. Ell, T.A., Sangwine, S.J.: Hypercomplex Wiener-Khintchine theorem with application to color image correlation. In: IEEE ICIP 2000, vol. II, pp. 792–795 (2000)
6. Ell, T.A., Sangwine, S.J.: Hypercomplex Fourier transform of color images. IEEE Trans. Image Proc. 16(1), 22–35 (2007)
7. Felsberg, M., Sommer, G.: The Monogenic Signal. IEEE Trans. Signal Proc. 49(12), 3136–3144 (2001)

8. Felsberg, M., Sommer, G.: The monogenic scale-space: A unifying approach to phase-based image processing in scale-space. J. Math. Imaging Vision 21, 5–26 (2004)

9. Heijmans, H.J.A.M.: Morphological Image Operators. Academic Press, Boston (1994)

10. Heijmans, H.J.A.M., Keshet, R.: Inf-Semilattice Approach to Self-Dual Morphology. J. Math. Imaging Vision 17(1), 55–80 (2002)

11. Matheron, G.: Les treillis compacts. Technical Report - Paris School of Mines, N-23/90/G (1990)

12. Meyer, F.: Constrast features extraction. In: Chermant (ed.) Quantitative Analysis of Microstructures in Materials Science, Biology and Medecine, pp. 374–380. Riederer Verlag (1977)

13. Meyer, F., Angulo, J.: Micro-viscous morphological operators. In: Mathematical Morphology and its applications to Signal and Image Processing (ISMM 2007), pp. 165–176 (2007)

14. Sangwine, S.J., Ell, T.A.: Mathematical approaches to linear vector filtering of colour images. In: Proc. CGIV 2002, pp. 348–351 (2002)

15. Serra, J.: Image Analysis and Mathematical Morphology, vol. I. Academic Press, London (1982)

16. Serra, J.: Image Analysis and Mathematical Morphology. Theoretical Advances, vol. II. Academic Press, London (1988)

17. Serra, J.: Anamorphoses and Function Lattices (Multivalued Morphology). In: Dougherty (ed.) Mathematical Morphology in Image Processing, pp. 483–523. Marcel-Dekker, New York (1992)

18. Shi, L., Funt, B.: Quaternion color texture segmentation. Computer Vision and Image Understanding 107(1-2), 88–96 (2007)

19. Zang, D., Sommer, G.: Signal modeling for two-dimensional image structures. J. Vis. Commun. Image R. 18, 81–99 (2007)

A Directional Rouy-Tourin Scheme
for Adaptive Matrix-Valued Morphology

Luis Pizarro, Bernhard Burgeth, Michael Breuß, and Joachim Weickert

Mathematical Image Analysis Group
Faculty for Mathematics and Computer Science
Saarland University, Building E11, 66041 Saarbrücken, Germany
{pizarro,burgeth,breuss,weickert}@mia.uni-saarland.de
http://www.mia.uni-saarland.de

Abstract. In order to describe anisotropy in image processing models or physical measurements, matrix fields are a suitable choice. In diffusion tensor magnetic resonance imaging (DT-MRI), for example, information about the diffusive properties of water molecules is captured in symmetric positive definite matrices. The corresponding matrix field reflects the structure of the tissue under examination. Recently, morphological partial differential equations (PDEs) for dilation and erosion known for grey scale images have been extended to matrix-valued data.

In this article we consider an adaptive, PDE-driven dilation process for matrix fields. The anisotropic morphological evolution is steered with a matrix constructed from a structure tensor for matrix valued data. An important novel ingredient is a directional variant of the matrix-valued Rouy-Tourin scheme that enables our method to complete or enhance anisotropic structures effectively. Experiments with synthetic and real-world data substantiate the gap-closing and line-completing properties of the proposed method.

1 Introduction

The enhancement and extraction of shape information from image objects is one of the principle tasks of mathematical morphology. Traditionally this task is successfully tackled with morphological operations based on the fundamental dilation process. Dilation and erosion can be realised in a set-theoretic or ordering based framework, see e.g. [1,2,3,4,5,6], but it may also be implemented within the context of partial differential equations (PDEs) [7,8,9,10,11] and their numerical solution schemes (see [12] as well as the extensive list of literature cited there). On a set-theoretic basis, locally adaptive linear structuring elements whose directions are inferred from a diffused squared gradient field have been introduced for binary images in [13]. The PDE-based approach is conceptually attractive since it allows for digital scalability and even adaptivity of the represented structuring element. This versatility was exploited, for example in [14] to create a adaptive, PDE-based dilation process for grey value images. In [15] the idea of morphological adaptivity has been transferred to the setting of matrix fields utilising the

M.H.F. Wilkinson and J.B.T.M. Roerdink (Eds.): ISMM 2009, LNCS 5720, pp. 250–260, 2009.

operator-algebraic framework proposed in [16]. The goal of [15] was to enhance line-like structures in diffusion tensor magnetic resonance imaging (DT-MRI), the main source of matrix fields consisting of positive semidefinite matrices.

In this* article we propose a concept for PDE-based adaptive morphology for matrix fields, involving directional derivatives in the formulation of the PDE-based dilation process. In contrast to the work in [15] the numerical realisation employed in this article takes advantage of the accurate calculation of directional derivatives that relies on bi-linear interpolation.

We will start from a scalar adaptive formulation for d-dimensional data u in form of the dilation PDE

$$\partial_t u = \|M(u) \cdot \nabla u\| \tag{1}$$

with a data dependent, symmetric, positive semidefinite $d \times d$-matrix $M = M(u)$. Let us consider greyvalue images $(d = 2)$: Then one has $M = \begin{pmatrix} a & b \\ b & c \end{pmatrix} = \begin{pmatrix} \|(a,b)\|\nu^\top \\ \|(b,c)\|\eta^\top \end{pmatrix}$ with unit vectors $\nu = \frac{1}{\|(a,b)\|}\begin{pmatrix} a \\ b \end{pmatrix}$ and $\eta = \frac{1}{\|(b,c)\|}\begin{pmatrix} b \\ c \end{pmatrix}$. This turns (1) into

$$\partial_t u = \sqrt{(a\partial_x u + b\partial_y u)^2 + (b\partial_x u + c\partial_y u)^2}\,, \tag{2}$$

$$= \sqrt{\|(a,b)\|^2(\partial_\nu u)^2 + \|(b,c)\|^2(\partial_\eta u)^2}\,. \tag{3}$$

In [15] the partial derivatives $\partial_x u$ and $\partial_y u$ in (2) were approximated with the standard Rouy-Tourin scheme [17] to obtain a directional derivative, which might lead to numerical artifacts. Now, however, we calculate the directional derivatives necessary for the steering process directly by means of equation (3). Hence it is decisive for our approach to implement the directional derivatives $\partial_\nu u$ and $\partial_\eta u$ in (3) via a directional version of the Rouy-Tourin scheme as will be explained in Section 4.

Equation (1) describes a dilation with an ellipsoidal structuring element since an application of the mapping $(x,y)^\top \mapsto M(x,y)^\top$ transforms a sphere centered around the origin into an ellipse. The necessary directional information of the evolving u contained in the matrix M may be derived from the so-called structure tensor. The structure tensor, dating back to [18,19], allows to extract directional information from an image. It is given by

$$S_\rho(u(x)) := G_\rho * \left(\nabla u(x) \cdot (\nabla u(x))^\top\right) = \left(G_\rho * \left(\partial_{x_i} u(x) \cdot \partial_{x_j} u(x)\right)\right)_{i,j=1,\ldots,d} \tag{4}$$

Here $G_\rho*$ indicates a convolution with a Gaussian of standard deviation ρ. For more details the reader is referred to [20] and the literature cited there. In [21,22] Di Zenzo's approach [23] to construct a structure tensor for multi-channel images has been extended to matrix fields yielding a standard structure tensor

$$J_\rho(U(x)) := \sum_{i,j=1}^{m} S_\rho(U_{i,j}(x)) \tag{5}$$

with matrix entries $U_{i,j}$, $i, j = 1, \ldots, m$. This tensor is a special case of the *full structure tensor concept* for matrix fields as proposed in [24]. For our purpose it suffices to use the standard tensor $J_\rho(U(x))$ to infer directional information from matrix fields.

The article is structured as follows: In Section 2 we will briefly give an account of basic notions of matrix analysis needed to establish a matrix-valued PDE for an adaptively steered morphological dilation process. We introduce the steering tensor that guides the dilation process adaptively in Section 3. It is explained how the numerical scheme of Rouy and Tourin is turned into a directional variant that can be used on matrix fields in Section 4. An evaluation of the performance of our approach to adaptive morphology for matrix fields is the subject of Section 5. The remarks in Section 6 conclude this article.

2   Elements of Matrix Analysis

This section provides the essential notions for the formulation of matrix-valued PDEs. For a more detailed exposition the reader is referred to [16].

A matrix field is considered as a mapping $F : \Omega \subset \mathbb{R}^d \longrightarrow \mathrm{Sym}_m(\mathbb{R})$ from a d-dimensional image domain into the set of symmetric $m \times m$-matrices with real entries, $F(x) = (F_{p,q}(x))_{p,q=1,\ldots,m}$. The set of positive semi-definite matrices, denoted by $\mathrm{Sym}_m^+(\mathbb{R})$, consists of all symmetric matrices A with $\langle v, Av \rangle := v^\top Av \geq 0$ for $v \in \mathbb{R}^m$. DT-MRI produces matrix fields with this property. Note that at each point x the matrix $F(x)$ of a field of symmetric matrices can be diagonalised yielding $F(x) = V(x)^\top D(x)V(x)$, where $V(x)$ is a orthogonal matrix, while $D(x) = \mathrm{diag}(\lambda_1, \ldots, \lambda_m)$ represents a diagonal matrix with the eigenvalues $\lambda_1 \geq \lambda_2 \geq \cdots \geq \lambda_m \in \mathbb{R}$ of $F(x)$ as entries.

The extension of a function $h : \mathbb{R} \longrightarrow \mathbb{R}$ to $\mathrm{Sym}_m(\mathbb{R})$ is standard [25]: We set $h(U) := V^\top \mathrm{diag}(h(\lambda_1), \ldots, h(\lambda_m))V \in \mathrm{Sym}_m^+(\mathbb{R})$. Specifying $h(s) = |s|, s \in \mathbb{R}$ as the absolute value function leads to the absolute value $|A| \in \mathrm{Sym}_m^+(\mathbb{R})$ of a matrix A. The *partial derivative* for matrix fields at ω_0 is handled *componentwise*: $\overline{\partial}_\omega U(\omega_0) = (\partial_\omega U_{p,q}(\omega_0))_{p,q}$ where $\overline{\partial}_\omega$ stands for a spatial or temporal derivative. We define the *generalised gradient* $\overline{\nabla}U(x)$ at a voxel $x = (x_1, \ldots, x_d)$ by

$$\overline{\nabla}U(x) := (\overline{\partial}_{x_1}U(x), \ldots, \overline{\partial}_{x_d}U(x))^\top \qquad (6)$$

which is an element of $(\mathrm{Sym}_m(\mathbb{R}))^d$, in close analogy to the scalar setting where $\nabla u(x) \in \mathbb{R}^d$. For (extended) vectors $W \in (\mathrm{Sym}_m(\mathbb{R}))^d$ with matrix components we set $|W|_p := \sqrt[p]{|W_1|^p + \cdots + |W_d|^p}$ for $p \in]0, +\infty[$. It results in a positive semidefinite matrix from $\mathrm{Sym}_m^+(\mathbb{R})$, the direct counterpart of a nonnegative real number as the length of a vector in $\mathbb{R}^d$. Since the product of two symmetric matrices is in general not symmetric we employ the so-called *Jordan product*

$$A \bullet B := \frac{1}{2}(AB + BA). \qquad (7)$$

It produces a symmetric matrix, and it is commutative but neither associative nor distributive. In the proposed numerical scheme we will use the maximum and minimum of two symmetric matrices A, B. In direct analogy with relations known to be valid for real numbers one defines [26]:

$$\max(A, B) = \frac{1}{2}(A + B + |A - B|), \quad \min(A, B) = \frac{1}{2}(A + B - |A - B|), \quad (8)$$

where $|F|$ stands for the absolute value of the matrix F. Now we are in the position to formulate the matrix-valued counterpart of (1) as follows:

$$\overline{\partial}_t U = |\overline{M}(U) \bullet \overline{\nabla} U|_2 \tag{9}$$

with an initial matrix field $F(x) = U(x, 0)$. Here $\overline{M}(U)$ denotes a symmetric $md \times md$-block matrix with d^2 blocks of size $m \times m$ that is multiplied block-wise with $\overline{\nabla} U$ employing the symmetrised product "$\bullet$". Note that $| \cdot |_2$ stands for the length of $\overline{M}(U) \bullet \overline{\nabla} U$ in the matrix valued sense. The construction of $\overline{M}(U)$ is detailed in the next section.

3 Steering Matrix $\overline{M}(U)$ for Matrix Fields

With these notions at our disposal we propose the following construction of the steering matrix $\overline{M}$ in the adaptive dilation process for matrix fields.

First, the directional information is deduced from the standard structure tensor $J_\rho(U)$ in (5); this symmetric $d \times d$-matrix $J_\rho(U)$ is spectrally decomposed, and the following mapping is applied:

$$H : \left\{ \begin{array}{c} \mathbb{R}_+^d \longrightarrow \mathbb{R}_+^d \\ (\lambda_1, \ldots, \lambda_d) \longmapsto \frac{c}{\lambda_1 + \cdots + \lambda_d}(\lambda_d, \lambda_{d-1}, \ldots, \frac{K}{c} \cdot \lambda_1) \end{array} \right. , \tag{10}$$

with constants $c, K > 0$. H applied to $J_\rho(U)$ yields the steering matrix M,

$$M := H\big(J_\rho(U)\big).$$

The eigenvalues of $J_\rho(U)$ fulfil $\lambda_1 \geq \cdots \geq \lambda_d$. Hence, the ellipsoid associated with the quadratic form of M is flipped, and, depending on the choice of K, more excentric if compared with $J_\rho(U)$. In this way we enforce dilation towards the direction of least contrast, i. e. along structures.

Second, in order to enable a proper matrix-vector-multiplication in (9) we enlarge the $d \times d$-matrix M to a $md \times md$-matrix $\overline{M}$ by an extension operation utilising the $m \times m$-identity matrix I_m and the so-called Kronecker product [25]:

$$\overline{M} = M \otimes \begin{pmatrix} I_m & \cdots & I_m \\ \vdots & \ddots & \vdots \\ I_m & \cdots & I_m \end{pmatrix} = \begin{pmatrix} M_{11} I_m & \cdots & M_{1d} I_m \\ \vdots & \ddots & \vdots \\ M_{d1} I_m & \cdots & M_{dd} I_m \end{pmatrix} \tag{11}$$

which yields a suitably sized (block-) matrix.

4   Matrix-Valued Directional Numerical Scheme

The first-order finite difference method of *Rouy and Tourin* [17] may be used to solve the scalar PDE (3) in the *isotropic* case with $M = I_d$. Let us denote by u_{ij}^n the grey value of a scalar 2D image u at the pixel centred in $(ih_x, jh_y) \in \mathbb{R}^2$ at the time-level $n\tau$ of the evolution, with n the iteration number and τ the time step size. Furthermore, we employ standard forward and backward difference operators, i.e., $D_+^x u_{i,j}^n := u_{i+1,j}^n - u_{i,j}^n$ and $D_-^x u_{i,j}^n := u_{i,j}^n - u_{i-1,j}^n$ with spatial grid size h_x, h_y in $x-$ and $y-$direction, respectively. The Rouy-Tourin method utilises an *upwind approximation* in the pixel (ih_x, jh_y) of the partial derivative u_x (and analogously u_y):

$$u_x \approx \max\left(\frac{1}{h_x}\max\left(-D_-^x u_{i,j}^n, 0\right), \frac{1}{h_x}\max\left(D_+^x u_{i,j}^n, 0\right)\right). \qquad (12)$$

For a unit vector $\nu = (\nu_1, \nu_2)^\top$ the directional derivative $\partial_\nu u$ of u may be approximated by $\partial_\nu u = \langle \nu, \nabla u \rangle = \nu_1 \partial_x u + \nu_2 \partial_y u$. Hence it is close at hand to approximate numerically equation (2) directly. However, this favours mass transport along the directions of the x- and y-axis leading to a poor representation of the directional derivative. Instead we take advantage of equation (3) in this article and propose an alternative involving an interpolated function value $u_{i+\nu_1, j+\nu_2}$ defined by the subsequent bi-linear[1] approximation (13).

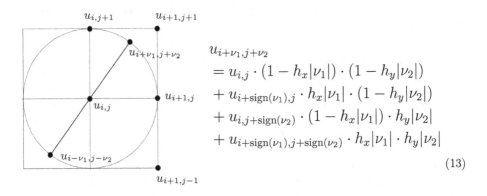

$$u_{i+\nu_1, j+\nu_2}$$
$$= u_{i,j} \cdot (1 - h_x|\nu_1|) \cdot (1 - h_y|\nu_2|)$$
$$+ u_{i+\text{sign}(\nu_1), j} \cdot h_x|\nu_1| \cdot (1 - h_y|\nu_2|)$$
$$+ u_{i, j+\text{sign}(\nu_2)} \cdot (1 - h_x|\nu_1|) \cdot h_y|\nu_2|$$
$$+ u_{i+\text{sign}(\nu_1), j+\text{sign}(\nu_2)} \cdot h_x|\nu_1| \cdot h_y|\nu_2|$$
$$(13)$$

Fig. 1. Interpolated image value $u_{i+\nu_1, j+\nu_2}$ with $\sqrt{\nu_1^2 + \nu_2^2} = 1$. It allows for backward and forward finite differences in the direction of $(\nu_1, \nu_2)^\top$.

This leads to forward and backward difference operators in the direction of ν with $\|\nu\| = \|(\nu_1, \nu_2)\| = \sqrt{\nu_1^2 + \nu_2^2} = 1$:

$$D_+^\nu u_{i,j}^n := u_{i+\nu_1, j+\nu_2}^n - u_{i,j}^n \quad \text{and} \quad D_-^\nu u_{i,j}^n := u_{i,j}^n - u_{i-\nu_1, j-\nu_2}^n \qquad (14)$$

---

[1] For the sake of efficiency we use bi-linear interpolation, although higher order alternatives such as bi-cubic or spline interpolation can be employed as well.

and to a direct approximation of the directional derivative

$$\partial_\nu u = u_\nu \approx \max\left(\frac{1}{h}\max\left(-D_-^\nu u_{i,j}^n, 0\right), \frac{1}{h}\max\left(D_+^\nu u_{i,j}^n, 0\right)\right) \tag{15}$$

where $h := \min(h_x, h_y)$. The extension to higher dimensions poses no problem. Furthermore, the resulting approximation of the directional derivatives is also consistent: Note that bi-linear approximation implies $u_{i+\nu_1,j+\nu_2} = u((i + \nu_1)h_x, (j + \nu_2)h_y) + \mathcal{O}(h_x \cdot h_y)$, and hence

$$\frac{1}{h}D_+^\nu u_{i,j} = \frac{1}{h}\left(u((i + \nu_1)h_x, (j + \nu_2)h_y) - u(ih_x, jh_y)\right) + \mathcal{O}(\max(h_x, h_y))$$
$$= u_\nu + \mathcal{O}(\max(h_x, h_y)). \tag{16}$$

Analogous reasoning applies to $D_-^\nu u_{i,j}$. With the calculus concept presented in Section 2 it is now straightforward to define one-sided directional differences in ν-direction for matrix fields of $m \times m$-matrices:

$$D_+^\nu U^n(ih_x, jh_y) := U^n((i+\nu_1)h_x, (j+\nu_2)h_y) - U^n(ih_x, jh_y) \in \mathrm{Sym}_m(\mathbb{R}), \tag{17}$$

$$D_-^\nu U^n(ih_x, jh_y) := U^n(ih_x, jh_y) - U^n((i-\nu_1)h_x, (j-\nu_2)h_y) \in \mathrm{Sym}_m(\mathbb{R}). \tag{18}$$

In order to avoid confusion with the subscript notation for matrix components we wrote $U(ih_x, jh_y)$ to indicate the (matrix-) value of the matrix field evaluated at the voxel centred at $(ih_x, jh_y) \in \mathbb{R}^2$. The η-direction is treated accordingly. The notion of supremum and infimum of two matrices – as needed in a matrix variant of Rouy-Tourin – has been provided in Section 2 as well. Hence, having these generalisations at our disposal a directionally adaptive version of the Rouy-Tourin scheme is available now in the setting of matrix fields simply by replacing grey values u_{ij}^n by matrices $U^n(ih_x, jh_y)$ and utilising the directional derivative approximations.

5   Experiments

Each matrix of the field is represented and visualised as an ellipsoid resulting from the level set of the quadratic form $\{x^\top A^{-2}x = const. : x \in \mathbb{R}^3\}$ associated with a matrix $A \in \mathrm{Sym}_3^+(\mathbb{R})$. By employing A^{-2} the length of the semi-axes of the ellipsoid correspond directly with the three eigenvalues of the matrix. We apply our PDE-driven adaptive dilation process to synthetic 2D data as well as to real DT-MRI data. For the explicit numerical scheme we used a time step size of 0.1, grid size $h = h_x = h_y = 1$, and $c = 0.01 \cdot K$ in (10).

Figure 2(a) exhibits a 32×32 synthetic matrix field used for testing. It is composed of two interrupted diagonal stripes with different thickness, built from cigar-shaped ellipsoids of equal size but different orientation. The lines intersect the x-axis with an angle of about -63 degrees. Figure 2(b) shows the result of applying the proposed adaptive dilation process using a directional Rouy-Tourin

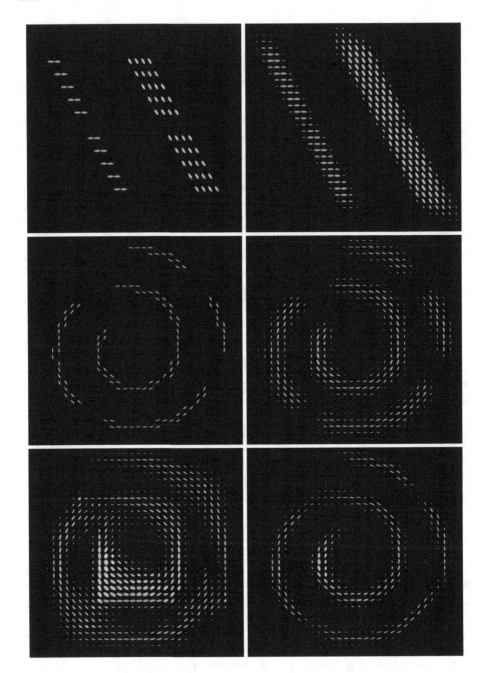

Fig. 2. Synthetic data sets. **(a) Top left:** Ellipsoids in line-like arrangement. **(b) Top right:** Proposed adaptive dilation with D-RT scheme; $K = 20, \rho = 4, t = 1$. **(c) Middle left:** Ellipsoids in spiral arrangement. **(d) Middle right:** Isotropic dilation; $t = 1$. **(e) Bottom left:** Adaptive dilation with RT scheme; $K = 20, \rho = 2, t = 1$. **(f) Bottom right:** Proposed adaptive dilation with D-RT scheme; $K = 20, \rho = 2, t = 1$.

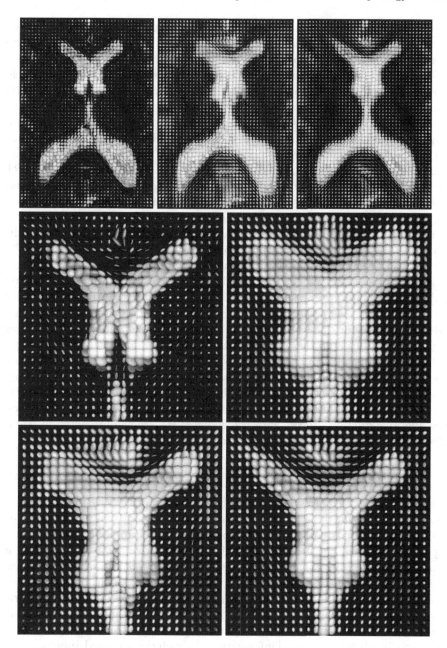

Fig. 3. Real data, 2D-slice of a 3D DT-MRI matrix field, and enlarged regions.
(a) Top left: Original data set. **(b) Top center:** Adaptive dilation with RT scheme;
$K = 10$, $\rho = 1$, $t = 1.5$. **(c) Top right:** Proposed adaptive dilation with D-RT scheme;
$K = 10$, $\rho = 1$, $t = 1.5$. **(d) Middle left:** Zoomed original data set. **(e) Middle right:**
Zoomed isotropic dilation after $t = 1.5$. **(f) Bottom left:** Zoomed adaptive dilation
with RT scheme. **(g) Bottom right:** Zoomed adaptive dilation with D-RT scheme.

(D-RT) scheme (9). Note that the direction and amount of anisotropic dilation does not depend on the orientation of the ellipsoids, but on the orientation and strength of the structural conformations.

Figure 2(c) displays another 32×32 testing image, namely a spiral-like field where large portions of the spiral have been removed. Figures 2(d), 2(e) and 2(f) depict the results of applying isotropic dilation [26], adaptive dilation with the classical Rouy-Tourin (RT) scheme [15], and the proposed adaptive dilation employing a directional Rouy-Tourin (D-RT) scheme (9). Comparatively, classical isotropic dilation requires much more time to fill in the missing ellipsoids and it also broadens the structures in all directions. Adaptive dilation with the RT scheme as in [15] does close the gaps in an anisotropic manner. However, numerous artifacts appear due to the numerical scheme bias to the coordinate axes. This problem is successfully solved in our PDE-based adaptive dilation process by utilising a D-RT scheme for approximating the partial derivatives. Evidently, relying on the D-RT scheme is much more accurate for longer dilation times.

We also tested the proposed method on a real DT-MRI data set of a human head consisting of a $128 \times 128 \times 38$-field of positive definite matrices. Figure 3(a) shows the lateral ventricles in a 40×55 2D section. Adaptive dilation with the classical RT scheme [15] and the proposed adaptive dilation process with a D-RT scheme (9) are shown in Figures 3(b) and 3(c), respectively. For a better comparison we scale-up these images around the *genu* area in Figures 3(d)-(g), including isotropic dilation [26] in Fig.3(e). Due to measurement errors the fibre tracts are interrupted in the original data. These holes are quickly and anisotropically filled by our directional-adaptive dilation process while enhancing slightly the directional structure of the fibres and preserving the shape of the ventricles. The adaptive dilation process with the classical RT scheme presented in [15] is affected by numerical artifacts and isotropic dilation [26] is too dissipative.

6   Conclusion

In this article we have presented a method for an adaptive, PDE-based dilation process in the setting of matrix fields. The evolution governed by a matrix-valued PDE is guided by a steering tensor. In order to enable proper directional steering we extended the classical Rouy-Tourin method in two ways: First, turning it into a directional Rouy-Tourin scheme based on directional finite differences via interpolation. Second, by means of matrix calculus, extending this directional scheme to matrix fields solving the matrix valued adaptive dilation PDE. Preliminary tests on synthetic and real DT-MRI data reveal a good performance of the method when it comes to filling in of missing data and segmentation of image structures involving directional information. As such the proposed approach may have its merits, for example, as a preprocessing step for fiber tracking algorithms.

Clearly, it is within our reach to formulate the anisotropic counterparts of other morphological operations such as erosion, opening, closing, top hats, gradients etc., which can be employed in more advanced image processing tasks for tensor fields, e.g. filtering and segmentation. In addition, the extension to the 3D setting is straightforward.

Acknowledgements. We gratefully acknowledge partial funding by the *Deutscher Akademischer Austauschdienst* (DAAD), grant A/05/21715.

DT-MRI data set by courtesy of Anna Vilanova i Bartoli, Eindhoven University of Technology.

References

1. Matheron, G.: Eléments pour une théorie des milieux poreux. Masson, Paris (1967)
2. Serra, J.: Echantillonnage et estimation des phénomènes de transition minier. PhD thesis, University of Nancy, France (1967)
3. Matheron, G.: Random Sets and Integral Geometry. Wiley, New York (1975)
4. Serra, J.: Image Analysis and Mathematical Morphology, vol. 1. Academic Press, London (1982)
5. Serra, J.: Image Analysis and Mathematical Morphology, vol. 2. Academic Press, London (1988)
6. Soille, P.: Morphological Image Analysis, 2nd edn. Springer, Berlin (2003)
7. Alvarez, L., Guichard, F., Lions, P.L., Morel, J.M.: Axioms and fundamental equations in image processing. Archive for Rational Mechanics and Analysis 123, 199–257 (1993)
8. Arehart, A.B., Vincent, L., Kimia, B.B.: Mathematical morphology: The Hamilton–Jacobi connection. In: Proc. Fourth International Conference on Computer Vision, Berlin, pp. 215–219. IEEE Computer Society Press, Los Alamitos (1993)
9. Brockett, R.W., Maragos, P.: Evolution equations for continuous-scale morphological filtering. IEEE Transactions on Signal Processing 42, 3377–3386 (1994)
10. Sapiro, G., Kimmel, R., Shaked, D., Kimia, B.B., Bruckstein, A.M.: Implementing continuous-scale morphology via curve evolution. Pattern Recognition 26, 1363–1372 (1993)
11. van den Boomgaard, R.: Mathematical Morphology: Extensions Towards Computer Vision. PhD thesis, University of Amsterdam, The Netherlands (1992)
12. Breuß, M., Weickert, J.: A shock-capturing algorithm for the differential equations of dilation and erosion. Journal of Mathematical Imaging and Vision 25(2), 187–201 (2006)
13. Verdú-Monedero, R., Angulo, J.: Spatially-variant directional mathematical morphology operators based on a diffused average squared gradient field. In: Blanc-Talon, J., Bourennane, S., Philips, W., Popescu, D., Scheunders, P. (eds.) ACIVS 2008. LNCS, vol. 5259, pp. 542–553. Springer, Heidelberg (2008)
14. Breuß, M., Burgeth, B., Weickert, J.: Anisotropic continuous-scale morphology. In: Martí, J., Benedí, J.M., Mendonça, A., Serrat, J. (eds.) IbPRIA 2007. LNCS, vol. 4478, pp. 515–522. Springer, Heidelberg (2007)
15. Burgeth, B., Breuß, M., Pizarro, L., Weickert, J.: PDE-driven adaptive morphology for matrix fields. In: Tai, X.-C., et al. (eds.) SSVM 2009. LNCS, vol. 5567, pp. 247–258. Springer, Heidelberg (2009)
16. Burgeth, B., Didas, S., Florack, L., Weickert, J.: A generic approach to diffusion filtering of matrix-fields. Computing 81, 179–197 (2007)
17. Rouy, E., Tourin, A.: A viscosity solutions approach to shape-from-shading. SIAM Journal on Numerical Analysis 29, 867–884 (1992)

18. Förstner, W., Gülch, E.: A fast operator for detection and precise location of distinct points, corners and centres of circular features. In: Proc. ISPRS Intercommission Conference on Fast Processing of Photogrammetric Data, Interlaken, Switzerland, June 1987, pp. 281–305 (1987)
19. Bigün, J., Granlund, G.H., Wiklund, J.: Multidimensional orientation estimation with applications to texture analysis and optical flow. IEEE Transactions on Pattern Analysis and Machine Intelligence 13(8), 775–790 (1991)
20. Bigün, J.: Vision with Direction. Springer, Berlin (2006)
21. Brox, T., Weickert, J., Burgeth, B., Mrázek, P.: Nonlinear structure tensors. Image and Vision Computing 24(1), 41–55 (2006)
22. Feddern, C., Weickert, J., Burgeth, B., Welk, M.: Curvature-driven PDE methods for matrix-valued images. International Journal of Computer Vision 69(1), 91–103 (2006)
23. Di Zenzo, S.: A note on the gradient of a multi-image. Computer Vision, Graphics and Image Processing 33, 116–125 (1986)
24. Burgeth, B., Didas, S., Weickert, J.: A general structure tensor concept and coherence-enhancing diffusion filtering for matrix fields. In: Laidlaw, D., Weickert, J. (eds.): Visualization and Processing of Tensor Fields. Mathematics and Visualization, pp. 305–323. Springer, Heidelberg (2009)
25. Horn, R.A., Johnson, C.R.: Matrix Analysis. Cambridge University Press, Cambridge (1990)
26. Burgeth, B., Bruhn, A., Didas, S., Weickert, J., Welk, M.: Morphology for matrix-data: Ordering versus PDE-based approach. Image and Vision Computing 25(4), 496–511 (2007)

Component-Trees and Multi-value Images: A Comparative Study

Benoît Naegel[1] and Nicolas Passat[2]

[1] Université Nancy 1, LORIA, UMR CNRS 7503, France
[2] Université de Strasbourg, LSIIT, UMR CNRS 7005, France
benoit.naegel@loria.fr, passat@unistra.fr

Abstract. In this article, we discuss the way to derive connected operators based on the component-tree concept and devoted to multi-value images. In order to do so, we first extend the grey-level definition of the component-tree to the multi-value case. Then, we compare some possible strategies for colour image processing based on component-trees in two application fields: colour image filtering and colour document binarisation.

Keywords: Component-trees, multi-value images, connected operators.

1 Introduction

Connected operators can be defined from various ways (for instance by region-adjacency graph merging [9], levelings [14], geodesic reconstruction, etc.). One possibility is to consider an image *via* its component-tree structure. Component-trees [19] (also known under different denominations [20,7,13]) have been devoted to several image processing tasks (segmentation, filtering, coding, etc.) and have, until now, exclusively involved grey-level images. Since there is an increasing need to process colour - and more generally multi-value - images, we propose to explore some solutions to use component-tree-based operators with such images. These solutions are experimentally assessed in the context of colour image filtering and colour document binarisation.

In Sec. 2, we briefly recall previous work and usual concepts related to multi-value image processing and component-trees. In Sec. 3, we propose an extension of the "grey-level" definition of the component-tree structure in order to use it with multi-value images. In Sec. 4, processing strategies allowing to use component-trees with multi-value images are described. In Sec. 5, we present two examples using the component-trees with colour images. A discussion and perspectives are given in Sec. 6.

2 Related Work

2.1 Multi-value Image Processing

The extension of mathematical morphology to the case of colour/multi-value images is an important task, which has potential applications in multiple areas. For decades, a significant amount of work has been devoted to this specific purpose (see *e.g.* [3] for a recent survey).

M.H.F. Wilkinson and J.B.T.M. Roerdink (Eds.): ISMM 2009, LNCS 5720, pp. 261–271, 2009.

Several attempts have been made to extend connected operators to colour images. Some of them are based on the contraction of a region adjacency graph structure [21]. Some others consider colour extrema using specific vectorial orderings [10,8]. However until now, no attempts have been made to use the component-tree data structure in combination with multi-value images.

In the mathematical morphology framework, two main ways are usually proposed to perform colour image processing. The first one, called *marginal processing*, consists in processing separately the different channels of a multi-value image, thus reducing the problem to the processing of mono-value images and their fusion to recover a multi-value result. This approach is straightforward, unfortunately it may also induce several drawbacks such as the generation of false colours, for instance.

The second one, called *vectorial processing*, consists in defining a total order (or preorder) relation on the set of multi-value components. To this end, various vector-based orderings have been proposed [4]: conditional ordering (C-ordering, including lexicographic ordering), reduced ordering (R-ordering, which implies to reduce a vector value to a scalar one) which has been extensively studied in [11], "partial ordering" (P-ordering, where vectors are gathered into equivalence classes as in [22]). Recently, usual morphological operators for colour images have been derived from a total ordering based on a reduced ordering (leading to a preorder) completed by a lexicographic ordering to obtain a total order [2].

2.2   Component-Trees

The component-tree structure provides a rich, scale-invariant, description for grey-level images [20,19]. It has been involved, in particular, in the development of attribute filtering [6,20], object identification [12,16,18], and image retrieval [15,1]. In the context of segmentation or recognition tasks, it enables to perform object detection without having to precompute a specific threshold (which is usually an error-prone task).

Another advantage of this structure lies in its low algorithmic cost: efficient algorithms have been designed to compute it [20,19,5]. Moreover, the component-tree computation can be done offline, therefore leading to very fast (real-time) and interactive processing [24].

Until now, component-tree-based processing has always involved binary or grey-level images. We now propose to investigate the multi-value case.

3   Component-Trees and Multi-value Images

3.1   Multi-value Images

Let $n \in \mathbb{N}^*$. Let $\{(T_i, \leq_i)\}_{i=1}^{n}$ be a family of (finite) totally-ordered sets (namely the sets of *values*). For any $i \in [1..n]$, the infimum of T_i is denoted by $\perp_i$. Let $\mathbf{T}$ be the set defined by $\mathbf{T} = \prod_{i=1}^{n} T_i = T_1 \times T_2 \times \ldots \times T_n$. A value $\mathbf{t} \in \mathbf{T}$ is then a vector composed of n scalar values: $\mathbf{t} \in \mathbf{T} \Leftrightarrow \mathbf{t} = (t_i)_{i=1}^{n} = (t_1, t_2, \ldots, t_n)$ with $t_i \in T_i$ for any $i \in [1..n]$. Let $\leq$ be the binary relation on $\mathbf{T}$ defined by: $\forall \mathbf{t}, \mathbf{u} \in \mathbf{T}, (\mathbf{t} \leq \mathbf{u} \Leftrightarrow \forall i \in [1..n], t_i \leq_i u_i)$. Then $(\mathbf{T}, \leq)$ is a complete lattice, the infimum $\perp$ of which is defined by $\perp = (\perp_i)_{i=1}^{n}$.

Let $d \in \mathbb{N}^*$. A (discrete) multi-value image is defined as a function $\mathbf{F} : \mathbb{Z}^d \to \mathbf{T}$. For all $i \in [1..n]$, the mappings $F_i : \mathbb{Z}^d \to T_i$ defined such that $\forall x \in \mathbb{Z}^d, \mathbf{F}(x) = (F_i(x))_{i=1}^n$ are called the *channels* (or the *bands*) of the multi-value image F. The support of $\mathbf{F}$ is defined by $\mathrm{supp}(\mathbf{F}) = \{x \in \mathbb{Z}^d \mid \mathbf{F}(x) \neq \perp\}$ and we note $\mathrm{supp}(\mathbf{F}) = E$. In the sequel, we will assume that for any image $\mathbf{F}$, $\mathrm{supp}(\mathbf{F})$ is finite. We will then assimilate an image $\mathbf{F}$ to its (finite) restriction $\mathbf{F}_{|E} : E \to \mathbf{T}$.

3.2   Component-Trees

Let $X \subseteq E$. The connected components of X are the subsets of X of maximal extent. The set of all the connected components of X is noted $C[X]$.

Let $\mathcal{R}$ be a total preorder on $\mathbf{T}$, *i.e.* a binary relation verifying (*i*) reflexivity ($\forall \mathbf{t} \in \mathbf{T}, \mathbf{t} \, \mathcal{R} \, \mathbf{t}$), (*ii*) transitivity ($\forall \mathbf{t}, \mathbf{u}, \mathbf{v} \in \mathbf{T}, (\mathbf{t} \, \mathcal{R} \, \mathbf{u}) \wedge (\mathbf{u} \, \mathcal{R} \, \mathbf{v}) \Rightarrow (\mathbf{t} \, \mathcal{R} \, \mathbf{v})$) and (*iii*) totality ($\forall \mathbf{t}, \mathbf{u} \in \mathbf{T}, (\mathbf{t} \, \mathcal{R} \, \mathbf{u}) \vee (\mathbf{u} \, \mathcal{R} \, \mathbf{t})$).

We set $\mathcal{P}(E) = \{X \mid X \subseteq E\}$. For all $\mathbf{t} \in \mathbf{T}$, let $X_{\mathbf{t}} : \mathbf{T}^E \to \mathcal{P}(E)$ be the thresholding function defined by $X_{\mathbf{t}}(\mathbf{F}) = \{x \in E \mid \mathbf{t} \, \mathcal{R} \, \mathbf{F}(x)\}$, for all $\mathbf{F} : E \to \mathbf{T}$. Since $\mathcal{R}$ is transitive, we have $\forall \mathbf{F} : E \to \mathbf{T}, \forall \mathbf{t}, \mathbf{t}' \in \mathbf{T}, \mathbf{t} \, \mathcal{R} \, \mathbf{t}' \Leftrightarrow X_{\mathbf{t}'}(\mathbf{F}) \subseteq X_{\mathbf{t}}(\mathbf{F})$.

Let $\mathbf{F} : E \to \mathbf{T}$ be a multi-value image. Let $\mathcal{K} = \bigcup_{\mathbf{t} \in \mathbf{T}} C[X_{\mathbf{t}}(\mathbf{F})]$. Then the relation $\subseteq$ is a partial order on $\mathcal{K}$, and the Hasse diagram $(\mathcal{K}, L)$ of the partially-ordered set $(\mathcal{K}, \subseteq)$ is a tree (*i.e.* a connected acyclic graph), the root of which is the supremum $R = \mathrm{sup}(\mathcal{K}, \subseteq) = E$. This rooted tree $(\mathcal{K}, L, R)$ is called the *component-tree of* $\mathbf{F}$. The elements $\mathcal{K}$, R and L are the set of the *nodes*, the *root* and the set of the *edges* of the tree, respectively.

Note that if $\mathbf{F} : E \to \mathbf{T}$ is monovalued (*i.e.* if $n = 1$ or, equivalently, $\mathbf{T} = (T_1)$) and equipped with a total order relation, then $\mathbf{F}$ can be assimilated to a function taking its values in a totally-ordered set $([0..|T_1| - 1], \leq_{\mathbb{Z}})$, and we actually retrieve the "usual" component-tree definition for grey-level images.

3.3   Tree Pruning

The nodes $\mathcal{K}$ of a component-tree store information, also called *attributes*, on the associated connected components of an input image $\mathbf{F}$. Practically, to each node $N \in \mathcal{K}$, we then associate an attribute $\sigma(N) \in K$ (where K is a set of knowledge).

Let $Q : K \to \mathbb{B}$ be a criterion (*i.e.* a predicate on K). By setting $\mathcal{K}_Q \subseteq \mathcal{K}$ as $\mathcal{K}_Q = \{X \in \mathcal{K} \mid Q(X)\}$, we generate the subset of the nodes satisfying the criterion Q. Such a selection process enables to perform pruning on the component-tree of an image $\mathbf{F}$ according to a given criterion, leading to a filtering process which generates an image reconstructed from $\mathbf{F}$ with respect to $\mathcal{K}_Q$.

As it will be illustrated in the next section, for any given criterion Q, it is generally possible to define a connected operator $\Psi : \mathbf{T}^E \to \mathbf{T}^E$ which associates to a multi-value image $\mathbf{F}$ the image $\Psi(\mathbf{F})$ generated from $\mathcal{K}_Q$. Similarly, a segmentation operator $\Gamma : \mathbf{T}^E \to \mathcal{P}(E)$ can be derived by setting $\Gamma(\mathbf{F}) = \bigcup_{X \in \mathcal{K}_Q} X$.

4   Processing Strategies for Multi-value Images

In this section, we explore some original processing strategies allowing to define connected operators for multi-value images based on the component-tree structure. These different approaches are illustrated in Fig. 1 on a synthetic image, considering an example of area opening.

4.1 Marginal Processing

Based on the marginal approach, an operator Ψ can be used to handle multi-value images by processing separately each one of the n component-trees associated to each channel of the image.

For all $i \in [1..n]$, let $\mathcal{K}^i$ be the set of nodes of the individual channel F_i, according to the total order relation $\leq_i$. Then, an operator Ψ can be defined by

$$\Psi(\mathbf{F}) = \bigvee_{i \in [1..n]} \bigvee_{X \in \mathcal{K}^i_Q} C_{X, \mathbf{v}_i(X)} , \tag{1}$$

where, $C_{X, \mathbf{v}_i(X)} : E \to \mathbf{T}$ is the cylinder function defined by $C_{X, \mathbf{v}_i(X)}(x) = \mathbf{v}_i(X)$ if $x \in X$ and $\perp$ otherwise, $\mathbf{v}_i(X) = (\perp_1, \ldots, \perp_{i-1}, m_i(X), \perp_{i+1}, \ldots, \perp_n)$ and $m_i(X)$ is the "value" of the component X in the channel F_i: $m_i(X) = \min\{F_i(x) \mid x \in X\}$.

A well-known drawback of marginal processing lies in the possible appearance of "false" values, *i.e.* values that are not present in the original image. Note however that Ψ is a connected operator, ensuring that no false contours are introduced.

4.2 Vectorial Processing

By contrast to the marginal case, considering a vectorial approach for processing multi-value images can ensure that no false values will be introduced in the result image. However, the nature of the relation $\mathcal{R}$ influences the way $\Psi(\mathbf{F})$ is reconstructed from the filtered set of nodes $\mathcal{K}_C \subseteq \mathcal{K}$. In particular, the image reconstruction is different whether $\mathcal{R}$ is a total *preorder* or a total *order* relation.

Total order. If $\mathcal{R}$ is a total order on $\mathbf{T}$ (*i.e.* if $\mathcal{R}$ is anti-symmetric: $\forall \mathbf{t}, \mathbf{u} \in \mathbf{T}, (\mathbf{t} \mathcal{R} \mathbf{u}) \wedge (\mathbf{u} \mathcal{R} \mathbf{t}) \Rightarrow (\mathbf{t} = \mathbf{u}))$, an operator Ψ can be defined by

$$\Psi(\mathbf{F}) = \bigvee_{\substack{\mathcal{R} \\ X \in \mathcal{K}_Q}} C_{X, \mathbf{v}(X)} , \tag{2}$$

where $\mathbf{v}(X) = \min_{\mathcal{R}}\{\mathbf{F}(x) \mid x \in X\}$.

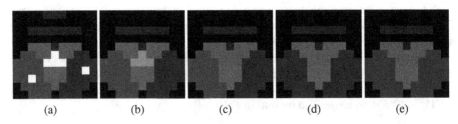

(a)   (b)   (c)   (d)   (e)

Fig. 1. Area opening of size 6 on a synthetic RGB image $\mathbf{F} : [0, 10]^2 \to [0..255]^3$. (a) Original image, (b) marginal processing, (c) vectorial processing based on a total lexicographic order (L), (d, e) vectorial processings relying on a distance-based reduced ordering using $\mathbf{r} = (255, 255, 255)$ as reference, with median reconstruction (P_{median}) (d) and mean reconstruction (P_{mean}) (e).

This is the case, for instance, when considering the lexicographic ordering, a case of C-ordering, or a total ordering based on the distance w.r.t. a reference vector and completed by a lexicographic ordering, as described in [2].

Total preorder. If $\mathcal{R}$ is a total preorder, there may possibly exist some $\mathbf{t}, \mathbf{u} \in \mathbf{T}$ such that $\mathbf{t} \neq \mathbf{u}$ while $(\mathbf{t} \, \mathcal{R} \, \mathbf{u}) \wedge (\mathbf{u} \, \mathcal{R} \, \mathbf{t})$. For any $U \subseteq \mathbf{T}$, let $\mathrm{Min}_{\mathcal{R}} U = \{\mathbf{u} \in U \mid \forall \mathbf{u}' \in U, \mathbf{u} \, \mathcal{R} \, \mathbf{u}'\}$. Then we can derive an operator Ψ by

$$\Psi(\mathbf{F}) = \bigvee_{\substack{\mathcal{R} \\ X \in \mathcal{K}_Q}} C_{X, \mathbf{v}(X)} \, , \tag{3}$$

where $\mathbf{v}(X)$ is a value computed from the set $\mathbf{V}(X) = \mathrm{Min}_{\mathcal{R}}\{\mathbf{F}(x) \mid x \in X\}$ (note that $\mathrm{Min}_{\mathcal{R}}\{\mathbf{F}(x) \mid x \in X\} = \{\min_{\mathcal{R}}\{\mathbf{F}(x) \mid x \in X\}\}$ whenever $\mathcal{R}$ is a total order). The value $\mathbf{v}(X)$ may belong to the set $\mathbf{V}(X)$ (for example by ranking the vectors using a total ordering and taking a representant value, as the median vector), or may be computed from the values of $\mathbf{V}(X)$ (by taking, for example, the mean vector of $\mathbf{V}(X)$).

Using component-trees based on a total preorder allow us, in particular, to merge in a same component several flat-zones taking distinct values. In particular, it can lead to highlight structures of interest according to some predetermined knowledge. In order to illustrate this assertion, let us consider the reduced ordering $\leq_r$ defined w.r.t. a function $r : \mathbf{T} \to \mathbb{R}$ by $\mathbf{t} \leq_r \mathbf{u} \Leftrightarrow r(\mathbf{t}) \leq r(\mathbf{u})$ for all $\mathbf{t}, \mathbf{u} \in \mathbf{T}$. If r is non-injective, $\leq_r$ induces a preorder relation on $\mathbf{T}$. For instance, we can consider the distance to a reference value $\mathbf{r} \in \mathbf{T}$, *i.e.* $r(\mathbf{t}) = d(\mathbf{r}, \mathbf{t})$. By choosing $\mathbf{r}$ close to the "colour" of some given structures of interest (*i.e.* to be removed, to be segmented, etc.), such structures may probably appear as single nodes close to the leaves of the tree. More generally, one could define a set of reference values $S = \{\mathbf{r}_i\}_{i=1}^{k}$ ($k \in \mathbb{N}^*$). A reduced ordering could then be induced by the function $\phi_S : \mathbf{T} \to \mathbb{R}$ defined by $\phi_S(\mathbf{t}) = \min_{i \in [1..k]}\{d(\mathbf{r}_i, \mathbf{t})\}$ for all $\mathbf{t} \in \mathbf{T}$.

5 Experiments

In this section we illustrate the proposed approaches in the context of two application fields. Let $\mathbf{F} : E \to \mathbf{T}$ be a colour image defined in the RGB space $\mathbf{T} = T_r \times T_g \times T_b = [0..255]^3$, with $E \subseteq \mathbb{Z}^2$.

The following methods have been considered. The first one (M) is based on marginal processing. The other ones (L, dL, P_{mean}, P_{median}) are based on vectorial processing. Two of them rely on a total ordering of the image values: lexicographic ordering (L)

$$\mathbf{t} \leq_L \mathbf{t}' \Leftrightarrow (t_r < t_r') \vee ((t_r = t_r') \wedge (t_g < t_g')) \vee ((t_r = t_r') \wedge (t_g = t_g') \wedge (t_b \leq t_b')) \, , \tag{4}$$

and a distance-based reduced ordering followed by a lexicographic ordering (dL)

$$\mathbf{t} \leq_{dL} \mathbf{t}' \Leftrightarrow (r(\mathbf{t}) < r(\mathbf{t}')) \vee ((r(\mathbf{t}) = r(\mathbf{t}')) \wedge (\mathbf{t} \leq_L \mathbf{t}')) \, . \tag{5}$$

The two other ones are based on a total distance-based preordering

$$\mathbf{t} \leq_P \mathbf{t}' \Leftrightarrow r(\mathbf{t}) \leq r(\mathbf{t}') \, , \tag{6}$$

with $r(\mathbf{t}) = \|\mathbf{t} - \mathbf{r}\|$ ($\|.\|$ being the Euclidean norm). One (P_{mean}) uses the *mean* vector to reconstruct the filtered image (see Sec. 4.2), the other one (P_{median}) uses the *median* vector based on a total lexicographic ordering $\leq_L$.

5.1   Colour Image Filtering

The proposed strategies have been evaluated from their performance in the context of image filtering. For this purpose, an image processing scheme based on area opening [23] using a component-tree implementation has been assessed. A filtering operator Ψ was defined by: $\Psi = \mathsf{C} \circ \gamma_\lambda \circ \mathsf{C} \circ \gamma_\lambda$, where C denotes the colour image complement defined for all $p \in E$ by: $\mathsf{C}\mathbf{F}(p) = (255 - F_r(p), 255 - F_g(p), 255 - F_b(p))$ and γ_λ is the area opening of parameter λ.

The experiments have been carried out by processing colour images corrupted by random noise and Gaussian noise.
The methods dL, P_{mean}, P_{median} used $\mathbf{r} = (255, 255, 255)$ as reference vector. From a qualitative point of view, the operator based on marginal processing (M) outperforms the other ones, since it is the one that visually preserves at best the image (see Fig. 2). The filtering operator based on preordering using a mean reconstruction (P_{mean}) preserves correctly image details, however it tends to reduce the quantisation of the image values, therefore reducing the overall image saturation. Note that all the other methods introduce undesired coloured artifacts.

As in [3], quantitative denoising comparisons based on the Normalised Mean Square Error (NMSE) measure have been performed. Given a reference image $\mathbf{F}$, the NMSE measure of the denoised image $\mathbf{F}'$ is defined by

$$\text{NMSE} = \frac{\sum_{p \in E} \|\mathbf{F}(p) - \mathbf{F}'(p)\|^2}{\sum_{p \in E} \|\mathbf{F}(p)\|^2} . \tag{7}$$

For the sake of comparison, the filtering operator Ψ has been compared with the OCCO filter [3] (using a marginal processing) defined by

$$\text{OCCO}_B = \frac{1}{2}\gamma_B[\phi_B(\mathbf{F})] + \frac{1}{2}\phi_B[\gamma_B(\mathbf{F})] , \tag{8}$$

where γ_B and ϕ_B denote the opening and closing operators, respectively. In these experiments B is the elementary ball. The results are summarised in Table 1.

Table 1. NMSE results of the proposed denoising operators based on different strategies (Lenna image, see Fig. 2)

	M	L	dL	P_{mean}	P_{median}	OCCO
Random noise (15%)	**0.33**	0.68	1.39	0.99	1.27	2.32
Gaussian noise ($\mu = 0, \sigma = 32$)	1.13	4.07	4.08	1.34	3.49	**0.84**

In the case of random noise, the proposed filtering approach based on marginal processing leads to the lowest NMSE, emphasising the efficiency of connected based operators in this particular context.

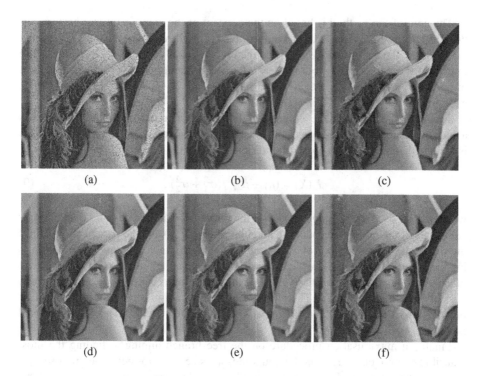

Fig. 2. Comparison of filtering operators Ψ based on area opening using different processing strategies. (a) Lenna image corrupted by 15% random noise. (b) Marginal processing (M). (c) Lexicographic ordering (L). (d) Distance-based total ordering (dL). (e) Preordering (P_{mean}). (f) Preordering (P_{median}).

In the case of Gaussian noise, the marginal strategy still performs better than the other ones based on the filtering Ψ operator. We notice however that the OCCO operator achieves the lowest NMSE in this context. Indeed, in the case of Gaussian noise, the original image values are not necessarily present in the corrupted version. Therefore, a denoising operator should enable to choose values that are not present in the image in order to restore the original ones. This is the case of the M, P_{mean} and OCCO methods, that are the best ones in this context. However, in a denoising context, the P_{mean} strategy suffers from the quantisation of the image values that results from the non-injective distance-based reduced ordering.

5.2 Colour Document Binarisation

Connected operators can be efficiently involved in object detection tasks. Based on the proposed strategies, an object extraction scheme relying on connected operators was experimented and applied to the case of colour document binarisation. This binarisation method was initially designed for grey-level document images and is fully described in [17]. We summarise it hereafter.

The core of the method is based on the concept of the component-tree *branch*. Let $(\mathcal{K}, L, R)$ be the component-tree of a monovalued (grey-level) image. The set of regional maxima (*i.e.* the set of tree *leaves*) is defined by $M = \{X \in \mathcal{K} \mid \forall Y \in \mathcal{K}, Y \not\subset X\}$. The branch of the tree starting from the leaf $M \in \mathcal{M}$ is defined by the (unique) sequence of nodes $\mathcal{B}_\mathcal{K}(M) = (X_k)_{k=1}^t \in \mathcal{K}$, such that $X_1 = M$, $X_t = R$, $\forall k \in [1, t-1], X_k \subset X_{k+1} \wedge \forall Y \in \mathcal{K}, X_k \subseteq Y \subset X_{k+1} \Rightarrow Y = X_k$.

The method is based on the assumption that, for each branch of the tree, there exists a node corresponding to an object of interest. In the considered application, such a node is the one that maximises a *contrast* criterion based on the Fisher discriminant

$$J_\lambda(X) = (\mu_1 - \mu_2)^2 / (\sigma_1^2 + \sigma_2^2) , \tag{9}$$

where μ_1 and σ_1 (resp. μ_2 and σ_2) denote the mean and standard deviation of the original values of the node X (resp. of the neighbourhood of X) and the parameter λ defines the size of the neighbourhood of X. Therefore, for each branch, a unique node maximising the criterion is preserved which allows to filter the component-tree without the use of any threshold parameter. However, using this procedure, each regional maximum can possibly create a component. Therefore, prior to the maximisation procedure, a rough binarisation based on the image grey-levels is first applied to discard irrelevant regional maxima.

Finally, a maximisation procedure on the tree branch aiming at finding the most plausible components based on the bounding-box size is performed. The proposed approach is then composed of three steps, each devoted to preserve relevant components according to a chosen criterion.

1. Rough binarisation based on a K-Means classifier applied to the pixel values.
2. For each branch of the tree, selection of the node X maximising the contrast measure $J_\lambda(X)$ (X marked as *active*).
3. For each branch of the tree, preservation of the active node maximising a size criterion (related to the bounding box of the component).

This method was applied on a set of colour documents from the MediaTeam Oulu Document Database[1]. It was implemented following the proposed strategies. Note that in order to extract the relevant components, prior knowledge related to the values of the objects of interest is necessary. As a consequence, marginal processing and total orderings (L and dL) - logically - led to unsatisfactory results.

The reduced ordering based on multiple reference vector provided interesting results, since it enabled to highlight the objects of interest w.r.t. the background. However it was not appropriate in this application: for example, in Fig. 3(a), the letter "G" in black in the image upper left became connected to the river in blue, therefore preventing the correct extraction of this component. The best result was obtained by using multiple distance-based reduced ordering, each aiming at extracting characters of a given colour and by taking the supremum of the results (see Fig. 3).

---

[1] http://www.mediateam.oulu.fi/downloads/MTDB

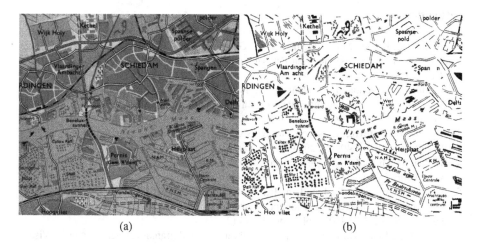

(a)   (b)

Fig. 3. Binarisation method of colour documents. (a) Original colour document, (b) Binarisation method based on a processing strategy involving two distance-based reduced orderings (dL) aiming at extracting black and dark blue characters.

6 Conclusion

In this article we have proposed an extension of the definition of the component-tree to the case of multi-value images equipped (at least) with a total preorder. Some solutions have also been explored to define connected operators based on the component-tree structure in the case of such images. The interest of component-tree-based operators in combination with colour images has, in particular, been illustrated in the context of denoising and binarisation.

In the case of denoising applications, the marginal processing approach remains the most efficient one, as pointed out in other works [3]. As far as object detection is considered, prior knowledge related to the object values is necessary, and therefore distance-based reduced ordering become more suited to such issues. We also believe that the usefulness of colour connected operators based on component-tree could be greatly increased by considering other - more perceptual - colour spaces. This possibility has been considered in other works [3,2] and has not been developed here for the sake of generality.

In terms of computational efficiency, marginal processing and strategies based on reduced ordering are the fastest in the case of colour images, since the number of different values remains limited (less than 255 for 24 bits colour images in marginal processing). Approaches based on total colour ordering lead to the construction of component-trees having a large depth (around 100 000 different values in the case of the Lenna image) and huge number of nodes, therefore implying longer processing times (although specific algorithms have been designed for this case [5]).

In this paper we have not explored the case in which the thresholding function X_t is defined w.r.t. the partial order $\leq$ of $\mathbf{T}$. In this case the Hasse diagram $(\mathcal{K}, L)$ obtained from the set $\mathcal{K} = \bigcup_{t \in \mathbf{T}} C[X_t(F)]$ is not a tree anymore. This leads to a, more general, graph structure, which will be investigated in future works.

References

1. Alajlan, N., Kamel, M.S., Freeman, G.H.: Geometry-based image retrieval in binary image databases. IEEE Transactions on Pattern Analysis and Machine Intelligence 30(6), 1003–1013 (2008)
2. Angulo, J.: Morphological colour operators in totally ordered lattices based on distances: Application to image filtering, enhancement and analysis. Computer Vision and Image Understanding 107(1-2), 56–73 (2007)
3. Aptoula, E., Lefèvre, S.: A comparative study on multivariate mathematical morphology. Pattern Recognition 40(11), 2914–2929 (2007)
4. Barnett, V.: The ordering of multivariate data. Journal of the Royal Statistical Society: Series A (Statistics in Society) 139(3), 318–354 (1976)
5. Berger, C., Géraud, T., Levillain, R., Widynski, N., Baillard, A., Bertin, E.: Effective component-tree computation with application to pattern recognition in astronomical imaging. In: Proc. of ICIP 2007, vol. 4, pp. 41–44 (2007)
6. Breen, E.J., Jones, R.: Attribute openings, thinnings, and granulometries. Computer Vision and Image Understanding 64(3), 377–389 (1996)
7. Chen, L., Berry, M.W., Hargrove, W.W.: Using dendronal signatures for feature extraction and retrieval. International Journal of Imaging Systems and Technology 11(4), 243–253 (2000)
8. Evans, A.N., Gimenez, D.: Extending connected operators to colour images. In: Proc. of ICIP 2008, pp. 2184–2187 (2008)
9. Garrido, L., Salembier, P., Garcia, D.: Extensive operators in partition lattices for image sequence analysis. Signal Processing: Special issue on Video Sequence Segmentation 66(2), 157–180 (1998)
10. Gimenez, D., Evans, A.N.: An evaluation of area morphology scale-spaces for colour images. Computer Vision and Image Understanding 110(1), 32–42 (2008)
11. Goutsias, J., Heijmans, H.J.A.M., Sivakumar, K.: Morphological operators for image sequences. Computer Vision and Image Understanding 62(3), 326–346 (1995)
12. Jones, R.: Connected filtering and segmentation using component trees. Computer Vision and Image Understanding 75(3), 215–228 (1999)
13. Mattes, J., Demongeot, J.: Efficient algorithms to implement the confinement tree. In: Nyström, I., Sanniti di Baja, G., Borgefors, G. (eds.) DGCI 2000. LNCS, vol. 1953, pp. 392–405. Springer, Heidelberg (2000)
14. Meyer, F.: From connected operators to levelings. In: Mathematical Morphology and its Applications to Image and Signal Processing (Proc. of ISMM 1998), pp. 191–198. Kluwer, Dordrecht (1998)
15. Mosorov, V.: A main stem concept for image matching. Pattern Recognition Letters 26(8), 1105–1117 (2005)
16. Naegel, B., Passat, N., Boch, N., Kocher, M.: Segmentation using vector-attribute filters: methodology and application to dermatological imaging. In: Proc. ISMM 2007, pp. 239–250 (2007)
17. Naegel, B., Wendling, L.: Document binarization based on connected operators. In: Proc. ICDAR 2009 (to appear, 2009)
18. Naegel, B., Wendling, L.: Combining shape descriptors and component-tree for recognition of ancient graphical drop caps. In: VISAPP 2009, vol. 2, pp. 297–302 (2009)
19. Najman, L., Couprie, M.: Building the component tree in quasi-linear time. IEEE Trans. Image Proc. 15(11), 3531–3539 (2006)
20. Salembier, P., Oliveras, A., Garrido, L.: Anti-extensive connected operators for image and sequence processing. IEEE Trans. Image Proc. 7, 555–570 (1998)

21. Salembier, P., Garrido, L.: Binary partition tree as an efficient representation for image processing, segmentation and information retrieval. IEEE Trans. Image Proc. 9, 561–576 (2000)
22. Titterington, D.M.: Estimation of correlation coefficients by ellipsoid trimming. Appl. Stat. 27(3), 227–234 (1978)
23. Vincent, L.: Grayscale area openings and closings, their efficient implementations and applications. In: Proc. EURASIP Workshop on Mathematical Morphology and its Applications to Signal Processing, pp. 22–27 (1993)
24. Westenberg, M.A., Roerdink, J.B.T.M., Wilkinson, M.H.F.: Volumetric attribute filtering and interactive visualization using the Max-Tree representation. IEEE Trans. Image Proc. 16(12), 2943–2952 (2007)

Fast Implementation of the Ultimate Opening

Jonathan Fabrizio[1,2] and Beatriz Marcotegui[1]

[1] MINES Paristech, CMM- Centre de morphologie mathématique,
Mathématiques et Systèmes, 35 rue Saint Honoré - 77305 Fontainebleau cedex, France
[2] UPMC Univ Paris 06
Laboratoire d'informatique de Paris 6, 75016 Paris, France
{jonathan.fabrizio}@lip6.fr, marcoteg@cmm.ensmp.fr
http://www-poleia.lip6.fr/~fabrizioj/

Abstract. We present an efficient implementation of the ultimate attribute opening operator. In this implementation, the ultimate opening is computed by processing the image maxtree representation. To show the efficiency of this implementation, execution time is given for various images at different scales. A quasi-linear dependency with the number of pixels is observed. This new implementation makes the ultimate attribute opening usable in real time. Moreover, the use of the maxtree allows us to process specific zones of the image independently, with a negligible additional computation time.

1   Introduction

The ultimate opening ($U.O.$) is a powerful morphological operator that highlights the highest contrasted areas in an image without needing any parameters. It has been used, for example, as a segmentation tool for text extraction [1]. We offer, in this paper, a fast implementation of the $U.O.$ based on a maxtree.

The article is organized as follows: in the first part, the $U.O.$ operator is briefly explained. In the second part, the fast implementation is explained and an adaptation of the algorithm is made, to implement the iterated U.O., which we introduce. In the last part the speed of the algorithm is measured and then comes the conclusion.

2   The Ultimate Opening Operator

Introduced by Beucher [2], the $U.O.$ is a residual operator that highlights patterns with the highest contrast. The operator successively applies increasing openings γ_i (of size i) and selects the maximum residue r_i computed between two successive results of opening, γ_i and γ_{i+1}, applied to the image (i.e. the difference $\gamma_i - \gamma_{i+1}$). The examples of Figs. 1 and 2, show that whatever small variations and noise, the operator only keeps the strongest patterns of the image: small variations on a contrasted structure are considered as noise and are automatically eliminated. For each pixel, the operator gives two pieces of information, the maximal residue ν and the size q of the opening leading to this residue:

M.H.F. Wilkinson and J.B.T.M. Roerdink (Eds.): ISMM 2009, LNCS 5720, pp. 272–281, 2009.
© Springer-Verlag Berlin Heidelberg 2009

$$\nu = sup(r_i) = sup(\gamma_i - \gamma_{i+1}) \tag{1}$$

$$q = \max(i) + 1; r_i = \nu(\neq 0). \tag{2}$$

ν, called the transformation in the literature, gives indications on the contrast of pattern and q, usually called residual or associated function, gives an estimation of the pattern size (a sort of granulometric function). It is also possible to use an attribute opening [3]. In this case, the associated function provides information linked with the given attribute. In this article, all tests and comments have been made on ultimate attribute openings. The attribute we have chosen is the height of the connected component, defined as the maximum difference of vertical co-ordinate among pixels that belong to the connected component. We have chosen this attribute according to our application. Height attribute is easily computed during maxtree creation. Other attributes could have been used instead.

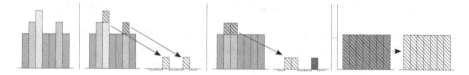

Fig. 1. The computation of the ultimate opening. From left to right: 1. the profile of the image is given, then 2. an opening of size 1 (γ_1) does not change the image and γ_2 removes two maxima. The residue is recorded in the transformation. 3. γ_3 generates a residue with the result of the previous opening, this residue is recorded in the transformation. Following openings (γ_4 to γ_7) do not generate residues. 4. At the end, γ_8 generates a bigger residue than the previous ones and is recorded in the transformation. As this residue is high, it erases all previous residues.

Fig. 2. On the left hand-side, an image, on the right hand-side the result (the transformation) of the ultimate height opening. Image copyright *institut géographique national (IGN)*.

3   The Implementation Based on Maxtree

The *U.O.* operator is a very powerful non-parametric operator. Its direct implementation (by applying successive openings) leads to a high computation time. A faster implementation has been proposed by Retornaz [4,5] but it is still time consuming because many regions of the image are processed several times.

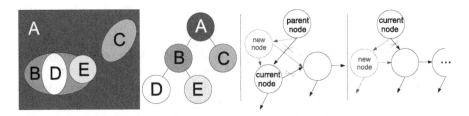

Fig. 3. From left to right: The image and the corresponding maxtree. The insertion of a node before in the branch (function *next_label_lower*()). The insertion of a node after in the branch (function *next_label_higher*()).

We propose a much faster implementation of the $U.O.$ operator. This implementation is based on the use of *maxtree* [6] and has multiple advantages:

- the image is processed only twice: once to create the tree and the other to deduce the transformation and the associated function,
- a region may be processed independently from the rest of the image by processing the corresponding branch of the tree,
- once the maxtree is created, many other filters [7] can be applied to the image directly on the tree without processing the image (which is much faster [8]).

The maxtree is a data structure that represents the image by a tree (Fig. 3). A node of level l is linked with a subset of pixels at level l in the image. Those pixels must be connected together by pixels laying at a level higher or equal to l. This means that a node and all its sub-nodes represent a region of the image which is the union of connected regions of level equal or higher than l. Two nodes at the same level l are separated in the image by pixels at a level lower than l. The mintree is the dual structure. Our algorithm is divided into three parts: 1. building the maxtree according to the image, 2. processing the tree and 3. generating the output of the operator. We will study each step in details.

3.1   Building the Maxtree

Algorithms to compute the tree already exist [9,6,10] and [11]. Any of them could have been used. In order to have the description of all the process steps and because the creation of the tree is time consuming, the tree creation is revisited.

Before writing the algorithm, the first thing to do is to choose data structures. Huang et al. in [10] offer to use a combination of a linked list and *a hash map*. The linked list is used to chain the tree nodes and the hash map is used to access, in a constant and small time period, each node of the tree. They also propose to store, in each node, a large amount of data and particularly: a unique identifier (id), its parent id and its children id list. Thanks to an adapted data structure, a lot of memory may be saved. Firstly, we do not keep the node id in the node. This id is implicitly given by the offset from the root node in the data structure. Next, as we do not need to go backward, we do not record the parent id. Moreover the list of children proposed by Huang implies the use of another linked list (one for each node) which increases the complexity and is memory consuming. These lists make us lose time during the tree creation. We replace these lists by two (integer) fields in the node: *son* the first child id and *brother* the first brother id. Then finding every children of a node i is easy and fast (all brothers may not have the same level):

```
01 child=node[i].son;              //first child
02 while(child !=0 ) child=node[child].brother; //next child
```

As the id gives us the index of a node in the data structure, we do not need the hash table any more and we save memory again. The underlying structure is also a bit different. Huang says that an array should not be used and uses a linked list instead. But a linked list has some drawbacks: it consumes time to allocate each node and does not offer a simple way to access a node (that is why they are obliged to use a hash table). We use instead an intermediate paginated structure: we do not link nodes but a group of nodes (a page). Each group of nodes is recorded in an array (of N elements). This simplifies the allocation because nodes are allocated by blocks. Elements can also be simply accessed: to get the ith element we compute the page number $page = i \,/\, N$ and the offset in the page $offset = i\, modulo\, N$.

Now that we have seen data structures, let us see the algorithm itself which is strongly inspired from [9]. To build the tree, a priority *lifo* structure is needed with standard functions: $push(e, p)$ which pushes an element e with the specified priority p, $pop()$ which provides the last element with the highest priority level and $get\_higher\_level()$ which gives the highest priority available.

The creation of the maxtree is based on a flooding process and starts on a pixel that has the minimum value. During the flooding process, from pixel p, when a new pixel p_next is flooded, three different situations may occur (figure 3):

- A) p_next has a level higher than p (line 14). Then, a new branch of the tree is created starting from branch p (function $next\_label\_higher()$),
- B) p_next has a level lower than p and this level is unknown (line 17). Then a new node is created at the correct level in the same branch of node p and adds this pixel to this node (function $next\_label\_lower()$).
- C) p_next has a level lower than p and this level is known (line 19). Then p_next is added to the node ancestor to node p at the correct level,

The tree is built by the following algorithm called with the min value of the pixel in the lifo. To simplify the readability some variables are used as if they were globally declared. Particularly, the tree, which is the main output of the function $flood$, is split into two indexed variables: *son* and *brother* ($son[i]$ gives the first child of node i and $brother[son[i]]$ gives his second and so on...). $attributes[i]$ and $level[i]$ give respectively the value of the attribute and the gray level of node i. Notice that the index of a node is the label used to fill in a region in lab_img, allowing to link the tree node with the corresponding region. Finally, the variable $next\_label$ always gives the next index available.

```
01 void flood(img, lab_img, labs_branch, level)
02 in
03  img: the input image,
04  labs_branch: labs_branch[l] gives the label of level l in the current branch of the tree
05  level: the level of the starting point of the flooding process
06 out
07  the labeled image (lab_img) and the tree (son, brother, attributes and level)
08 {
09  index: the label;
10  p: the current pixel;
11  while( (get_higher_level()>=level) && (p=pop())!=-1)
12  {
13    for all pixels p_next, unlabeled neighbor of p {
14      if (img[p_next]>img[p]) {                        // CASE A
15        for(j=img[p]+1;j<img[p_next];j++) labs_branch[j]=0;
16        index=lab_img[p_next]=labs_branch[j]=
                    next_label_higher(img[p], img[p_next], labs_branch);
17      } else if (labs_branch[img[p_next]]==0) {        // CASE B
18        index=lab_img[p_next]=
                    labs_branch[img[p_next]]=next_label_lower(img[p_next], labs_branch);
19      } else index=lab_img[p_next]=labs_branch[img[p_next]];    // CASE C
20      // *** update here attributes[index] according to p_next ***
21      push(p_next, img[p_next]);
22      if (img[p_next]>img[p]) flood(img, lab_img, labs_branch, img[p]);
23    }
24  }
25 }
26
27 label next_label_lower(value, labs_branch)
28 in:
29  value: the level of the pixel,
30  labs_branch: labs_branch[l] gives the label of level l in the current branch of the tree
31 {
32  level[next_label]=value;
33  l=level_of_parent_of(labs_branch[value]);
34  son[next_label]=son[labs_branch[l]];
35  son[labs_branch[l]]=next_label;
36  brother[next_label]=brother[son[next_label]];
37  brother[son[next_label]]=NO_BROTHER;
38  // *** init here attributes[next_label] ***
39  return next_label++;
40 }
41
42 label next_label_higher(parent_value, value, labs_branch)
43 in:
44  parent_value: the level of the ancestor pixel,
45  value: the level of the pixel,
46  labs_branch: labs_branch[l] gives the label of level l in the current branch of the tree
47 {
48  level[next_label]=value;
49  brother[next_label]=son[labs_branch[parent_value]];
50  son[labs_branch[parent_value]]=next_label;
51  // *** init here attributes[next_label] ***
52  return next_label++;
53 }
```

A comparison with a non optimized version of the algorithm proposed in [9] (provided by the authors) shows that our implementation is in average 33% faster. This difference seems to be mainly due to memory management.

3.2 Computing Ultimate Opening

An attribute opening γ_i discards regions with attributes j smaller than i. γ_i can easily be deduced from the previously computed maxtree by pruning branches

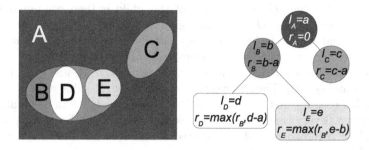

Fig. 4. The computation of the residue in the tree according to an attribute height opening

with an attribute smaller than i [7] (each node is valuated with the attribute value of its corresponding region). A residue can be computed between two successive openings γ_i and γ_{i+1} by the difference of the resulting images of each opening. In the tree, this residue is computed (in every node removed by γ_{i+1}) by the difference between the gray level of each node and its first ancestor with a different attribute. A node with attribute different from i has no residue at step i (i.e $r_i = 0$). The $U.O.$ analyzes the residue of successive growing size openings and the highest residue is kept (Fig. 4, Eq. 1-2). During the process, an opening γ_i may generate a residue r_1 for a node n_1. Later, a bigger opening γ_j will generate a residue r_2 for a node n_2 ancestor of n_1 in the tree. As n_2 encompasses n_1, if r_2 is bigger than r_1, r_2 must also be assigned to n_1 (if not n_1 keeps r_1). To find the highest residue of all nodes, each ancestor n_k of each node must be checked. This is why the maxtree data structure is suitable: the highest residue will not be searched among ancestors but all ancestors will transmit their own maximum residue. The tree is recursively processed: on a given node, the computed residue is transmitted to every child (see l 22-25 of pseudo-code). Every child will compare its own residue (*var. contrast*) with the one transmitted by its ancestors (l 11-17), and will keep the maximum of them (*transform_node_LUT*) as well as the opening attribute size associated with the maximum residue (*associate_node_LUT*) (l 18-19). This maximum is transmitted again and so on. This process is in o(n), with n the number of nodes.

Figure 4 illustrates an example of ultimate opening computation on the max-tree. The height attribute is chosen for the example (we note H_k the height of region k). First, the max-tree is created and each node is given the gray level and the attribute of the corresponding region. Note that regions B and D have the same attribute. Both regions are then removed by the same opening, of size $H_B + 1$ ($= H_D + 1$). The residue generated on region D by γ_{H_B+1} is not $l_D - l_B$ (the gray level difference with its parent node) but $l_D - l_A$, A been its first ancestor with a different attribute.

Let us process the tree to compute the ultimate height opening. The process follows a depth-first traversal starting from the root node. The first computed residue is residue r_B of region B. This residue is equal to the difference of the level of region B and the level of region A ($l_B - l_A$). Then, this residue r_B

is transmitted to child D. The residue of this child is then computed as the difference between region D level and region A level ($l_D - l_A$ and not $l_D - l_B$ as region B and region D have the same attribute and are then removed by the same opening). r_D value is the highest value between r_B and $l_D - l_A$. The next child E is processed. The residue of this child is $l_E - l_B$ and r_E is the max between r_B and $l_E - l_B$. The last region C is processed. Residue r_C is simply the difference between region C level and region A level ($l_C - l_A$). r value for all nodes is then known. Every time a residue is selected, the size of the opening leading to the residue is recorded (and transmitted to children) in order to generate the associated function q of equation 1.

```
01 void compute_uo(node, max_tr, max_in, parent_attribute, parent_value, previous_value)
02 In:
03 node: the processed node
04 max_tr: the maximum of contrast of previous nodes
05 max_in: the attribute that generate max_tr
06 parent_attribute: the attribute of the parent node
07 parent_value: the value of the parent node
08 previous_value: the value of the 1st ancestor with a
                                       different attribute: the parent of the branch
09 {
10   contrast=(attributes[node]==parent_attribute)?
                               level[node]-previous_value:level[node]-parent_value;
11   if (contrast>=max_tr) {
12     max_contrast=contrast;
13     linked_attributes=attributes[node];
14   } else {
15     max_contrast=max_tr;
16     linked_attributes=max_in;
17   }
18   transform_node_LUT[node]=max_contrast;
19   associated_node_LUT[node]=linked_attributes+1;
20   child=son[node];
21   if (attributes[node]==parent_attribute) pv=previous_value; else pv=parent_value;
22   while (child!=0) {
23     compute_uo(child, max_contrast, linked_attributes, attributes[node], level[node], pv);
24     child=brother[child];
25   }
26 }
```

3.3   Generating the Output Images

The last step consists in generating the two results (the transformation and the associated function). This is an easy step: by the use of the labeled image obtained by the flooding step and the two look-up tables computed in *compute_uo* (*transform_node_LUT* and *associated_node_LUT* which give, for a given label, respectively the maximum contrast and the associated attribute for this contrast), we can deduce the transformation and associated function images:

```
01 for all pixel p {
02   transform[p]=transform_node_LUT[lab_img[p]];
03   associate[p]=associated_node_LUT[lab_img[p]];
04 }
```

3.4   Iterated Ultimate Opening

The maxtree is well adapted to process $U.Os.$. Several filters may be deduced from the same tree structure. These filters can be applied on the entire image

Fig. 5. From left to right: input image, the result of the *U.O.* (some details such as the text "UNITED STATES" are masked), and the result of the iterated *U.O.* where the text is clearly visible (contrast of both images is enhanced to be visible). With the maxtree, negligible additional time is needed to iterate the *U.O.* onto some areas of the image. Here, the condition to re-start *U.O.* is based on feature size and contrast.

or on a part of it (which corresponds to a branch of the tree). In this section we illustrate this property. A major issue with the *U.O.* is that sometimes, when an interesting area of the image is surrounded by a highly contrasted border (text over a sign board for example), the content of this area is masked by the *U.O.* (Fig. 5). To solve this issue, we offer to iterate the *U.O.* on such an area (Fig. 5). Function *compute_uo* can simply be adapted to this improvement and is modified in *compute_uo_iterative* to process specific areas of the image (ie. a specific branch) from the rest of the image. A condition is checked before processing branches:

```
1 if (specific condition) max_tr_propag=0; else max_tr_propag=max_contrast;
2 while (child!=0) {
3    compute_uo_iterative(child, max_tr_propag, linked_attributes,
                                 attributes[node], level[node], pv);
4    child=brother[child];
5 }
```

Instead of transmitting the estimated contrast deeper in the branch, we just transmit 0 as the previous contrast. The branch will be processed as a new tree without consuming additional time. Trying to perform this operation directly on the image is much harder and much more time consuming. The main difficulty with this approach is to determine when the *U.O.* will be restarted but this question is out of the scope of this article and depends on the application.

Format	128x128	256x256	512x512	1024x1024	2048x2048
Nb of pixels	16384	65536	262144	1048576	4194304
Time (ms)	0,18	2,39	12,01	52,04	235,53

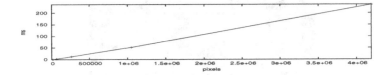

Fig. 6. Execution time of the *U.O.* according to the number of pixels

4 Results

To measure the efficiency of the implementation, we have tested it on a personal image database (about 570 various photos). We evaluate the average time consumed by our implementation by computing ten times the *U.O.* on every picture. We perform the test at different scales. The result is approximately linear according to the number of pixels in the image (Fig. 6) (except for very small images; the cache may introduce a bias). All tests have been carried out on a DELL D630 laptop with 2,4GHz T7700 processor and the implementation is in C. Given times include all the process: all allocations (lifo structure...) maxtree creation, *U.O.* process, result generation and memory cleaning (intermediary data such as the maxtree, the lifo structure...). Only I/O operations are not included in given times. Moreover, let us consider the execution time distribution: 72% of time in average is spent to build the tree, 9% and 19% of time is spent for processing the tree and generating the result respectively. This new implementation of the *U.O.* operator is a major improvement and makes this operator usable in various contexts and even in real time for rather large images.

5 Conclusion

We have presented a fast implementation of the *U.O.* based on a maxtree. Even if existing maxtree implementations may be used for our purpose, we propose a new implementation with a more efficient memory management.

As stated in the literature, several connected operators may be deduced from the same tree-structure. This is why, the maxtree is an efficient representation to implement an ultimate attribute opening operator (which definition involves a series of connected openings). The maxtree creation itself, remains the most time consuming task of our process. Moreover, we propose an iterated version of the ultimate opening. It provides more details in a given region, with negligible additional time, exploiting the re-usability of the maxtree representation.

At the end, we evaluate the executing time and show that the algorithm is quasi-linear according to the number of pixels. The *U.O.* is a powerful nonparametric tool. This fast implementation makes it very competitive and usable in real time.

Acknowledgments. We thank Pr. Salembier for providing us his maxtree implementation to perform comparisons. This work is supported by the French ANR (Agence Nationale de la Recherche) in the itowns project.

References

1. Retornaz, T., Marcotegui, B.: Scene text localization based on the ultimate opening. In: International Symposium on Mathematical Morphology, vol. 1, pp. 177–188 (2007)
2. Beucher, S.: Numerical residues. Image Vision Computing 25(4), 405–415 (2007)
3. Breen, E.J., Jones, R.: Attribute openings, thinnings, and granulometries. Computer Vision and Image Understanding 64(3), 377–389 (1996)
4. Retornaz, T.: Détection de textes enfouis dans des bases d'images généralistes. Un descripteur sémantique pour l'indexation. PhD thesis, Ecole Nationale Suprieure des Mines de Paris - C.M.M., Fontainebleau - France (October 2007)
5. Retornaz, T., Marcotegui, B.: Ultimate opening implementation based on a flooding process. In: The 12th International Congress for Stereology (September 2007)
6. Garrido, L.: Hierarchical Region Based Processing of Images and Video Sequences: Application to Filtering, Segmentation and Information Retrieval. PhD thesis, Universitat Politnica de Catalunya - Department of Signal theory and Communications, Barcelona, Spain (April 2002)
7. Salembier, P., Garrido, L.: Connected operators based on region-tree pruning strategies. In: International Conference on Pattern Recognition, ICPR 2000, September 2000, vol. 3, pp. 3371–3374 (2000)
8. Meijster, A., Wilkinson, M.H.: A comparison of algorithms for connected set openings and closings. IEEE Trans. on Pattern Analysis and Machine Intelligence 24(4) (April 2002)
9. Salembier, P., Oliveras, A., Garrido, L.: Anti-extensive connected operators for image and sequence processing. IEEE Trans. on Image Processing 7(4), 555–570 (1998)
10. Huang, X., Fisher, M., Smith, D.: An efficient implementation of max tree with linked list and hash table. In: Proc. VIIth Digital Image Computing: Techniques and Applications, pp. 299–308 (2003)
11. Najman, L., Couprie, M.: Building the component tree in quasi-linear time. IEEE Transactions on Image Processing 15(11), 3531–3539 (2006)

Stack Filter Classifiers

Reid Porter[1], G. Beate Zimmer[2], and Don Hush[1]

[1] Los Alamos National Laboratory, Los Alamos, NM 87544
rporter@lanl.gov, dhush@lanl.gov
[2] Department of Mathematics and Statistics,
Texas A&M University – Corpus Christi,
6300 Ocean Drive, Corpus Christi, TX 78412-5825
Beate.Zimmer@tamucc.edu

Abstract. Stack Filters define a large class of increasing filter that is used used widely in image and signal processing. The motivations for using an increasing filter instead of an unconstrained filter have been described as: 1) fast and efficient implementation, 2) the relationship to mathematical morphology and 3) more precise estimation with finite sample data. This last motivation is related to methods developed in machine learning and the relationship was explored in [1]. In this paper we investigate this relationship by applying Stack Filters directly to classification problems. This provides a new perspective on how monotonicity constraints can help control estimation errors, and also suggests new learning algorithms for Boolean function classifiers when they are applied to real-valued inputs.

1 Introduction

Just as linear models generalize the sample mean and weighted average, weighted order statistic models generalize the sample median and weighted median [2]. This analogy can be continued informally to generalized additive models in the case of the mean, and Stack Filters in the case of the median. Both of these model classes have been extensively studied for signal and image processing, but it is surprising to find that for pattern classification, their treatment has been significantly one sided. Generalized additive models are now a major tool in pattern classification and many different learning algorithms have been developed to fit model parameters to finite data. Several model classes related to Stack Filters have been suggested for classification including morphological networks [3], min-max networks [4], order statistics [5] and Positive Boolean Functions [6], [7]. However direct application of Stack Filters to classification problems is yet to be seen. One of the reasons why Stack Filters have not been directly applied to classification problems is because Stack Filter classifiers appear to reduce to a known problem: learning a Boolean function. In this paper we show that on closer inspection, optimizing

M.H.F. Wilkinson and J.B.T.M. Roerdink (Eds.): ISMM 2009, LNCS 5720, pp. 282–294, 2009.

Stack Filters for classification leads to a different Boolean function learning problem than has been traditionally considered.

Since Stack Filter classifiers reduce to Boolean function classifiers, they also share many properties with decision tree classifiers, including fast and simple implementation, and increased interpretability. Some of the difficulties encountered with these types of classifiers include high approximation error and combinatorial learning problems. Several important learning algorithms have been developed to address these difficulties in different ways. Traditionally tree models are built with a top-down greedy method, and then pruned to control over-fitting [8]. More recently theoretical results and increased computing resources have enabled the development of optimal learning algorithms over the class of dyadic decision trees [9]. These methods have been applied successfully to practical problems and provide an exact minimization of a complexity penalized loss function.

In this paper we propose an approach most similar to this second method, in that we suggest a global optimization problem for Boolean function classifiers that can be exactly minimized. We also show that by approaching the problem as a Stack Filter, we arrive at a new and unique method to control over-fitting.

2   Main Results

We consider two-class classification, where we are given a training set of N points, $x \in \mathbb{R}^D$, with labels, $y \in \{-1, 1\}$, drawn from a distribution $P_{X,Y}$. The task is to find a model (or function) $F : \mathbb{R}^D \to \mathbb{R}$ that has small error $e(F) = E_{X,Y}(\mathbf{1}_{\{sgn(F(x)) \neq y\}})$. Classification performance is measured by the excess error of the classifier $e(F)$ compared to the Bayes optimal classifier $e^* = \inf_{\forall F} e(F)$ and can be viewed as a combination of approximation and estimation errors (these quantities are related to bias and variance):

$$e(F) - e^* = \left(e(F) - \inf_{F' \in \mathcal{F}} e(F') \right) + \left(\inf_{F' \in \mathcal{F}} e(F') - e^* \right) \tag{1}$$

The first term is estimation error and is due to the fact that we only have a finite number of examples to select the best model from the model class $\mathcal{F}$. The second term is approximation error and is due to the fact that the Bayes classifier is not represented in the model class. These two errors have conflicting needs: a common way to reduce approximation error is to increase the capacity of the model class but this typically increases the estimation error. The learning algorithm must balance these needs and the most common approach is to choose a function F that minimizes a training set error:

$$\hat{F} \in \arg\min_{F \in \mathcal{F}} \hat{e}(F, L) \tag{2}$$

$$\hat{F} \in \arg\min_{F \in \mathcal{F}} \frac{1}{N} \sum_{i=1}^{N} L(F(x(i)), y(i)) \tag{3}$$

where $L : (\mathbf{R} \times \{-1, 1\}) \to \mathbf{R}$ is a loss function. The choice of loss function affects both the estimation and approximation errors of $\hat{F}$ and must be carefully chosen. A popular approach is to define a very rich model class and then parameterize the loss function in a way that allows the tradeoff to be easily tuned to the application: $L_\gamma(F(x), y)$. At one extreme of γ, the loss function would define a classifier with zero approximation error and at the other extreme, a classifier with zero estimation error. We would also like both errors to decrease as N increases. It would also be desirable if the value of γ was well behaved, or in some way easy to tune e.g. it is a smooth (convex) function of the excess error, and/or it is constrained to a small, finite number of values.

Support vector machines provide one solution to this problem for Reproducing Kernel Hilbert space model classes, and in this case the loss function includes a regularization parameter. In this paper we suggest a loss function and calibration parameter for Stack Filter classifiers with several desirable properties. In particular, for misclassification loss:

$$L(F(x), y) = \mathbf{1}_{\{F(x) \neq y\}} \tag{4}$$

a Stack Filter minimizer can be found via a linear program of $O(N)$ variables. For large-margin misclassification loss:

$$L_\gamma(F(x), y) = \mathbf{1}_{\{yF(x) < \gamma\}} \tag{5}$$

a Stack Filter minimizer

$$\hat{F}_\gamma(x) \in \arg\min_{F \in \mathcal{F}} \hat{e}(F, L_\gamma) \tag{6}$$

is equivalent to minimizing misclassification loss with a Stack Filter from a restricted function class:

$$\hat{F}_\gamma(x) \in \arg\min_{F \in \mathcal{F}_\gamma} \hat{e}(F, L) \tag{7}$$

where $\mathcal{F}_\gamma \subseteq \ldots \subseteq \mathcal{F}_1 \subseteq \mathcal{F}$. This margin parameter is monotonically related to the size of the Stack Filter function class and is also discrete and bounded. For large margin hinge loss:

$$L_\gamma^h(F(x), y) = (\gamma - yF(x))_+ \tag{8}$$

a Stack Filter minimizer also minimizes the sum of large-margin misclassification loss functions:

$$\hat{F}_\gamma^h(x) \in \arg\min_{F \in \mathcal{F}} \sum_{\gamma'=-\gamma}^{\gamma} \hat{e}(F, L_{\gamma'}) \tag{9}$$

This result implies that large-margin hinge loss is a good choice for optimizing stack filter classifiers. It has one parameter, which determines the size of the model class considered during optimization, and it minimizes the dependence on that parameter, which should make it easier to tune. The size of the model class $\mathcal{F}$, although finite, can be made arbitrarily large and minimization over $\mathcal{F}$ is exact with a linear program of $O(2\gamma D)$ variables.

3   Stack Filter Classifiers

Stack Filters are defined using threshold decomposition and monotonicity constraints. Given a real valued input vector $x = [x_1, x_2, \ldots, x_D]$ we define a thresholding function $u = x \succcurlyeq c$, parameterized by a scalar c, that produces a binary vector with components $u_i = \mathbf{1}_{\{x_i \geqslant c\}}$. We then define a Stack Index Filter, $SI : \mathbb{R}^D \to \{1, \ldots, D\}$ as:

$$SI(x) = \sum_{d=1}^{D} f(x \succcurlyeq x_{(d)}) \tag{10}$$

where $x_{(d)}$ is the dth smallest component of x and $f : \{0,1\}^D \to \{0,1\}$ is a positive Boolean function (PBF). A Boolean function is positive (or monotone, non-decreasing) if it satisfies the stacking constraint that $u_i \geqslant v_i, \forall i$ implies $f(u) \geqslant f(v)$. A Boolean function that is defined using 'and' and 'or', but no negations, satisfies this constraint. A Stack Filter, $S : \mathbb{R}^D \to \mathbb{R}$, is related to a Stack Index Filter by the relationship:

$$F(x) = x_{(SI(x))} \tag{11}$$

In classification problems we typically threshold a real valued function to produce an indicator for class labels. However the output from a Stack Filter is always one of the inputs, which means choosing a sensible threshold is non-trivial. We suggest extending the input space using the mirror-map $\mathcal{M} : \mathbb{R}^D \to \mathbb{R}^{2D}$ given by $\mathcal{M}(x) = [x, -x]$ [10]. This means zero is guaranteed to lie between the Dth and $(D + 1)^{th}$ order statistics, and we can use the sign of the Stack Filter as a class indicator. Figure 1a provides an example of a Stack Filter classifier predicting $y = 1$ for a mirrored input sample $x = [3, 1, 2, -3, -1, 2]$. The monotonicity constraints mean that the the output column is always a solid *stack* of ones, and the height corresponds to the Stack Index Filter output. In addition, monotonicity also means that:

$$\mathbf{1}_{\{F(x) \geqslant t\}} = \mathbf{1}_{\{f(x \succcurlyeq t)\}} \tag{12}$$

In Figure 1a we see a Stack Filter thresholded at zero is equivalent to a positive Boolean function applied to an abstract middle row between D^{th} and

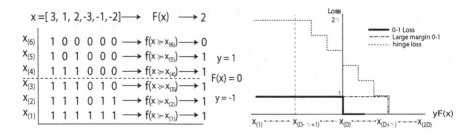

Fig. 1. a) Stack Filter Classifier. b) Classification loss functions investigated.

$(D+1)^{th}$ thresholds. A topic of interest in this paper are learning algorithms that require the Stack Filter output to be further from the decision boundary. This distance can be measured in terms of the number of threshold levels and is called rank-order margin [11]. For example, in Figure 1a the sample has been predicted with rank-order margin $\gamma = 2$.

There are several loss functions we might consider for Stack Filter classifiers. We investigate three, which are illustrated in Figure 1b, and described in the next few sub-sections.

3.1   0-1 Loss

Finding the Stack Filter which minimizes 0-1 loss, is equivalent to finding the positive boolean function that minimizes 0-1 loss. From Equation 12 it follows that:

$$L(F(x), y) = \mathbf{1}_{\{yF(x)<0\}} \tag{13}$$
$$= \mathbf{1}_{\{-yf(x \succeq 0)\}}$$

where we redefine the Boolean function output labels to simplify notation: $f : \{0,1\}^D \to \{1,-1\}$. We first consider the related problem of finding a Boolean function that minimizes 0-1 loss. We define a partially specified Boolean function where we assign class labels to the rows of a look-up table that appear in the training set thresholded at zero: $u = x \succeq 0$. The same row can appear multiple times in the training set and so we identify the unique set by $Q = \{q(1), q(2), \ldots, q(M)\}$. A straightforward solution is to implement a plug-in type classifier and estimate the class conditional probability for each $q(i)$ independently:

$$\hat{P}_{q(i)} = \frac{\sum_{n=1}^{N} \mathbf{1}_{\{u(n)=q(i), y(n)=1\}}}{\sum_{i=1}^{N} \mathbf{1}_{\{u(n)=q(i)\}}}.$$

We assign class labels z_i for each q_i with the rule:

$$z_i = \begin{cases} 1 & \text{if} \quad \hat{P}_{q(i)} > 0.5 \\ -1 & \text{otherwise} \end{cases} \tag{14}$$

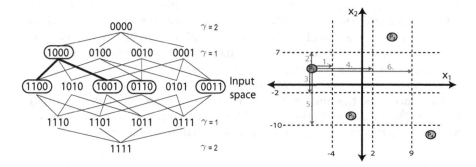

Fig. 2. a) Lattice diagram of the mirrored input space. b) Example of the input expansion described in Section 4.

If we restrict the Boolean function to be positive, then we must introduce monotonicity constraints. This means the plug-in rule of Equation 14 is replaced by an integer linear program:

$$\begin{aligned}
minimize \quad & c.z \\
subj \quad & z_i \geqslant z_j \quad when \ q_i \geqslant q_j \\
and \quad & z_i \in \{0,1\} \ \forall i,j
\end{aligned} \tag{15}$$

where the cost for variable z_i is $c_i = 0.5 - \hat{P}_{q(i)}$. Note, to simplify notation we switched to class labels $\{0,1\}$. This linear program was first suggested for Stack Filter optimization under mean absolute error [12].

3.2 Large Margin 0-1 Loss

We now consider large-margin loss functions and define margin, γ, as the number of thresholds above (and below) zero in Figure 1. This leads to the large margin 0-1 loss:

$$\begin{aligned}
L_\gamma(F(x),y) &= \mathbf{1}_{\{yF(x) < x_{(D+\gamma)}\}} \\
&= \mathbf{1}_{\{-yf(x \succeq x_{(D+y\gamma)})\}}
\end{aligned} \tag{16}$$

where to simplify notation: $f : \{0,1\}^D \to \{1,-1\}$ and we have omitted a class dependent offset. For class 1 samples, x is thresholded by $x_{(D+\gamma)}$, which is larger than $x_{(D)}$, which means there are less ones. In a similar way, for class -1 samples, x is thresholded by $x_{(D-\gamma+1)}$, which is smaller than $x_{(D)}$, which means there are more ones. The problem has the same form as the 0-1 loss problem, however the binary input samples are different.

In Figure 2a the monotonicity constraints of positive Boolean functions are illustrated as a lattice where links between two Boolean values u and v implies an ordering $u \geqslant v$ $(u_i \geqslant v_i, \forall i)$. The mirrored representation means that the original input space is a subset of entries in the middle row of the lattice where $[u, \bar{u}]$. As rank order margin is increased, samples move higher (for class 1) and lower (for class -1) in the lattice, which produces increasing numbers of constraints. In Figure 2 a sample $u = [1100]$ moves to $u' = [1000]$ at margin 1, which places an additional constraint on $v = [1001]$.

As γ increases, the number of positive boolean functions that can satisfy the additional constraints decreases. The large margin 0-1 loss functions for Stack Filters therefore define reduced sets of PBF function classes.

3.3   Hinge Loss

Hinge loss is typically defined as $(1 - F(x))_+$, but for Stack Filters, the loss function is discrete and bounded. Furthermore, as shown in Figure 1b, the maximum loss incurred is 2γ at threshold level $(D - \gamma + 1)$. This is because for class 1, threshold levels $1 \ldots (D - \gamma)$ do not introduce any additional constraints, i.e., all samples at threshold $(D-\gamma)$ are below the class -1 samples at $(D - \gamma + 1)$ and therefore can be trivially satisfied. The same reasoning applies to class -1 samples above $(D+\gamma)$. Given this reduced set of thresholds, we can write Stack Filter hinge loss as:

$$L_\gamma^h(F(x), y) = \sum_{\gamma'=-\gamma}^{\gamma} \mathbf{1}_{\{-yf(x \succcurlyeq x_{(D+y\gamma')})\}} \tag{17}$$

By reordering summations we see minimizing hinge loss is equivalent to minimizing the sum of large margin 0-1 loss functions as described in Equation 9. The solution has the same form as Equation 15, but with more variables (2γ times more) and more constraints. Note, that this decomposition of hinge loss to a sum of misclassification loss functions follows directly from the original results for Stack Filters under mean absolute error [12]. For classification, this decomposition suggests that the optimal Stack Filter classifier will have some degree of invariance to the rank order margin parameter. This is useful in practice since we need to choose this parameter for the application. Put another way, optimizing Stack Filters with hinge loss smoothes the error estimate as a function of margin, which should help methods like cross-validation converge.

4   Input Expansion

Direct application of Stack Filters typically leads to significant approximation error, e.g., in two dimensions, the Stack Filter function class has only two functions (maximum and minimum). The solution is to map the input space into a higher-dimensional feature space. This is typically an application specific problem, but here we consider a general purpose expansion that work wells with Stack Filter learning algorithms. First, we map each input independently using a set of constant thresholds:

$$xx_d = [x_d - t_d(1), x_d - t_d(2), \ldots, x_d - t_d(T_d)] \tag{18}$$

The two main ways we choose thresholds are: 1) evenly spaced across the input range, and 2) midpoints between consecutive samples in the training set. When applying Stack Filter classifiers we threshold the expanded input at zero. This means we can calculate and represent the thresholded expansion by ranks, i.e., each dimension is replaced by an integer which simply counts how many thresholds are below the given sample:

$$r_d = \sum_{i=0}^{T_d} \mathbf{1}_{\{x_d - t_{1,i} > 0\}} \tag{19}$$

We call r the rank expansion and it allows us to manipulate a $(2D * T_D)$ dimensional Stack Filter with $2D$ integers. In Figure 2b we provide an example of this input expansion in two-dimensions with 4 points: $\{P_1 = (-6, 4), P_2 = (-2, -8), P_3 = (6, 10), P_4 = (12, -12)\}$ and 3 data dependent thresholds defined per component $t_1 = \{-4, 2, 9\}$ and $t_2 = \{-10, -2, 7\}$. Point $P_1 = [-6, 4]$ would be expanded to $[\{-2, -8, -15\}, \{14, 6, -3\}]$. We then threshold the expanded input at zero to produce a binary string $[\{000\}, \{110\}]$ which we represent with integer ranks $r = [0, 2]$.

The final step in the input expansion, is to apply the mirror map. We use the same threshold constants for both original and mirrored input components. This allows us to assign any class label to any partition with a PBF. That is, for any two partitions a and b, it is not true that $a_i \leqslant b_i \forall i$, and hence there is always a PBF that can assign arbitrary class labels to a and b. Note that partitions, r, were described in Section 3.1 as rows of a look-up-table, u, but that the two terms are equivalent.

The rank expansion has a simple geometric interpretation. Misclassification loss minimization is a tiling problem where we maximize training sample coverage with γ-sized partitions. At zero margin training samples have equal numbers of ones and zeros and define non-overlapping partitions i.e., $q(i) \nleqslant q(j) \nleqslant q(i)$. This means that there are no monotonicity constraints and a pbf can be found using Equation 14. As we increase margin, partitions grow in size, one threshold at a time. Eventually partitions overlap and this means that monotonicity constraints must be satisfied using Equation 15.

For real valued inputs, the order in which components of $r(n)$ are reduced (or increased) as margin increases, depends on the distance between the sample and the threshold constants. In Figure 2b we show an example for P_1 which we will assume has a class label 1. As margin is increased from 1 through to 6, we subtract 1 from r_d in the following order $d = \{0, 1, 3, 0, 3, 0\}$. The corresponding thresholds are numbered in Figure 2b.

For other types of inputs, e.g. categorical or binary, the distances to thresholds are less meaningful, and often equal. In this case, the Stack Filter approach does not suggest which thresholds should be relaxed first. In this paper we use a simple heuristic to resolve tied distances: we select the threshold which produces the smallest number of conflicts.

5   Rank-Distance Classifier

The hinge loss classifier can be found via a Linear Program of $O(2\gamma N)$ variables. One way to view the optimization is shown on the left in Figure 3. The monotonicity constraints of positive (crosses) and negative (circle) margin samples define local contours of a margin function and the Linear Program selects a continuous path from these contours that maximizes the sum of sample margins. The solid gray line in Figure 3 is a hypothetical solution that misclassifies one negative sample. One of the main problems with the hinge loss solution is computational cost.

The main objective in optimizing hinge loss is to assign class labels to input partitions that are poorly represented in the training data. As we have seen, Stack Filter minimizers of hinge loss have attractive properties for this problem, however we now consider an alternative, which is to directly optimize class labels for the input partitions independently. This is illustrated on the right in Figure 3. We define the rank-order distance classifier as a function of r (the mirrored, rank expansion of an input x) as:

$$\hat{f}(r) = sgn\Big(\sum_{n \in C1} \sum_{m=0}^{\gamma} \mathbf{1}_{\{r \geqslant r_m(n)\}} - \sum_{n \in C0} \sum_{m=0}^{\gamma} \mathbf{1}_{\{r \leqslant r_m(n)\}} \Big) \qquad (20)$$

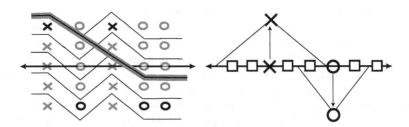

Fig. 3. A one-dimensional representation of samples (zeros and crosses), and monotonicity constraints. a) Hinge loss minimization and b) direct estimation of input partitions (squares) with rank-order distance.

where $r_m(n)$ is a margin modified version of the nth training sample. In geometric terms, this classifier is defined by counting the number of positive and negative partitions that overlap a given point r. In practice this classifier is implemented by constructing a rank-order distance matrix, and we add (and subtract) the distances from a given point r to each training sample. We call the distance function rank-order distance and it is defined as the value of margin where the point is covered by a training sample. In contrast to the Linear program, this approach is memory-based and appears similar to Parzen window or nearest neighbor methods.

The rank-order distance approach assumes we really only care about the statistics of the thresholded hinge-loss Stack Filter. By estimating these statistics independently for each partition we obtain significant computational savings but also reduce approximation error. That is, the partitions used by the hinge loss are larger than those estimated with the rank-order distance approach. The price one pays is the density of solution and the interpretability. The hinge-loss solution typically produces a small number of terms and each term directly dictates class labels for large partitions of the input space. This model is both fast to implement and easy to interpret in a decision tree like fashion. With the rank-order distance classifier we no longer have this simple partitioning of the input space. Instead we derive class labels for a given point by accumulating many terms.

6   Experiments

We investigate the relationship between the different loss functions and learning algorithms with synthetic experiments. For the first experiment samples for two classes are drawn from 4-dimensional symmetric Gaussians. The parameters for the Gaussians are $\mu_{-1} = 0, \sigma_{-1} = I$ and $\mu_1 = 1.5I, \sigma_1 = 1.5I$. The training sample size is fixed at 50 and performance evaluated with 5000 test samples. The number of data dependent thresholds is fixed at 8 for each dimension. In Figure 4a we show the performance of zero-one, hinge loss classifiers as well the rank-order distance classifier as a function of margin, averaged over 20 trials. The rank-order distance classifier clearly outperforms hinge loss which clearly outperforms zero-one loss. The rank-order distance classifier obtained the best performance at maximum margin, which we attribute to the limited capacity of the model class defined by the small number of thresholds.

To investigate this further we apply the rank-order distance classifier to a multi-modal 2-dimensional xor problem where samples are drawn from Gaussian distributions with equal variance $\sigma = 2$, and class means centered on $\mu = \pm 2$. We compare 3 classifiers in Figure 4b. RankDistance8 and rankDistance500 are the rank-order distance classifier with 8 and 500 thresholds/dimension respectively. We also compare the performance of an

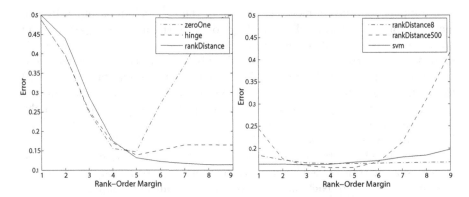

Fig. 4. a) Test error versus rank-order margin for different learning algorithms. b) Test error of rank-distance classifiers with different numbers of thresholds compared to an SVM. Note that the x-axis scale applies only to the rankDistance8 result.

SVM rbf classifier as the regularization parameter is varied: $C = [1e-3, 1e-2, 1e-1, 1, 5, 10, 50, 100, 500]$. The SVM rbf parameter is set at $\sigma = 0.1$, the best value found with $C = 1$. With the increased model capacity, we see that the rank-order margin parameter behaves as we would expect, and that its performance appears competitive with the SVM.

The rank-order distance classifier is applied to the UCI benchmark datasets described in [9]. Each problem is provided as 100 pre-partitioned training and test set pairs and the reported percentage is the average test set error over the 100 trials. During these experiments a simple cross-validation scheme is used to choose the value of rank-order margin for each trial independently. 75% of the training set is used to train the classifier and the remaining 25% is used as a validation set. We choose the value of margin with the minimal average validation error over 10 folds. Table 1 summarizes results reported in [9] and the results obtained with the rank-order distance method (SFC: Stack Filter Classifier).

In all problems, the SFC approach outperformed C4.5 and in two of the problems it outperformed ODT. We observed that the SFC had difficulty with purely categorical, or binary inputs such as the Flare-Solar and Titanic datasets. As discussed in Section 4 the best way to expand partitions for binary, or categorical, inputs is not well defined with our approach. Future work will need to address this problem and we suggest incorporating techniques from the decision tree literature may be useful. For the Titanic problem, we also observed that an error rate of 22.3 could be obtained by simply memorizing the data (zeroOne loss classifier at 0 margin). This error rate is in fact lower than the best reported score for this problem and indicates how important the choice of margin (or regularization) parameter is for learning algorithms. In

Table 1. Classification accuracies on selected benchmarks. *Results reproduced from [9].

DATA SET	$C4.5^*$	ODT^*	SFC
BANANA	15.2± 1.3	14.9 ± 1.2	11.03 ± 0.6
BREAST CANCER	30.8± 4.9	28.7 ± 4.2	29.4 ± 4.2
DIABETES	27.9± 2.6	26.0 ± 2.3	26.7 ± 1.9
FLARE-SOLAR	34.5± 2.1	32.6 ± 1.9	34.4 ± 2.2
THYROID	8.4± 3.5	8.2 ± 3.4	4.9 ± 2.3
TITANIC	23.0± 1.1	22.5 ± 1.2	22.9 ± 1.9

fact, we observed that better performance could often be achieved for several of the problems, by simply choosing a fixed margin for the SFC.

7 Discussion

Stack Filter classifiers and decision tree classifiers produce similar decision boundaries. The two approaches place different constraints on how partitions, induced by thresholds, can be assigned, but both approaches produce a unique rule for each partition. In contrast, the rank-order distance method does not produce a rule based representation for input partitions. In this regard, the rank-order distance method is perhaps better compared to a non-rule based classifier such as an SVM. The best results obtained on the benchmark datasets using this larger class of methods can be found in [9]. The results reported here have higher error than these methods, which we attribute to the fact that the SVM has a well motivated method to optimize weights associated with each training sample. An open question is whether the Stack Filter approach suggests methods to introduce and optimize weights for the rank-order distance classifier. In summary, we have proposed two complementary and related methods for designing Stack Filter Classifiers: one that produces a decision tree like model and one that produces a Parzen window like model. This relationship appears unique to Stack Filter classifiers and could lead to new methods for maximizing the benefit of both approaches for a given application.

References

1. Dougherty, E.R., Barrera, J.: Nonlinear filtering and pattern recognition: Are they the same? In: Proc. SPIE, vol. 4304, pp. 1–6 (2001)
2. Arce, G.: A general weighted median filter structure admitting negative weights. In: Proc. 11th Int. Joint Conf. on Artifical Intelligence, vol. 46, pp. 3195–3205 (1998)

3. Ritter, G.X., Sussner, P.: An introduction to morphological neural networks. In: 13th Int. Conf. on Pattern Recognition, vol. 4, pp. 709–717 (1996)
4. Yang, P., Maragos, P.: Min-max classifiers: Learnability, design and application. Pattern Recognition 28, 879–899 (1995)
5. Tumer, K., Ghosh, J.: Linear and order statistics combiners for pattern classification. Combining Artificial Neural Nets, 127–162 (1999)
6. Han, C.C.: A supervised classification scheme using positive boolean function. In: 16th Int. Conf. on Pattern Recognition, vol. 2, pp. 100–103 (2002)
7. Muselli, M.: Switching neural networks: A new connectionist model for classification. In: Apolloni, B., Marinaro, M., Nicosia, G., Tagliaferri, R. (eds.) WIRN 2005 and NAIS 2005. LNCS, vol. 3931, pp. 23–30. Springer, Heidelberg (2006)
8. Quinlan, J.R.: C4.5: Programs for Machine Learning. Morgan Kaufmann, San Francisco (1993)
9. Blanchard, G., Schafer, C., Rozenholc, Y., Muller, K.R.: Optimal dyadic decision trees. Machine Learning 66(2-3), 709–717 (2007)
10. Paredes, J.L., Arce, G.R.: Optimization of stack filters based on mirrored threshold decomposition. IEEE Trans. on Signal Processing 49, 1179–1188 (2001)
11. Porter, R., Eads, D., Hush, D., Theiler, J.: Weighted order statistic classifiers with large rank-oder margin. In: Proc. 20th Int. Conf. on Machine Learning (2003)
12. Wendt, P.D., Coyle, E.J., Gallagher, N.C.: Stack filters. IEEE Trans. on Acoustics, Speech, and Signal Processing 34, 898–910 (1986)

Milena: Write Generic Morphological Algorithms Once, Run on Many Kinds of Images

Roland Levillain[1,2], Thierry Géraud[1,2], and Laurent Najman[2]

[1] EPITA Research and Development Laboratory (LRDE)
14-16, rue Voltaire, FR-94276 Le Kremlin-Bicêtre Cedex, France
[2] Université Paris-Est, Laboratoire d'Informatique Gaspard-Monge, Équipe A3SI,
ESIEE Paris, Cité Descartes, BP 99, FR-93162 Noisy-le-Grand Cedex, France
{roland.levillain,thierry.geraud}@lrde.epita.fr, l.najman@esiee.fr

Abstract. We present a programming framework for discrete mathematical morphology centered on the concept of genericity. We show that formal definitions of morphological algorithms can be translated into actual code, usable on virtually any kind of compatible images, provided a general definition of the concept of image is given. This work is implemented in Milena, a generic, efficient, and user-friendly image processing library.

1 Introduction

Software for mathematical morphology targets several audiences: *end users*, *designers* and *providers*. *End users* of morphological tools want to apply and assemble algorithms to solve image processing, pattern recognition or computer vision problems. *Designers* of morphological operators build new algorithms by using constructs from their software framework (language, libraries, toolboxes, programs, etc.). Finally, *providers* of data structures are interested in extending their framework with new data types (images, values, structuring elements, etc.).

The size of the population of these categories is decreasing: there are more end users of morphological software than designers of algorithms, and the latter themselves outnumber providers of data structures. Morphological frameworks usually address the needs of their clients in this order, and even sometimes ignore the third or second categories. However, a full morphological framework should suit all groups of users so that structures of providers and algorithms of designers can be used by every actor. In this article, we present a software framework for mathematical morphology designed with two major goals in mind:

1. Be as simple as calling C routines for end users.
2. Be modular enough to be extended w.r.t. algorithms and data structures;

and four minor:

3. Be generic: if a morphological operator admits a general definition whatever the context (topology of the image, structuring element, etc.), then this algorithm should have a corresponding single implementation.

M.H.F. Wilkinson and J.B.T.M. Roerdink (Eds.): ISMM 2009, LNCS 5720, pp. 295–306, 2009.

4. Be close to theory: reading (and writing) algorithms should eventually become natural to scientists used to mathematical morphology notations.
5. Retain efficiency (with respect to run time speed and memory usage) when it is possible. Dedicated and efficient implementations of morphological algorithms for certain cases are known and should be selected whenever possible.
6. Be user-friendly: users should not have to address memory-related issues or deal with a program silently failing because of an arithmetic overflow. The tool should handle these situations, and help the user diagnose any problem.

The paradigm of Generic Programming (GP) [1] which is at the heart of many modern C++ libraries [2,3] and its application to the C++ language address many of these concerns. Developing a software library in the context of GP requires some effort. One of the key ideas is that such a library should be based on *abstractions* of the domain (mathematical morphology in this case). The above requirements will not be fully satisfied if we fail to reify intrinsic concepts of the domain as abstractions. Several image processing libraries relying on the GP paradigm exist (ITK [4], VIGRA [5], Morph-M [6]) but as far as the authors know, none of them seem to meet all of the above requirements.

From the general lattice theory on which is built mathematical morphology, many authors have proposed derived theoretical frameworks. The first ones are graphs [7,8], later extended to store information both *on* vertices and *between* vertices (on edges) [9,10]. The notion of *complex* (see Section 3.2) has also been used to express topological and geometrical attributes of images beyond the scope of graphs [11,12]. Generic programming frameworks to implement algorithms on complexes and grid data structures have been proposed [13,14,15]. Other possible frameworks include combinatorial maps [16] and orders [17].

Let us for instance consider the framework of graphs as the basis of morphological image processing in order to express definitions and properties as general as possible (and meet requirement 3). We could then use a graph-related library like the Boost Graph Library (BGL) [3]. However, such a design suffers from limitations, as mathematical morphology, despite having many intersections with graph theory, has its own definitions, idioms, notations, and issues. Therefore, adapting morphological algorithms to a graph software framework would distort their definitions, which is contrary to requirements 4 and 6. Moreover, we would probably lose efficiency (requirement 5) for restricted use cases in image processing (but at the same time, the most common ones): regular 2D or 3D images on grids, classical structuring elements, etc. Finally, setting graphs as the ultimate representation of images in mathematical morphology once and for all might prevent future extensions. For example the notion of complex mentioned previously, which extends the notion of graph, can be considered to form the basis of a morphological framework.

Therefore, instead of using a fixed system, we propose to rethink mathematical morphology under the light of generic programming [18]. The first step is to define software abstractions matching morphological entities (topology, sets, functions, lattices, structuring elements, geometry, etc.), starting with the concept of *discrete image*. Then, it will be possible to express algorithms in terms of these concepts on the one hand, and provide actual data structure implementing these abstractions on the other hand.

In this paper, we present a generic and efficient C++ programming library, Milena, a part of the Olena image processing platform [19,20]. Milena uses and extends the idea of GP [21]. It implements the abstractions for mathematical morphology software mentioned previously.

This article mainly targets end users of the library and designers of algorithms. It is structured as follows: in Section 2, we study how morphological algorithms are commonly implemented and what are the issues of classical yet restrictive designs. Section 3 proposes a generic definition of an image and shows how this genericity is expressed through the image's traits. As an illustration, a small generic image processing chain is given in Section 4 and applied to various images.

2 Software Implementation of Mathematical Morphology

Translating mathematical morphology methods and objects into readable and usable algorithms is often biased either to satisfy constraints of actual data or meet software and hardware requirements. An example of the first circumstance is the prominent case of a 2-dimensional single-valued image, set on a rectangular (boxed) domain with integer coordinates (a discrete grid $D \subseteq \mathbb{Z}^2$). Many morphological algorithms are solely expressed with this framework in mind. The second bias is computer-dependent: for the sake of efficiency or simplicity of implementation, algorithms sometimes include language- or hardware-related constructs: buffers, loops, dimension decomposition, out-of-bounds behavior, etc.

Let us consider a simple example: the elementary morphological dilation of a gray-level image `ima` with a (flat) structuring element. A shortened definition in the framework of complete lattices [22] would be:

$$\delta_B(I)(x) = \sup_{h \in B} I(x + h)$$

where I (the image to process) is a function $D \to V$ associating a point from the domain D to a value from the set V; and B the structuring element associated to, e.g., the usual 4-connectivity neighborhood. A simple implementation in C++ could be as the one from Algorithm 1. However, this solution makes extra hypotheses that were not contained in the definition of the operation, e.g.:

1. The image is 2-dimensional, since it is accessed using a (row, col) notation.
2. Sites are points with nonnegative integers coordinates starting at 0.
3. The values of the image are compatible with the 8-bit **unsigned char** type.
4. The values of the image form a totally ordered set; hence the operator $<$ can be used to compute the supremum.
5. The structuring element is based on the 4-connectivity.

Each of the previous hypotheses is an actual limitation on the generality of Algorithm 1. It cannot be reused as-is if for instance one or several of the following conditions are expected:

```
image dilation(const image& input) {
  image output (input.nrows(), input.ncols()); // Initialize an output image.
  for (unsigned int r = 0; r < input.nrows(); ++r) // Iterate on rows.
    for (unsigned int c = 0; c < input.ncols(); ++c) { // Iterate on columns.
      unsigned char sup = input(r, c);
      if (r != 0              && input(r−1, c) > sup) sup = input(r−1, c);
      if (r != input.nrows() − 1  && input(r+1, c) > sup) sup = input(r+1, c);
      if (c != 0              && input(r, c−1) > sup) sup = input(r, c−1);
      if (c != input.ncols() − 1  && input(r, c+1) > sup) sup = input(r, c+1);
      output(r, c) = sup;
    }
  return output;
}
```

Algorithm 1. Non generic implementation of a morphological dilation of an 8-bit gray-level image on a regular 2D grid using a 4-c flat structuring element.

1. The input is a 3-dimensional image.
2. Its points are located on a box subset of a floating-point grid, that does not necessarily include the origin.
3. The values are encoded as 12-bit integers or as floating-point numbers.
4. The image is multivalued (e.g., a 3-channel color image).
5. The structuring element represents an 8-connectivity.

Even if the class of images accepted by Algorithm 1 covers day-to-day needs of numerous image processing practitioners, image with features from the previous list are also quite common in fields like biomedical imaging, astronomy, document image analysis or arts. Algorithm 1 also highlights less common restrictions. As is, it is unable to process images with the following features:

- A domain
  - which is not an hyperrectangle (or "box");
  - which is not a set of *points* located in a geometrical space, e.g., given a 3D triangle mesh, one can build an image by mapping each triangle to a set of values;
  - which is a restriction (subset) of another image's domain, still preserving essential properties, like the adjacency of the sites.
- A neighborhood where neighbors of a site are not expressed with a fixed-set structuring element, but through a function associating a set of sites to any site of the image. This is the case when the domain of the image is a graph, where values are attached to vertices [8].
- Non scalar image values, like color values.

Furthermore, the style used in Algorithm 1 does not allow for optimizations. An optimized code (taking advantage, for example, of a totally ordered domain of values, with an attainable upper bound), requires a whole new algorithm per compatible data structure.

```
template <typename I, typename W>
mln_concrete(I)  dilation (const I& input, const W& win) {
  mln_concrete(I) output;  initialize (output, input); // Initialize  output.
  mln_piter(I)  p(input.domain()); // Iterator on sites of the domain of 'input'.
  mln_qiter(W)  q(win, p); // Iterator on the neighbors of 'p' w.r.t. 'win'.
  for_all(p) {
    accu::supremum sup = input(p); // Accumulator computing the supremum.
    for_all(q)  if  (input.has(q))
      sup.take(input(q));
    output(p) = sup.to_result();
    return output;
  }
}
```

Algorithm 2. Generic implementation of a morphological dilation.

In the remainder of this paper, we show how the programming framework of Milena allows programmers to easily write generic and reusable [23] image processing chains using mathematical morphology tools. For instance a Milena equivalent of Algorithm 1 could be Algorithm 2. In this algorithm I is a generic image type, while W is the type of a generic structuring element (also named *window*). p and q are objects traversing respectively the domain of ima and the sites of the structuring element win centered on p. The predicate input.has(q) ensures that q is a valid site of input (this property may not be verified e.g. when p is on the border of the image). sup iteratively computes the supremum of the values under win for each site p. An example of use similar to Algorithm 1 would be:

```
image2d<unsigned char> ima_dil = dilation(ima, win_c4p());
```

where win_c4p() represents the set of neighboring sites in the sense of the 4-connectivity plus the center of the structuring element.

Algorithm 2 is a small yet readable routine and is no longer specific to the aforementioned 2-dimensional 8-bit gray-level image case of Algorithm 1. It is generic with respect to its inputs, and no longer restricted by the limitations we mentioned previously. For instance it can be applied to an image defined on a Region Adjacency Graph (RAG) where each site is a region of an image, associated to an n-dimensional vector expressing features from each underlying region, provided a supremum is well defined on such a value type.

3   Genericity in Mathematical Morphology

3.1   A Generic Definition of the Concept of Image

The previous considerations about the polymorphic nature of a discrete image require a clear definition of the concept of image. To embrace the whole set of aforementioned aspects, we propose the following general definition.

Definition. An image I is a function from a domain D to a set of values V. The elements of D are called the *sites* of I, while the elements of V are its *values*.

For the sake of generality, we use the term *site* instead of *point*: if the domain of I were a RAG, it would be awkward to refer to its elements (the regions) as "points". This definition forms the central paradigm of Milena's construction. However, an actual implementation of an image object cannot rely only on this definition. It is too general as is, and mathematical morphology algorithms expect some more information from their inputs, like whether V is a complete lattice, how the neighboring relation between sites is defined, etc. Therefore, we define additional notions to supplement the definition of an image. These notions are designed to address orthogonal concerns in image processing and mathematical morphology, so that actual definitions (*implementations*) of images can be changed along one axis (e.g., the topology of D) while preserving another (e.g., the existence of a supremum for each subset $X \subseteq V$).

Algorithms are then no longer defined in terms of specific image characteristics (e.g., a domain defined as two ranges of integers representing the coordinates of each of its points) but using *abstractions* (e.g., a *site iterator* object, providing successive accesses to each site of the image, that can be deduced from the image itself). This paradigm based on Generic Programming promotes "Write Once, Reuse Everywhere Applicable" design of algorithms by introducing abstract entities (akin to mathematical objects) in software defined by their *properties*.

The *genericity* of our approach resides in both the organization of the library around entities dedicated to morphological image processing (images, sites, site sets, neighborhoods, value sets, etc.) and in the possibility to extend Milena with new structures and algorithms, while preserving and reusing existing material.

The next section presents the main entities upon which we define morphological algorithms in Milena, and how they provide genericity in mathematical morphology.

3.2   Genericity Traits

We define actual images as models of the previous definition of an image, with extra *properties* on I, D or V. These traits express the generic nature of this definition, and are related to the notions of this section. Each of them is as much orthogonal (or loosely coupled) to the others as possible, so that an actual implementation of one of these concepts can be defined and used with many algorithms regardless of the other features of the input(s). In the rest of this section, we illustrate how the limitations of Algorithm 1 mentioned in Section 2 are lifted by the generic implementation of Algorithm 2.

Restriction of the Domain. It is possible to express the restriction of an image **ima** to a subset **s** of its domain using the dedicated operator **|**; the result can then be used as input of an algorithm:

```
image2d<int> ima_dil = morpho::dilation(ima | s, win);
```

The subset s can either be a comprehensive collection of sites (array, set, etc.) or a predicate. A classical example is the use of a "mask" to restrict the domain of an image. This mask can for instance be a watershed line previously computed on ima; the dilation above would act as a reconstruction of the pixels of ima belonging to this watershed line.

Structuring Elements, Neighborhoods and Windows. Structuring elements of mathematical morphology can be generalized with the notion of *windows*: functions from D to $\mathcal{P}(D)$. A special case of window is a *neighborhood*: a non-reflexive symmetric binary relation on D. In the case of images set on n-dimensional regular grids (as in the previous example of dilation of a 2D image), D is a subset of $\mathbb{Z}^n$ and is expressed as an n-dimensional bounding box. Windows' members can be expressed regardless of the considered site, using a (fixed or variable) set of vectors, called *delta-sites*, as they encode a difference between two sites. For instance a 4-connectivity window is the set of 2D vectors $\{(-1,0),(0,-1),(0,0),(0,1),(1,0)\}$.

In more general cases, windows are implemented as domain-dependent functions. For instance, the natural neighbors of a site p (called the center of the window) of a graph-based image, where D is restricted to the set of vertices, are its adjacent vertices, according to the underlying graph. Such a window is implemented by an object of type adjacent_vertices_window_p in Milena (see below). This window does not contain delta-sites; instead, it encodes the definition of its member sites as a function of p. Using an iterator q to iterate over this window (as in Algorithm 2) successively returns each of its members.

Topological Structure. The structure of D defines relations between its elements. Classical images types are set on the structure of a regular graph, where each vertex is a site of I. More general images can be defined on general graphs, where sites can be either the vertices of the graph, its edges or even both.

An example of dilation on regular 2D image was given in Section 2. In the case of an image associating 8-bit integer values to the elements (vertices and edges) of graph, computing an elementary dilation with respect to the adjacent vertices would be written as this:

```
graph_image<int_u8> ima_dil =
  morpho::dilation(ima | vertices, adjacent_vertices_window_p());
```

ima | vertices creates an image based on the subset of vertices on-the-fly, while adjacent_vertices_window_p() returns a window mapping each vertex to the set of its neighbors plus the vertex itself.

We can generalize this idea by using *simplicial complexes*. An informal definition of a simplicial complex (or simplicial d-complex) is "a set of simplices" (plural of simplex), where a simplex or n-simplex is the simplest manifold that can be created using n points (with $0 \leq n \leq d$). A 0-simplex is a point, a 1-simplex a line segment, a 2-simplex a triangle, a 3-simplex a tetrahedron. Simplicial complexes extends the notion of graphs; a graph is indeed a 1-complex. They can be used to define topological spaces, and therefore serve as supports for images. Figure 1 shows an example of simplicial 3-complex.

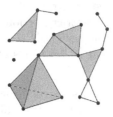

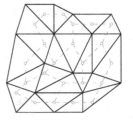

Fig. 1. A simplicial 3-complex   **Fig. 2.** A mesh seen as a simplicial 2-complex

Let us consider an image **ima** based on a simplicial 2-complex (Figure 2) where each element is located in space according to a geometry G (the notion of site location and geometry is addressed later) with 8-bit integer values. The domain D of this image is composed of points, segments and triangles. We consider a neighboring relation among triangles (also known as 2-faces) where two triangles are neighbors iff they share a common edge (1-face). The code to compute the dilation of the values associated to the triangles of D with respect to this relation is as follows:

```
complex_image<2, G, int_u8> ima_dil =
    dilation (ima | faces (2), complex_lower_dim_connected_n_face_window_p<2, G>());
```

As in the example of the graph-based image, **ima | faces(2)** is a restriction of the domain of **ima** to the set of 2-faces (triangles). The expression **complex_lower_dim_connected_n_face_window_p<2, G>()** creates the neighboring relation given earlier (for a site p of dimension n, this window is the set of n-faces sharing an $(n-1)$-face, plus p itself).

Site Location and Geometry. In many context, the location of the sites of an image can be independent from the structure of D. For instance if the domain of I is built on the vertices of a graph, these sites can be located in $\mathbb{Z}^n$ or $\mathbb{R}^n$ with $n \in \mathbb{N}^*$. In some cases, the location of sites is polymorphic. E.g., if D is a 3-dimensional simplicial complex located in a 3D space (as in Figure 1), the location of site p can be a 3D point (if p is a vertex), a pair of points (if p is an edge), a triplet of points (if it is a triangle) or a quadruplet (if it is a tetrahedron). We encode such information as a set of locations called a *geometry*. For instance, the term **G** from the previous code is a shortcut for **complex_geometry<2, point2d>**.

Value Set. Almost all framework support several (fixed) value sets representing mathematical entities such as $\mathbb{B}$, $\mathbb{N}$, $\mathbb{Z}$, $\mathbb{Q}$, $\mathbb{R}$ or subsets of them. Some of them also support Cartesian products of these sets. Not so many support user-defined value types. To be able to process any kind of values, properties should be attached to these sets: quantification, existence of an order relation, existence of a supremum or infimum, etc. Then it is possible to implement algorithms with expected constraints on V. For instance, one can perform a dilation of a color image with 8-bit R, G, B channels by defining a supremum on the rgb8 type:

```
rgb8 sup (const rgb8& x, const rgb8& y) {
  return rgb8(max(x.r(),y.r()), max(x.g(),y.g()), max(x.b(),y.b()));
}
image2d<rgb8> ima_dil = morpho::dilation(ima, win_c4p());
```

3.3 Design and Implementation

Milena aims at genericity (broad applicability to various inputs, reusability) and efficiency (fast execution times, minimum memory footprint). The design of the library focuses on the following features, that we can only sketch here.

Ease of Use. The interface of Milena is akin to classical C code to users, minus the idiosyncratic difficulties of the language (pointers, manual memory allocation and reclaim, weak typing, etc.). Users do not need to be C++ experts to use the library. Images and other data are allocated and released automatically and transparently with no actual performance penalty.

Efficiency. Milena handles non-trivial objects (images, graphs, etc.) through shared memory, managed automatically. The mechanism is efficient since it avoids copying data. As for algorithms, programmers can provide several versions of a routine in addition to the generic one. The selection mechanism is static (resolved at compile-time), and more powerful than function overloading: instead of dispatching with respect to *types*, it dispatches with respect to one or several *properties* attached to one or several types [21].

Usability. Milena targets both prototyping and effective image processing. In the case of very large images (1 GB), we cannot afford multiples copies of values or sometimes even loading a whole image (of e.g. several gigabytes). Therefore, the library provides alternative memory management policies to handle such inputs: in this case, memory-mapped image types which, by design, have no impact whatsoever on the way algorithms are written or called.

4   Illustrations

In this part, we consider a simple, classical image processing chain: from an image ima, compute an area closing c using criterion value lambda; then, perform a watershed transform by flooding on c to obtain a segmentation s. We apply this chain on different images ima. All of the following illustrations use the exact same Milena code corresponding to the processing chain above. Given an image ima (of type I), a neighborhood relation nbh, and a criterion value (threshold) lambda, this code can be written as this (nb is a placeholder receiving the number of catchment basins present in the watershed output image) :

```
template <typename L, typename I, typename N>
mln_ch_value(I, L) chain(const I& ima, const N& nbh, int lambda, L& nb) {
    return morpho::watershed::flooding(morpho::closing::area(ima, nbh, lambda),
                                        nbh, nb);
}
```

Regular 2-Dimensional Image. In the example of Figure 3(a), we first compute a morphological gradient used as an input for the processing chain. A 4-c window is used to compute both this gradient image and the output (Figure 3(d)), where basins have been labeled with random colors.

Graph-Based Image. Figure 3(b) shows an example of planar graph-based [7] gray-level image, from which a gradient is computed using the vertex adjacency as neighboring relation. The result shows four basins separated by a watershed line on pixels.

Simplicial Complex-Based Image. In this last example [24], a triangular mesh is viewed as a 2-simplicial complex, composed of triangles, edges and vertices (Figure 3(c)). From this image, we can compute maximum curvature values on each triangle of the complex, and compute an average curvature on edges.

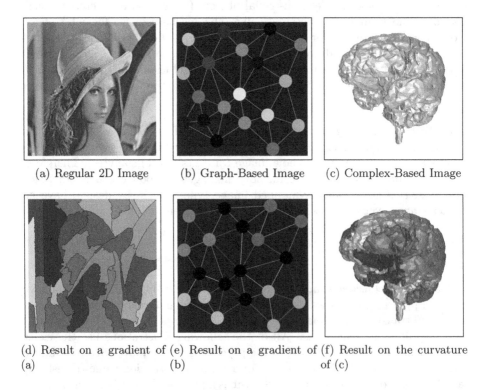

(a) Regular 2D Image   (b) Graph-Based Image   (c) Complex-Based Image

(d) Result on a gradient of (e) Result on a gradient of (f) Result on the curvature
(a)   (b)   of (c)

Fig. 3. Results of the image processing chain of Section 4 on various inputs

Finally, a watershed cut [25] on edges is computed, and basins are propagated to adjacent triangles and vertices for visualization purpose (Figure 3(f)).

All examples use Meyer's watershed algorithm [26], which has been proved to be equivalent to watershed cuts when used on the edges of a graph [27].

5 Conclusion

We have presented the fundamental concepts at the heart of Milena, a generic programming library for image processing and mathematical morphology, released as Free Software under the GNU General Public License. Milena allows users to write algorithms once and use them on various image types. The programming style of the library promotes simple, close-to-theory expressions.

As far as implementation is concerned, Milena extends the C++ language "from within", as a library extension dedicated to image processing. Though we designed the library to make it look familiar to image processing practitioners, it does not require a new programming language nor special tools: a standard C++ environment suffices. Moreover, as Generic Programming allows many optimizations from the compiler, the use of abstractions does not introduce actual run-time penalties.

We encourage practitioners of mathematical morphology interested in Milena to download the library at `http://olena.lrde.epita.fr/Download` and see if it can be useful to their research experiments.

Acknowledgments. The authors thank Guillaume Lazzara for his work on Milena as part of the SCRIBO project, and Alexandre Duret-Lutz for proofreading and commenting on the paper. This work has been conducted in the context of the SCRIBO project (`http://www.scribo.ws/`) of the Free Software Thematic Group, part of the "System@tic Paris-Région" Cluster (France). This project is partially funded by the French Government, its economic development agencies, and by the Paris-Région institutions.

References

1. Garcia, R., Järvi, J., Lumsdaine, A., Siek, J., Willcock, J.: A comparative study of language support for generic programming. In: Proc. of OOPSLA, pp. 115–134 (2003)
2. CGAL: Computational Geometry Algorithms Library (2008), `www.cgal.org`
3. Siek, J.G., Lee, L.Q., Lumsdaine, A.: The Boost Graph Library: User Guide and Reference Manual, 1st edn. Addison Wesley Professional, Reading (2001)
4. Yoo, T.S. (ed.): Insight into Images: Principles and Practice for Segmentation, Registration, and Image Analysis. AK Peters Ltd. (2004)
5. Köthe, U.: STL-style generic programming with images. C++ Report Magazine 12(1), 24–30 (2000)
6. Enficiaud, R.: Algorithmes multidimensionnels et multispectraux en Morphologie Mathématique: approche par méta-programmation. PhD thesis, CMM, ENSMP, Paris, France (February 2007)

7. Vincent, L.: Graphs and mathematical morphology. Signal Processing 16(4), 365–388 (1989)
8. Heijmans, H., Vincent, L.: Graph morphology in image analysis. In: Dougherty, E. (ed.) Mathematical Morphology in Image Processing, pp. 171–203. M. Dekker, New York (1992)
9. Meyer, F., Angulo, J.: Micro-viscous morphological operators. In: Proc. of ISMM, pp. 165–176 (2007)
10. Cousty, J., Najman, L., Serra, J.: Some morphological operators in graph spaces. In: Wilkinson, M.H.F., Roerdink, J.B.T.M. (eds.) ISMM 2009. LNCS, vol. 5720, pp. 149–160. Springer, Heidelberg (2009)
11. Bertrand, G., Couprie, M., Cousty, J., Najman, L.: Ligne de partage des eaux dans les espaces discrets. In: Najman, L., Talbot, H. (eds.) Morphologie mathématique: approches déterministes, pp. 123–149. Hermes Sciences (2008)
12. Loménie, N., Stamon, G.: Morphological mesh filtering and α-objects. Pattern Recognition Letters 29(10), 1571–1579 (2008)
13. Köthe, U.: Generic programming techniques that make planar cell complexes easy to use. In: Bertrand, G., Imiya, A., Klette, R. (eds.) Digital and Image Geometry. LNCS, vol. 2243, pp. 17–37. Springer, Heidelberg (2002)
14. Kettner, L.: Designing a data structure for polyhedral surfaces. In: Proc. of SCG, pp. 146–154. ACM, New York (1998)
15. Berti, G.: GrAL: the grid algorithms library. FGCS 22(1), 110–122 (2006)
16. Edmonds, J.: A combinatorial representation for polyhedral surfaces. Notices of the American Mathematical Society 7 (1960)
17. Bertrand, G., Couprie, M.: A model for digital topology. In: Bertrand, G., Couprie, M., Perroton, L. (eds.) DGCI 1999. LNCS, vol. 1568, pp. 229–241. Springer, Heidelberg (1999)
18. d'Ornellas, M.C., van den Boomgaard, R.: The state of art and future development of morphological software towards generic algorithms. International Journal of Pattern Recognition and Artificial Intelligence 17(2), 231–255 (2003)
19. LRDE: The Olena image processing library (2009), http://olena.lrde.epita.fr
20. Darbon, J., Géraud, T., Duret-Lutz, A.: Generic implementation of morphological image operators. In: Proc. of ISMM, Sydney, Australia, CSIRO, pp. 175–184 (2002)
21. Géraud, T., Levillain, R.: A sequel to the static C++ object-oriented programming paradigm (SCOOP 2). In: Proc. of MPOOL, Paphos, Cyprus (July 2008)
22. Goutsias, J., Heijmans, H.J.A.M.: Fundamenta morphologicae mathematicae. Fundamenta Informaticae 41(1-2), 1–31 (2000)
23. Köthe, U.: Reusable software in computer vision. In: Jähne, B., Haussecker, H., Geißler, P. (eds.) Handbook of Computer Vision and Applications. Systems and Applications, vol. 3, pp. 103–132. Academic Press, San Diego (1999)
24. Alcoverro, M., Philipp-Foliguet, S., Jordan, M., Najman, L., Cousty, J.: Region-based 3D artwork indexing and classification. In: 3DTV, pp. 393–396 (2008)
25. Cousty, J., Bertrand, G., Najman, L., Couprie, M.: Watershed cuts: minimum spanning forests and the drop of water principle. IEEE PAMI (to appear, 2009)
26. Meyer, F.: Un algorithme optimal de ligne de partage des eaux. In: Actes du 8e Congrès AFCET, Lyon-Villeurbanne, France, AFCET, pp. 847–857 (1991)
27. Cousty, J., Bertrand, G., Najman, L., Couprie, M.: On watershed cuts and thinnings. In: Coeurjolly, D., Sivignon, I., Tougne, L., Dupont, F. (eds.) DGCI 2008. LNCS, vol. 4992, pp. 434–445. Springer, Heidelberg (2008)

An Efficient Algorithm for Computing Multi-scale Connectivity Measures

Georgios K. Ouzounis

Second Department of Surgery
School of Medicine
Democritus University of Thrace,
University General Hospital of Alexandroupoli,
68100 Alexandroupoli, Greece
gouzoun@med.duth.gr

Abstract. Multi-scale connectivity measures have been introduced in the context of shape analysis and image segmentation. They are computed by progressive shape decomposition of binary images. This paper presents an efficient method to compute them based on the dual-input Max-Tree algorithm. Instead of handling a stack of binary images, one for each scale, the new method reads a single gray-level image, with each level associated to a unique scale. This reduces the component labeling iterations from a total number equal to the number of scales to just a single pass of the image. Moreover, it prevents the repetitive decomposition of each component under study, for the remaining scale range, since these information are already mapped from the input image to the tree hierarchy. Synthetic and real image examples are given and performance issues are discussed.

1 Introduction

Connected operators [1] are morphological functions that modify the intensity of image regions, known as *connected components*, instead of individual points. They rely on some notion of image connectivity, which in the lattice-theoretic framework of connected morphology [2], it is defined through set families known as *connectivity classes* [3].

A connected operator given a point on the image domain, extracts the connected component containing it in its entirety and without edge or shape modifications. This property, though particularly valuable in many applications, is responsible for what is known as the "leakage" problem [4, 5] of connected operators. Wide object regions linked by narrow, elongated bridging paths, which could be the result of background texture, are extracted as one object when it is often desirable to treat them separately. Contraction-based second-generation connectivity [6, 7] can in part resolve this issue by handling pixels in these paths as singleton sets. The result of connected operators configured with this type of connectivity is a heavy edge distortion and a severe blurring effect in gray-scale images, known as *oversegmentation* [8]. Moreover, computing connected pattern-spectra [9] from granulometries using this notion of connectivity, saturates the spectrum bin accounting for objects of size 1. That is, much of the structural

M.H.F. Wilkinson and J.B.T.M. Roerdink (Eds.): ISMM 2009, LNCS 5720, pp. 307–319, 2009.

information contained on the edges of the image objects is discarded by being placed to the spectrum entry accounting mostly for noise.

To counter this, a multi-scale connectivity analysis framework was introduced, based on the axiomatic definition of generalized morphological connectivity measures [5]. An example is the *adjunctional multi-scale connectivity function* employed for computing generalized granulometries. This function incorporates geometrical cues to quantify "how strongly connected" a set is. This is by evaluating the rate at which it can be partitioned into a set of disjoint non-empty subsets through the recursive application of an anti-extensive local operator such as an erosion or an opening. The adjunctional multi-scale connectivity functions can be computed using the *Connectivity Tree* (C-Tree) [5], which is a hierarchical binary image representation structure, encoding in each level how "strong" the image connections are. Fine-tuning of the tree allows the control of the leakage problem. Moreover, it allows the handling of structures that are inaccessible with operators configured with the regular topological connectivity.

In this paper a new method for computing the adjunctional multi-scale connectivity functions is presented. It is based on the dual-input Max-Tree algorithm [10] and handles gray-level images. Each level corresponds to a unique shape decomposition scale of the original binary image. The decomposition is computed through a sequence of anti-extensive erosions or openings at an off-line stage, with scale corresponding to the radius of the structuring element involved. After configuring the input image and the mask [10], the Max-Tree is computed. This process, described in Section 3.2, delivers a structure in which the entire image is fragmented to singleton sets. Following, the tree is re-partitioned with the aid of a wavefront expansion routine that groups sets of singletons to larger components, under the connectivity scheme of [5] (Section 4). In a step further - Section 5, the Max-Tree nodes satisfying the C-Tree node criteria, are pinpointed in the structure, the connectivity function parameters for these nodes are obtained by accumulating information from the remaining nodes and the adjunctional multi-scale connectivity function for each node visited is computed. The algorithm is tested on real and synthetic images of known structural characteristics - Section 6. The findings, together with the advantages of this method over the regular C-Tree algorithm and a short analysis into computational complexity issues are discussed in Section 7.

2   Connectivity Measures and the C-Tree

The concept of connectivity measures $\mu(X)$ (X being a binary set) was introduced in the context of multi-scale shape connectivity analysis, aiming at quantifying the strength of object connections. An example is given in Fig. 1 where the objective is to differentiate the three cases based on the structural characteristics of each path, i.e.

$$r_1 < r_2 < R_1 < R_2 \Rightarrow \mu(A_1) > \mu(A_2) > \mu(A_3) \tag{1}$$

Tzafestas et al. [5] proposed a multi-scale connectivity function based on adjunctions, given by:

Definition 1. *Let $\alpha = (\epsilon_B, \delta_B)$ denote an adjunction on $\mathcal{P}(E)$. A function $\mu_a : \mathcal{P}(E) \times \mathbb{R}_+ \to [0, 1]$ defined as:*

$$\mu_a(X, s) = e^{-\lambda r_a(X, s)}, \text{ with} \tag{2}$$

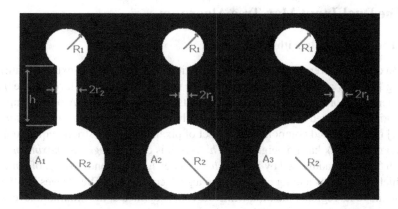

Fig. 1. An example of two circular objects linked with a path of different structural characteristics in each case

$$r_a(X, s) = \bigvee \{r \in \mathbb{N} : \delta_B^r(e_B^s(X)|X) \in \mathcal{C} \setminus \emptyset\}, \tag{3}$$

is called an adjunctional connectivity function and gives a measure of the connectivity of a set $X \subseteq E$, at scale s.

E is an arbitrary superset, $\mathcal{P}(E)$ is the powerset of E, $\mathcal{C}$ is a connectivity class [3], and ϵ_B and δ_B are Minkowski erosions and dilations with a structuring element B.

The adjunctional multi-scale connectivity functions can be computed using the Connectivity Tree. In [5], a detailed description of the algorithm is given. In brief, the C-Tree creation resides on a recursive procedure, called *create-CTree(child[j])*, that takes as input a C-Node structure, constructs the children C-Nodes and recursively calls itself to complete the remaining part of the C-Tree hierarchy. Following is a brief summary of this procedure (as given in [5]) in four steps.

1. Perform erosion $X_\epsilon = \epsilon_B^s(X)$ on the input image (X = C-node→Image), with progressively increasing scale s until X_ϵ is partitioned into a number ($nc > 1$) of disjoint connected components $Y_j (j = 1, ..., nc)$ (if X vanishes completely for a particular scale s without being partitioned into separated connected components, then the current C-Node is a leaf node.
2. Perform a conditional wavefront expansion on the partition $\{Y_j\}$ of X_ϵ, to reconstruct a partition $\{Z_j\}$ of X, that is, a new set of disjoint connected components Z_j such that: $\bigcup Z_j = X$.
3. Create children C-Node structures (child[j], for $j = 1, ..., nc$). Call recursively the *create-CTree(child[j])* procedure.
4. Compute the adjunctional multiscale connectivity function $\mu_a(X_j^k, s)$, where k is the tree level.

The Max-Tree node structure [4] discussed next, is enriched with some of the members of the C-Node structure, namely the *c_max_scale* which marks the maximum scale $s_{max} : \forall s > s_{max} \Rightarrow \epsilon_B^s(X_j^k) = \emptyset$, the *c_func[]* which is an array storing $\mu_a(X_j^k, s), \forall s : 0 \leq s \leq s_{max}$, and the number of C-Node equivalent Max-Tree children nodes *c_num_children*.

3 The Dual-Input Max-Tree Algorithm

3.1 The Original Algorithm

The Max-Tree is a versatile image representation structure for anti-extensive attribute filtering [4]. It is a rooted, unidirected tree in which, given a gray-level image f, the node hierarchy corresponds to the nesting of its *peak components*. A peak component P_h [11] is a connected component of the threshold set $T_h(f)$ at level h and a *flat-zone* F_h [11] is a connected component of the set of pixels with level strictly equal to h. If a peak component P_h has no neighbors at $h' > h$, it is called a *regional maximum*.

Each tree node, addressed by its level h and index i, corresponds to a set of flat-zones for which there exists a unique mapping to a peak component. The "leaves" of the tree correspond to its regional maxima while the root is defined at the minimum level h_{min} and represents the background. Each node except for the root points to its parent at level $h' < h$. The root node points to itself. An example is shown in Fig. 2.

Each tree node stores auxiliary data from the set of image flat-zones it associates with, and from all its descendant nodes. Node attributes can be computed directly and filtering is done on a node basis. Data inheritance from child to parent node is a simple

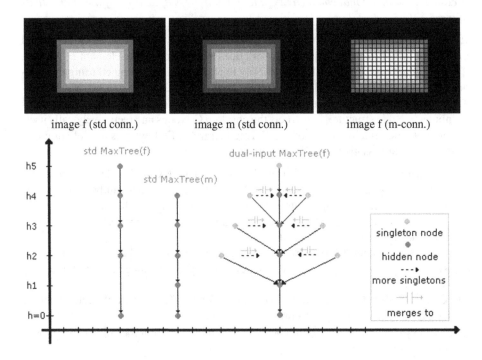

Fig. 2. Max-Tree examples: the first two trees on the left, are of the images f and m respectively, both associated to standard (std) connectivity, and with regular nodes only. The third Max-Tree, (*dimt* algorithm), is of the image f using m as a connectivity mask. f is associated to a contraction-like m-connectivity, in which all foreground regions are converted to singletons.

accumulation in the case of increasing attributes but for the more complicated case of shape descriptors, a more delicate handling is required [10].

The Max-Tree algorithm runs a three-stage process in which the construction of the tree and the collection of auxiliary data is separate from image filtering and restitution. The construction is based on a recursive flood-fill function fed by a set of hierarchical FIFO queues (see [4] for pseudo-code). Differences between the standard and dual-input mode are limited to the construction stage only. The dual-input mode (*dimt* algorithm) [10] is designed for operators associated to second-generation connectivity [6, 7, 10]. The algorithm reads two input images; the original f and the gray-scale mask image m. An inspection routine in the flooding function checks for intensity mismatches between the same pixel p in f and m. Depending on whether p is higher in m or in f, the algorithm assigns it to a local cluster or contraction of connected components according to $\mathcal{C}$. Fig. 2 shows an example of a contraction-like $m-$connectivity according to which, all foreground components of the image f are treated as singletons. This type of second-generation connectivity is only supported from the mask-based connectivity framework in [10].

3.2   Fragmenting the Image Domain

The Max-Tree structure is employed in this work to represent hierarchically the different shape decomposition scales s of a binary image. To build the Max-Tree structure, the input images are first created at an off-line stage. Starting from a binary image X at $s = 0$ and $X_\epsilon^0 = X$, $X_\epsilon^s = \epsilon_B^s(X_\epsilon^{s-1})$ is computed recursively until $X_\epsilon^{s_{max}+1} = \emptyset$. Each binary set X_ϵ^s, is given a unique gray-level with all levels being equally-spaced (separated by *lev_diff*). Superimposing the total of $s_{max}+1$ images (including the original) yields a gray-level image f in which any peak component at level h is strictly a subset of its parent peak component at $h' < h$. An example is given in Fig. 3. 8-bit images support up to $s = 253$ because the original binary image should be assigned a level at least 2 levels up from the background, which is at $h = 0$. If $s_{max} > 253$ then 16-bit images should be used. The mask image m is a replica of f reduced in intensity by *lev_diff*. With $m < f$, the *dimt* computes a structure in which f is characterized by a contraction-like mask-based second-generation connectivity [10]. This is a rather

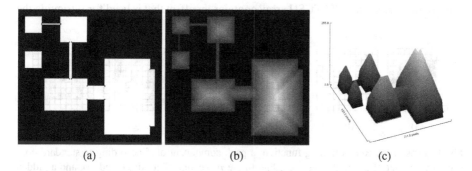

        (a)   (b)   (c)

Fig. 3. Input image (from left): original binary image (a); gray-scale encoding of 37 erosion-based decomposition scales (0-36) (b); the surface plot of the previous image (c)

unusual type of connectivity because essentially every point not belonging to the background is represented by a singleton set. Each singleton node marked by a point x at level h points to its parent, which is a replica of the connected component of f, according to $\mathcal{C}$, that is marked by x, only reduced in intensity by *lev_diff*. Its parent, and thus every parent of the tree other than the root node, is a "hidden node", i.e. it has no flat-zones. An example is shown in Fig. 2 - third tree from the left.

4 The Repartitioning Function

The Max-Tree of an image f computed as described in the previous section, is a structure with singleton and hidden nodes only. Using the hidden nodes as markers (Fig. 4(b)), the repartitioning function delivers sets of connected components by merging progressively all singletons of the same level to their closest marker. This is in a fashion similar to the wavefront expansion routine described in [5].

   Merging singletons to the appropriate hidden nodes is done on a scale basis, i.e. for each level of the mask image separately. This is an iterative procedure controlled in a *while()* loop for each given level h. The loop records the number of singletons merged at each iteration (stored in *new_count*), and exits if no more merges take place. In each iteration all nodes at level h are accessed sequentially. Each node *Tree[idx]* is tested to see if it is a singleton or not. If true, its coordinates (pixel p) and those of its neighbors are retrieved. For each neighbor, the node id to which it belongs to, is computed and if it is not a singleton, two further conditions are checked, otherwise it is ignored. If not a singleton, it is either a hidden node which remains "hidden" or it has already been expanded. In the first case, the neighbor's intensity *ORI_F[q]* (the array containing the intensity of each pixel in f, pixel q in this case) is higher than h. We proceed if *GLUE_STEP[q]* (an array storing the number of iterations before a pixel gets merged to a node) is less than the current number of iterations. In Fig. 4(c), the GLUE_STEP[] entry for the marker (the hidden node) is still 0 while the process is in its first iteration, thus all pixels adjacent to it will be merged at this pass. This is done to prevent the next neighbor of p which might not be adjacent to the current boundaries of the hidden node, to be merged at the iteration step in progress. Merging the singleton node containing p to a hidden node is by setting the *STATUS[p]* entry to the hidden node's index. Each entry of the image-size STATUS[] array contains an offset that is used for retrieving the

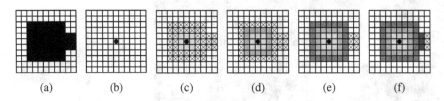

(a)   (b)   (c)   (d)   (e)   (f)

Fig. 4. Steps of the Repartitioning function: a peak component of f according to standard connectivity or a set of 55 singletons according to the $m-$connectivity discussed (a), and a hidden node used as marker (b). 1st expansion step (c). $2^{nd}, 3^{d}$ and 4th expansion steps in (d),(e) and (f) respectively. The gray shades and crosses indicate the expansion steps and pixels to be processed.

node to which each pixel belongs to. In the case of a marker that remains hidden, this is retrieved by accessing the parent id of the node above it.

In the second case, that is when a hidden node has already been expanded, we proceed if the neighbor is not a singleton node, if ORI_F[q]=h and if the GLUE_STEP[q] condition is true as before. Merging is simply by setting STATUS[p]=STATUS[q]. Examples of this case are shown in Fig. 4(d-e), i.e. iteration steps 2-4 respectively.

After a singleton node at p is processed, the GLUE_STEP[p] is updated to the current number of iterations. Moreover, the singleton node updates the auxiliary data of the expanded marker, found at q, and the hidden node's GLUE_STEP[q] is updated to the current iteration step. The singleton node needs to be invalidated so that in the next iteration though the nodes or in the case of wavefront collision, it will be ignored. That is by marking it as *ghost*. The *new_count* is incremented by one for every merge.

When all singleton nodes for a level h are processed and the *while()* loop exits, for each one of them including pixels p assigned to the original hidden nodes at h, the corresponding entry of GLUE_STEP[] is initialized to 0 for the next level to be processed. The C-style pseudo-code for the repartition function is given in Fig. 5. When finished with all levels, a loop through all nodes sets the number of children for each node and performs a final merge of all remaining singletons; these are the pixels corresponding to the regional maxima of f according to standard connectivity.

5   Computing the Connectivity Measures

After re-partitioning the Max-Tree to represent structures according to the definition of connectivity given in [5], the Max-Tree nodes (*MTNodes*) corresponding to the equivalent C-Nodes (*MTCnodes*) need to be identified. Additionally, a parent-to-child relation directly between them needs to be established. With each level associated to a shape decomposition at the corresponding scale, this becomes rather trivial since all that is needed is to check which MTNodes exist at the first scale $s = 0$ and for the rest, which ones are children nodes of a parent with more than one child.

Next, given any MTCnode at level h (or scale s_{ref}), the maximum number of expansion steps from a marker at scale $s : s_{ref} < s \leq s_{max}$, that are needed to reconstruct it, must be computed. This number is stored in a scale-size array called *MaxIterPerScale[]*. The process breaks in two loops going through the tree structure. In the first loop, from $s = 0$ to s_{max}, for each MTCnode found at scale s_{ref}, MaxIterPerScale[s_{ref}] is set to 0, i.e. the partition of X it represents needs 0 iterations to be reconstructed from itself. Moreover, if it is not a root MTCnode, the child count of its MTCnode parent is increased by one. The MTCnode parent updates its MaxIterPerScale[s_{ref}] too, with the maximum between the existing and a newly computed value. This is the sum of the MaxIterPerScale[$s_{ref} - 1$] of itself and the number of expansion steps registered to the MTCnode under inspection, during the re-partitioning phase. Note that this is not sufficient for computing the connectivity functions of each MTCnode because the contribution of regular MTnodes has not been considered to that stage.

This is addressed at the second loop that goes from s_{max} to $s = 0$. For each MTCnode detected (at s_{ref}) its MaxIterPerScale[s] entries corresponding to regular MTnodes along the same root-path, that are used as markers at $s > s_{ref}$, are computed and propagated to its MTCnode parent. Moreover, for each MTCnode its connectivity measures

```
PROCEDURE: MaxTreeRepartition(MaxTree Tree)
{
  /* general purpose variables */
  var: p, q, i, l, x, y, neighs, num_neighbors;
  /* node ID storing variables */
  var: idx, parent, root, neigh_par, neigh_idx;
  /* counting and iteration variables    */
  var: old_count, new_count, iterations;
  /* stores the 8-neighbors coordinates */
  var: *neighbors;

  Allocate(neighbors);

  for(h= h_max_m; h>=h_min_m; h--)
  {
   old_count=0; new_count=1; iterations = 0;
   while(new_count>old_count)
   {
      old_count = new_count;
      iterations++;
      for(i=0; i<numbernodes[h]; i++)
      {
          idx    : Get_Node_ID;
          parent : Get_Node_Parent_ID;
          if (Tree[idx]->singleton==true) /*"->" denotes member of...*/
          {
             p = Tree[idx]->singleton_coord;
             x : Get_X_Coord(p); y : Get_Y_Coord(p);
             /* retrieve its 8- neighbors */
             num_neighbors = Get8Neighbors(p, neighbors);
             for (neighs=0; neighs<num_neighbors; neighs++)
             {
                 q = neighbors[neighs];
                 neigh_idx : Get_Node_ID;
                 if(Tree[idx]->singleton==true)
                 {
                     neigh_par : /* get neighbor's parent id */
                     if((ORI_F[q]>h)&& (GLUE_STEP[q]<iterations))
                     {
                         /* Get the node's index to which p belongs to (singleton)*/
                         STATUS[p] : Get_Node_ID(p);
                         GLUE_STEP[p] = iterations;
                         AddToAuxiliaryData(Tree[neigh_idx]->Attribute,x,y);
                         Tree[neigh_par]->steps=
                            MAX(GLUE_STEP[p],Tree[neigh_par]->steps);
                         /* mark this singleton as ghost*/
                         Tree[idx]->singleton=ghost;
                         new_count++;
                     }
                     if((ORI_F[q]==h)&&(GLUE_STEP[q]<iterations)&&
                        (Tree[neigh_idx]->singleton==false))
                     {
                         STATUS[p] = STATUS[q];
                         GLUE_STEP[p] = iterations;
                         AddToAuxiliaryData(Tree[neigh_idx]->Attribute,x,y);
                         Tree[neigh_idx]->steps=
                            MAX(GLUE_STEP[p],Tree[neigh_idx]->steps);
                         Tree[idx]->singleton=ghost;
                         new_count++;
     } } } } } }  /* EO while */
     /* Need to reset all GLUE_STEP entries for that level */
     for(i=0; i<ImageSize; i++)
        if(ORI_F[i]==h) GLUE_STEP[i] = 0;
     } /* EO for */
}
```

Fig. 5. The Max-Tree Repartitioning function in C-style pseudo code

c_func[s], at each scale $s > s_{ref}$ are computed as in (2). The entry c_func[s_{ref}] of the MTCnode at s_{ref} is set to 1.0, since a partition of X prior to any decomposition is fully connected. The same applies for all $s : 0 \leq s \leq s_{ref}$, i.e. $\mu_a(X, s) = 1.0$ by definition.

The process is complemented with three attribute overrules. The first concerns MTC-nodes that remain connected after the reference scale. The value of $\mu_a(X, s)$ in this case, remains constant until the scale $s > s_{ref}$ that the MTCnode breaks in two or more components. The attribute overrules are enforced by overwriting the previously computed values for $\mu_a(X, s)$ wherever applicable. The second case concerns the MTCnodes which are regional maxima in the C-Tree sense. For such a node defined at scale s_{ref}, $\mu_a(X, s)$ is set to 1.0 for each scale $s_{ref} < s \leq s_{max}$ for which MaxIterPerScale[s]>0 and 0.0 otherwise. This is because, it doesn't split until the scale at which it vanishes, i.e. it is fully connected. The third attribute overrule is rather more complicated and concerns the case in which two or more regional maxima emerge from a single MTn-ode, independent of whether it is an MTCnode or not. Consider an example in which, at a given root-path there exist two regional maxima at scales s_{max1} and s_{max2} respectively, such that $s_{max1} > s_{max2}$. Moreover, let the two paths meet at an MTCnode at scale s_{ref}. At $s_{ref} + 1$, there exist two MTCnodes, i.e. the MTCnode parents of the regional maxima, one of which has descendants that span for $diff\_sc = s_{max1} - s_{max2}$ scales more than the other. This is called the "dominant component". Without considering an overrule, the values $\mu_a(X, s_{max1})$ and $\mu_a(X, s_{max2})$ for all MTCnodes at $s : 0 < s \leq s_{ref}$ are computed based on their MaxIterPerScale[] entries for these two scales. And though for any scale $s_{max2} < s \leq s_{max1}$ this is rather clear, for s_{max2} their MaxIterPerScale[s_{max2}] entries are computed based on the max between the expansion steps of the dominant and non-dominant components. This is in fact wrong. MTCnodes at scales smaller or equal to s_{ref} must have values of $\mu_a(X, s)$, with $s > s_{ref}$, computed from the dominant component only. Moreover, for the last $diff\_sc$ scales before the top, $\mu_a(X, s)$ must be forced to 1.0 because the regional maximum is the last remaining structure that cannot be further partitioned. In the case that all regional maxima have the same height, i.e. appear at the same scale, $diff\_sc$ is set to $s_{max1} - s_{ref}$. An in-depth analysis of these routines with pseudo-code samples are given at the corresponding page in http://www.georgios-ouzounis.info.

6 Experiments

The proposed algorithm for computing the adjunctional multi-scale connectivity measures was validated on a number of synthetic images with known structural characteristics. An example, similar to the one in [5], is shown in Fig. 3. The MTCnodes and their child to parent relations from the Max-Tree of Fig. 3(b) are shown in Fig. 6(a). Images (b-c) of the same figure show two MTCnode profiles, i.e. the adjunctional connectivity measures as functions of scale. The MTCnodes have C-Tree labeling with $s = 0$ for the original binary image. It is observed that the connectivity measure of MTCnodes increases as we move upwards in the Max-Tree hierarchy. This means that peak components at higher scales correspond to "stronger" structures compared to those at lower scales, and further that the connectivity measure $\mu_a(X, s)$ is non-increasing, i.e.

$$s_1 < s_2 : X_{s_1} \supset X_{s_2} \Rightarrow \mu_a(X_{s_1}, s_1) \subset \mu_a(X_{s_2}, s_2). \tag{4}$$

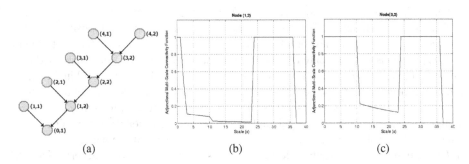

Fig. 6. The MTCnodes and child to parent relations from the Max-Tree of Fig.3(b) in (a); multi-scale connectivity function profiles of nodes (1,2) and (3,2), (b) and (c) respectively

Strong connections are evident at the surface plot (Fig. 3(c)), where for example, the path between the largest two subsets of X (if treated as disjoint), is considerably brighter than the rest, i.e. it requires more erosions to disappear compared to the others.

To analyze the two node profiles, consider first the original binary image in Fig. 3(a). It shows a single object made of five block regions that are linked with four bridges. The MTCnode at scale $s = 0$ corresponds to this object which remains connected until scale $s = 1$, i.e. SE side $r = 1$. For $s = 1$, the left most bridge of diameter 1 is removed, breaking the original object to two components. The smaller one labeled as $Node(1, 1)$ remains connected until it vanishes after 12 erosions on the original image. $Node(1, 2)$, shown in Fig. 6(b), remains connected only until $s = 1$. For $s = 2$ it is fragmented further, as is for $s = 3$. The dominant object remains connected from $s = 3$, to $s = 10$, and the decline of the values of $\mu_a(X_s, s)$ is rather smooth as expected from (2). At $s = 11$ the thickest, right-most bridge is removed fragmenting the dominant object further. Moreover, at $s = 24$, there remains just one component which cannot be further fragmented until the scale it is removed ($s = 37$). The value of $\mu_a(X_s, s)$, for $s = 24$ to $s = 36$, remains constant and equal to 1 since the component is fully connected. $Node(3, 2)$ in Fig. 6(c), refers to the dominant object that results after an erosion with $r = 3$ on the original. That is, the object containing the two largest blocks, which is fully connected from $s = 0$ to 10 and from $s = 24$ to 36 as before.

The adjunctional multi-scale connectivity measure as a function of scale, has been used to compute generalized pattern spectra for soil-section image analysis. For attribute filtering purposes however, it is preferable dealing with scalar values instead. In this case, the *average adjunctional connectivity measure* is used [5], defined as

$$\bar{\mu}_a(X) = \frac{\int_{s=0}^{s_{max}} \mu_a(X, s)ds}{s_{max}}. \tag{5}$$

A filter based on the $\bar{\mu}_a(.)$ measure as an attribute, can be computed efficiently using the proposed method thanks to the existing Max-Tree functionality. An example employing the *direct filtering rule* [4] is given in Fig. 7. Image (a) shows a diatom, of the *Skeletonema* species, courtesy of the Analytical Instruments and Field Research Laboratory, Texas A& M University-Kingsville, USA (http://www.tamuk.edu /chemistry/research/AnalyticalLab/ana_lab.htm) The diatom is first

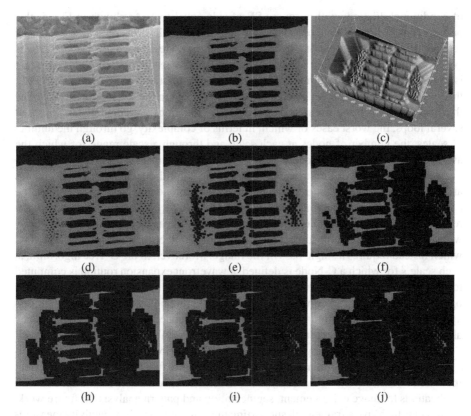

Fig. 7. Diatom filtering: the original image of a Skeletonema diatom (a); the segmented binary input, superimposed with 81 decomposition scales (b) and its surface plot (c). Images (d-j) show the filter output for attribute thresholds from 0.4 to 0.9 respectively.

segmented and the resulting binary image is eroded 81 times (SE side of 1) before everything disappears, i.e. $\forall s \geq 82 \Rightarrow \epsilon_B^s(X_\epsilon^{s-1}) = \emptyset$. Assigning the level $h = 100$ to the set X^0 and setting $diff\_lev = 1$, the superposition of all 81 shape decomposition scales with the original is shown in image (b) and its surface plot in (c). The mask is computed by subtracting the value 1 from the input image. Images (d-j) show the filter outputs for attribute thresholds starting from 0.4 to 0.9 (the step is 0.1), respectively.

7 Discussion

The proposed algorithm is based on the Max-Tree structure and consists of three separate stages. In the first stage, a Max-Tree of the input image is constructed, associated to a contraction-like $m-$connectivity. The complexity of this stage has been analyzed in [10] and is approximately $O(GN)$, where G is the number of gray levels and N the number of pixels. In the second stage, the tree consisting of hidden and singleton nodes only, is repartitioned based on a wavefront expansion routine that is equivalent to a set of conditional dilations with a square SE of side equal to 1. As shown in Fig. 5, this process runs through four loops. The first goes through all levels or scales G and the second, through the number of expansion steps of the nodes at each level (worst case is $N/2$ for nodes $n \simeq N/2$ in a 3-level image). If the average number of nodes per scale is N/G and the average number of merges in each iteration for all scales is $N/2G$, this yields an overall complexity equal to $O(N^2/G)$ and in the worst case $O(N^2)$. The third stage in which the multi-scale connectivity functions are computed, consists of several loops, the worst cases of which, in terms of complexity, go through the number of scales, the number of nodes at each scale and through a scale range for which the average value of $G/2$ is considered. This yields an approximate complexity of $O(GN)$.

Comparing this algorithm to the regular C-Tree algorithm, a number of advantages are observed. The C-Tree, at each scale s operates a component labeling routine on the corresponding binary input image resulting in total of s_{max} searches. If this is with the regular Max-Tree for example, which has an almost linear computational complexity with image size, the overhead becomes s_{max} times higher than the proposed method which instead requires a single pass of the image. Moreover, in the C-Tree algorithm, at each scale s for which a C-Node is defined, a wavefront expansion routine is computed to reconstruct the new partition of X from the marker. A number of erosions are then operated on the marker to compute $\mu_a(X, s)$ for the remaining scales before the it disappears. In the new algorithm no further erosions are needed to compute $\mu_a(X, s)$ since all partitions of X (one per scale) are already mapped into the Max-Tree structure.

The *dimt* is an efficient algorithm for computing attribute filters on sets characterized by second-generation connectivity. Using its existing functionality and the adjacent multi-scale connectivity measures $\mu_a(X, s)$ as attributes, new filters can be defined with applications to image enhancement, segmentation and pattern analysis. In future work, the target is to derive a framework supporting the computation of connectivity measures on gray-level images and extend the functionality of the current algorithm to support it.

References

1. Heijmans, H.J.A.M.: Connected morphological operators for binary images. Comp. Vis. Image Understand. 73(1), 99–120 (1999)
2. Serra, J.: Connectivity on complete lattices. Journal of Mathematical Imaging and Vision 9(3), 231–251 (1998)
3. Serra, J. (ed.): Image Analysis and Mathematical Morphology. II: Theoretical Advances. Academic Press, London (1988)
4. Salembier, P., Oliveras, A., Garrido, L.: Anti-extensive connected operators for image and sequence processing. IEEE Trans. Image Proc. 7(4), 555–570 (1998)
5. Tzafestas, C., Maragos, P.: Shape connectivity: Multiscale analysis and application to generalized granulometries. Journal Math. Imaging and Vision 17(2), 109–129 (2002)
6. Ronse, C.: Set-theoretical algebraic approaches to connectivity in continuous or digital spaces. Journal of Mathematical Imaging and Vision 8(1), 41–58 (1998)
7. Braga-Neto, U., Goutsias, J.: Connectivity on complete lattices: New results. Comp. Vis. Image Understand. 85(1), 22–53 (2002)
8. Ouzounis, G.K., Wilkinson, M.H.F.: Countering oversegmentation in partitioning-based connectivities. In: Proc. Int. Conf. Image Processing, pp. 844–847 (2005)

9. Urbach, E.R., Roerdink, J.B.T.M., Wilkinson, M.H.F.: Connected shape-size pattern spectra for rotation and scale-invariant classification of gray-scale images. IEEE Trans. Pattern Anal. Mach. Intell. 29(2), 272–285 (2007)
10. Ouzounis, G.K., Wilkinson, M.H.F.: Mask-based second-generation connectivity and attribute filters. IEEE Trans. Pattern Anal. Mach. Intell. 29(6), 990–1004 (2007)
11. Salembier, P., Serra, J.: Flat zones filtering, connected operators, and filters by reconstruction. IEEE Trans. Image Proc. 4(8), 1153–1160 (1995)

Author Index